# W. H. Freeman and Company's Laboratory Separates for Biology

W. H. Freeman's Laboratory Separates allow you to create your own customized laboratory program easily and economically. Simply choose from the more than 120 experiments we offer: each one is thoroughly introduced and completely self-contained-- 8 1/2" x 11", three-hole punched and ready for students to insert into their notebooks. They may be ordered directly from your bookstore. Orders of ten or more titles will be collated by the publisher at no charge. Examination copies of Laboratory Separates may be requested in writing to : *W. H. Freeman and Company, Professor Services Department, 4419 West 1980 South, Salt Lake City, UT 84104*

## Laboratory Separates from *Laboratory Outlines in Biology V* by Abramoff/Thomson

1000 Light Microscopy
1001 Cell Structure and Function
1002 Subcellular Structure and Function
1003 Cellular Reproduction
1004 Plant Anatomy: Roots and the Shoots
1005 Plant Anatomy: Leaves, Flowers, and Fruits
1006 Vertebrate Anatomy: External Anatomy, Skeleton, and Muscles
1007 Vertebrate Anatomy, Digestive, Respiratory, Circulatory, and Urogenital System
1008 Movement of Materials Through Cell Membranes
1009 Biologically Important Molecules: Proteins, Carbohydrates, Lipids, and Nucleic Acids
1010 Photosynthesis
1011 Enzyme Activity
1012 Cellular Respiration
1013 Transport in Biological Systems
1014 Biological Coordination in Plants
1015 Biological Coordination in Animals
1016 Mendelian Genetics
1017 Molecular Genetics
1018 Molecular Biology
1019 Fertilization and Early Development of the Sea Urchin
1020 Fertilization and Early Development of the Frog
1021 Early Development of the Chick
1022 Plant Growth and Development
1023 Plant Development: Hormonal Regulation
1024 Plant Development: Effect of Light
1025 Kingdom Monera: Divisions Schizophyta and Cyanobacteria
1026 Kingdom Protista I: Algae and Slime Molds
1027 Kingdom Protista II: Protozoa
1028 Kingdom Fungi
1029 Kingdom Plantae: Division Bryophyta
1030 Kingdom Plantae: The Vascular Plants
1031 Kingdom Animalia: Phyla Porifera, Cnidaria, and Ctenophora
1032 Kingdom Animalia: Phylum Platyhelminthes
1033 Kingdom Animalia: Phyla Nematoda and Rotifera; Parasites of the Frog
1034 Kingdom Animalia: Phylum Mollusca
1035 Kingdom Animalia: Phylum Annelida
1036 Kingdom Animalia: Phyla Onycophora and Arthropoda
1037 Kingdom Animalia: Phylum Echinodermata
1038 Kingdom Animalia: Phyla Hemichordata and Chordata

## Additional Separates by Abramoff/Thomson

978 Glossary of Anatomic Terms
606 Chemical Aspects of Life
609 Respiration
610 Biological Transport
664 Mitosis
768 The Role of Nucleic Acids in Inheritance
917 Cell Structure
919 Chromosome Morphology
920 Cellular Reproduction: Meiosis
921 Permeability of Cell Membranes
922 Extraction of Proteins
925 Extraction and Identification of Nucleic Acids
926 Energy Capture: Photosynthesis
931 Energy Transformation: Biological Oxidation
932 Energy Utilization: Biological Work
933 Human Genetics
934 Drosophila Genetics I
945 Environmental Biology: Bacterial Pollution in Surface Water Supplies
946 Algae in Water Supplies
949 Cellular Reproduction
955 Tissue Structure and Function: Muscle and Nervous Tissues
968 Kingdom Animalia: Phylum Arthropoda I
969 Kingdom Animalia: Phylum Arthropoda II
971 Kingdom Animalia: Protochordates
972 Kingdom Animalia: Phylum Vertebrata I
973 Kingdom Animalia: Phylum Vertebrata II
974 Adaptation

## Laboratory Separates by Newcomb/Gerloff/Whittingham

696 The Metric System
700 Enzymes: The Catalysts of Living Matter
705 Cellular Fine Structure
711 Genetics Problems

## Laboratory Separates by Walker
### Dissection of the Frog

770 External Anatomy and Skeleton
771 Muscles
772 Digestive and Respiratory Systems
773 Circulatory System
774 Urogenital System
775 Nervous System
776 Sense Organs

### Dissection of the Rat

840 External Anatomy
841 Muscles
842 Digestive and Respiratory Systems
843 Circulatory System
844 Urogenital System
845 Nervous System and Sense Organs

### Anatomy and Dissection of the Fetal Pig

882 A Guide to the Teacher; Preface; A Note for the Student References
883 External Anatomy, Skin, and Skeleton
884 Muscles
885 Digestive and Respiratory Systems
886 Circulatory System
887 Urogenital System
888 Nervous Coordination: Sense Organs
889 Nervous Coordination: Nervous System
980 Glossary of Vertebrate Anatomical Terms

## Laboratory Separates by Wischnitzer
### Anatomy of the Dogfish Shark

738 External Morphology
739 Skeletal System
740 Muscular System
741 Digestive and Respiratory System
742 Circulatory System
743 Urogenital System
744 Sense Organs
745 Nervous System
977 Sectional Morphology
981 Evolution and Function

### Anatomy of the Protochordates

737 Morphology

### Anatomy of the Lamprey

839 Morphology

### Anatomy of the Mud Puppy Necturus

746 External Morphology
747 Skeletal System
748 Muscular System
749 Digestive and Respiratory Systems
750 Circulatory System
751 Urogenital System
752 Sense Organs and Nervous System
983 Evolution and Function

### Anatomy of the Cat

753 External Morphology
754 Skeletal System
755 Muscular System
756 Digestive and Respiratory Systems
757 Circulatory System
758 Urogenital System
759 Endocrine System and Sense Organs
760 Nervous System
982 Sheep Brain

P9-CKB-210

# Atlas and Dissection Guide
## for Comparative Anatomy

Mrinal K. Das
08 August 1994

# ATLAS AND DISSECTION GUIDE FOR COMPARATIVE ANATOMY

SAUL WISCHNITZER
Long Island University

FIFTH EDITION

W. H. Freeman and Company
New York

Library of Congress Cataloging-in-Publication Data

Wischnitzer, Saul.
    Atlas and dissection guide for comparative anatomy / Saul
Wischnitzer.—5th ed.
        p.   cm.
    Includes bibliographical references and index.
    ISBN 0-7167-2374-3
    1. Anatomy, Comparative—Laboratory manuals.   2. Chordata—
Dissection—Laboratory manuals.   I. Title.
    QL812.5.W57   1993                                    92-30800
    596′.04′078—dc20                                         CIP

2 3 4 5 6 7 8 9 0   RRD   9 9 8 7 6 5 4 3

*This book is dedicated*
*to the memory of Dr. Meyer Atlas,*
*Professor of Biology, Yeshiva University,*
*who as my teacher and then colleague*
*inspired me by his devotion*
*to teaching, scholarship,*
*and science.*

# Contents

## ANATOMY OF THE MUD PUPPY *Necturus*

## ANATOMY OF THE CAT
### (INCLUDING THE SHEEP HEART, EYE, AND BRAIN)

# Preface

Comparative vertebrate anatomy is one of the first advanced biology courses selected for study by many students majoring in biology, as well as by many premedical and predental students. It is valuable preparation for the courses in human anatomy that are required in medical or dental school or for those in advanced vertebrate anatomy that are offered in graduate school. The laboratory sessions in comparative vertebrate anatomy usually give students their first major opportunity to carry out careful and comprehensive dissections of biological specimens and thus to acquire a critical scientific approach to such studies.

Both atlases and dissection guides have been long established as important aids for learning in the anatomy laboratories of professional and graduate schools. However, their use in the undergraduate comparative anatomy laboratory has become popular only during the past 20 years.

A course in comparative anatomy is intended to reveal the structural specialization in chordates that has resulted from the process of evolution. The need in such a course for an atlas–dissection guide is as important as it is in the study of human gross anatomy, which concentrates on pure morphology. Both courses use dissection as the main learning approach.

This manual is more comprehensive and flexible than many currently available separate atlases, dissec-tion guides, and unlabeled drawing books covering in-dividual animal forms. Functioning as both an atlas and a dissection guide, it covers all five animal forms commonly studied in comparative anatomy courses: protochordates, lamprey, dogfish shark, mud puppy (*Necturus*), and cat. Each exercise deals with a spe-cific animal system and is available as a loose-leaf sepa-rate. Thus, although the material in the book appears in the traditional sequence, with all of the systems for each form being given in its entirety, the individual exercises may be arranged into systemic units so that the instructor may use the comparative approach if he or she prefers this method for teaching the course.

This atlas–dissection guide is designed to be used by the student in four ways: (1) as a study guide to the anatomic material to be covered in a forthcoming lab-oratory dissection; (2) as an instruction guide to the dissection procedure; (3) as an aid in locating struc-tures that might otherwise be difficult to find; and (4) as a review to recall the original condition of the ani-mal after dissection has altered its normal structural relationships.

In the two- and three-dimensional illustrations used throughout the book, only the anatomic terms that are being discussed, usually on the facing page, are la-beled. For further study, students may want to label some of the other recognizable structures that appear in the art.

Students using this manual will be able to utilize laboratory time more efficiently as they acquire the ability to perform knowledgeable dissection and to identify structures rapidly and easily. The amount of material that can be covered effectively in a course will also depend on the amount of time allotted to laboratory work. My own experience has been with a one-semester course that involves two 2-hour laboratory sessions per week. A course of this length usually covers material on protochordates, either the dogfish shark or the mud puppy, and the cat. If more time is allotted to the course, all the forms described in this manual can be covered.

Since its introduction, this manual has received very positive responses from both students and instructors; these have encouraged me to prepare this fifth edition.

My aim in the fifth edition of *Atlas and Dissection Guide for Comparative Anatomy* has been to make the material more meaningful to students. Thus I have made four major changes to material from the pre-

vious editions: (1) the addition of a short section devoted to dissection of the sheep heart to Cat Exercise 5, Circulatory System, (2) the addition of a section discussing dissection of the sheep or the ox eye to the sensory organ portion of Cat Exercise 7, (3) the addition of a new *Necturus* exercise, Exercise 8, to cover evolutionary and functional aspects, (4) the addition of five tear-out illustrations supplementing the cat circulatory system. I have also made some minor improvements in layout of text and labeling of illustrations.

I am indebted to Dr. Edwin L. Bell of Albright College, Pennsylvania, for his long-standing interest and valued suggestions. I am also grateful to Dr. B. L. Allen of Ohio University for allowing me to use illustrations from her *Basic Anatomy: A Laboratory Manual* in the supplementary figure section.

S. Wischnitzer
October 1992

# Systemic Study Program

The instructor has the option of obtaining any of the exercises in this manual as individual Separates, which can then be arranged by system (see the list of Separates inside the front cover). Or the instructor may wish to use this bound manual and assign the exercises for systemic study as follows:

Anatomy of the Protochordates

Anatomy of the Lamprey

External Morphology
    Dogfish Shark
    Mud Puppy
    Cat
Skeletal System
    Dogfish Shark
    Mud Puppy
    Cat
Muscular System
    Dogfish Shark
    Mud Puppy
    Cat
Digestive and Respiratory Systems
    Dogfish Shark
    Mud Puppy
    Cat

Circulatory System
    Dogfish Shark
    Mud Puppy
    Cat
Urogenital System
    Dogfish Shark
    Mud Puppy
    Cat
Sense Organs
    Dogfish Shark
    Mud Puppy
    Cat
Nervous System
    Dogfish Shark
    Mud Puppy
    Cat
    Sheep
Endocrine System
    Cat

# To the Student

Courses in comparative anatomy traditionally include laboratory exercises in which various vertebrate forms are carefully dissected. The usefulness of dissection to the student is indisputable. No other approach can provide as deep an impact as the visual and tactile examination of the components that make up the organ systems of animals.

This atlas–dissection guide is intended to assist you in the laboratory. However, it cannot replace your own exploration by dissection and study of the parts that you have exposed.

To learn to dissect properly takes practice and patience. The objective of dissection is to expose the body parts in such a manner that their form and relations can be studied readily. To accomplish this type of exposure, separate structures from one another along natural lines of cleavage and attachment wherever possible. This separation can best be done by blunt dissection, using a finger, a scalpel handle, or forceps. When cutting is essential or unavoidable, do so carefully to avoid damaging adjacent structures. *Do not cut unless it is called for in the laboratory instructions.*

Successful dissection requires you to have a clear understanding of the dissection instructions before you begin work. You should endeavor to become skillful at carrying out a good dissection in a reason-able length of time. Preparation before the laboratory period and review afterward will make the dissection more meaningful.

At the conclusion of each laboratory period, you should moisten the specimen and place it in a suitable container, such as a plastic bag or a box, so that it does not dry out between sessions. The work area should then be cleaned up.

Remember, not all specimens are exactly alike. Variations in morphology may result from age and sex differences. Variations from the normal pattern are not uncommon, especially in the circulatory system. When such variations seem to exist in the specimen, you should first make sure that you have not misinterpreted the description or dissected the material incompletely or inaccurately before assuming that a variant has been uncovered.

To help visualize the elements of the circulatory system, the specimens' arterial and venous pathways may come injected with colored latex so that they stand out more clearly and are more readily indentifiable.

Terminology in this manual is expressed in either Latin or English, depending on common usage. Wherever possible, eponyms have been replaced by scientific terms. Because a formal standard nomenclature has not yet been adopted by comparative anatomists,

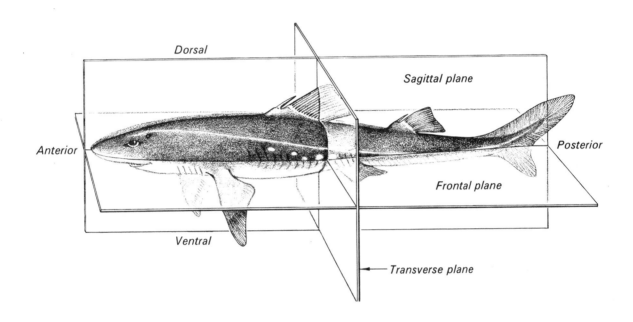

*Dorsal*

*Sagittal plane*

*Anterior*

*Posterior*

*Frontal plane*

*Ventral*

*Transverse plane*

some structures are known by several names. The alternative terms for these structures may be included in the text.

In this manual, the terms *right* and *left* refer to the side of an animal positioned on its back with its head forward. However, if the animal is on its abdomen, head forward, the right and left sides of the specimen will correspond to the dissector's.

The following anatomic terms are those most frequently used in the text.

**anterior** or **cranial:**   toward the head end.
**posterior** or **caudal:**   toward the tail end.
**dorsal:**   toward or near the back.
**ventral:**   pertaining to the belly.
**median:**   in or near the longitudinal plane in the middle of the body.
**medial:**   pertaining to the median plane.
**proximal:**   lying near the base or site of attachment.

**distal:**   lying near the tip.

Sections through the body are called:

**sagittal:**   if they run parallel to the median longitudinal plane to divide the specimen into right and left sides.
**frontal:**   if they run horizontally to divide the specimen into dorsal and ventral parts.
**transverse:**   if they run at right angles to the longitudinal plane.

To facilitate rapid identification of the anatomic structures described at a particular point in dissection, the number of labels on each illustration has been restricted to the terms mentioned in the related text. As an additional study aid, it may be beneficial to add labels for those structures previously identified in the specimen.

# Classification of the Chordates

*Branchiostoma = Amphioxus*

Studies have shown that there is a hierarchy among the many animals that are known, both existing and extinct. That hierarchy can be outlined in a reasonably logical order of categories, with each category having a different rank: phylum, class, order, family, genus, and species. (Using these categories every animal can be classified by category in order from the most inclusive rank to the least inclusive rank.) How an animal is classified depends on the degree of relationships it has with an associated animal located at some particular level in the hierarchy.

In this manual seven animals are discussed, three of which belong to the protochordates and the other four to the craniates (vertebrates). The major categories of chordates and where these seven animals fit into the scheme of things is shown in boldface in the following abbreviated, traditional classification system. (An asterisk preceding a category denotes a totally extinct group.)

## PHYLUM HEMICHORDATA

Forms with little or no development of notochord or dorsal nerve cord, which are closely related to chordates.

*Class Pterobranchia.* Simple, sessile, plant-like animal.

*Class Enteropneusta.* **Acorn worms.**

## PHYLUM CHORDATA

*Subphylum Urochordata.* **Tunicates** and related forms.

*Subphylum Cephalochordata.* **Amphioxus.**

*Subphylum Vertebrata.* Forms with developed backbone, skull, brain, and kidneys.

*SuperClass Agnatha.* Jawless fishes.

*\*Subclass Ostracodermi.* Extinct jawless fishes.

*Subclass Cyclostomata.* Living jawless fishes.

*Order Petromyzontia.* **Lampreys.**

*Order Myxinoidea.* **Hagfishes.**

*Class Elasmobranchiomorphi.* Cartilaginous fishes and certain relatives.

*\*Subclass Placodermi.* Armored, jawed fishes.

*Subclass Chondrichthyes.* Cartilaginous fishes.

*Infraclass Elasmobranchii.* Sharks and related forms, including the **dogfish shark.**

*Infraclass Bradyodonti.* Chimeras and related forms.

*Class Osteichthyes.* Bony fishes.

*\*Subclass Acanthodii.* Paleozoic "spiny sharks."

*Sublclass Actinopterygii.* Ray-finned fishes.

*Superorder Chondrostei.* Primitive ray-finned fishes of the Paleozoic era represented by paddlefishes and sturgeons.

*Superorder Holostei.* Dominant ray-finned fishes of the Mesozoic era, represented by the gar and bow-fin. *Superorder Teleostei.* Dominant fishes of the Cenozoic and Recent eras.

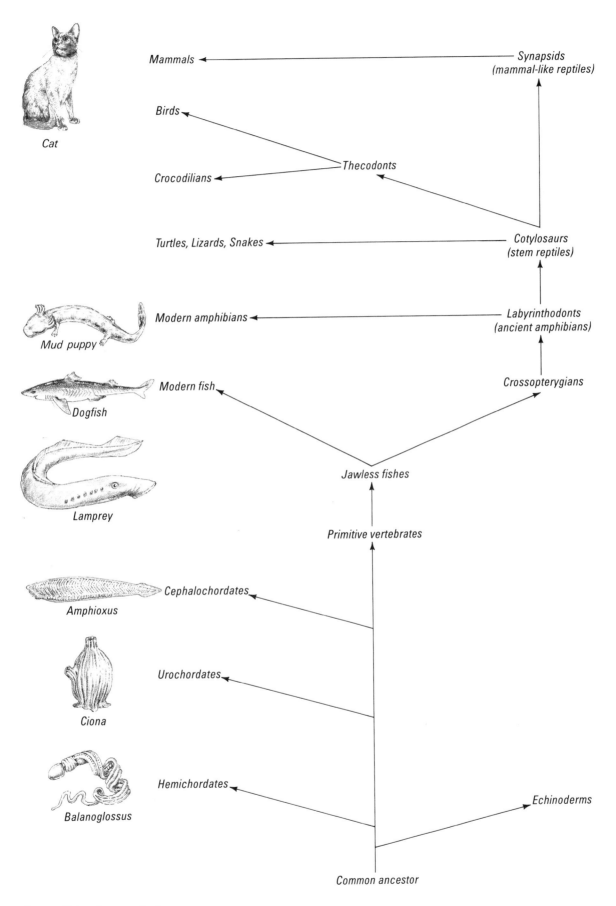

Mammals ← Synapsids (mammal-like reptiles)

Birds ←

Crocodilians ← Thecodonts

Cat

Turtles, Lizards, Snakes ← Cotylosaurs (stem reptiles)

Modern amphibians ← Labyrinthodonts (ancient amphibians)

Mud puppy

Modern fish ← Crossopterygians

Dogfish

Jawless fishes

Lamprey

Primitive vertebrates

Cephalochordates

Amphioxus

Urochordates

Ciona

Hemichordates ← Echinoderms

Balanoglossus

Common ancestor

Scheme of chordate evolution.

*Subclass Sarcopterygii.* Fleshy-finned fishes.
  *Order Crossopterygii.* Predaceous forms.
  *Order Dipnoi.* Lungfishes.
*Class Amphibia.* Tetrapods.
 *\*Subclass Labyrinthodontia.* Ancient amphibians.
 *\*Subclass Lepospondyli.* Ancient snake-like, and salamander-like forms.
  *Subclass Lissamphibia.* Modern amphibians.
   *Order Anura.* Tailless amphibians, such as frogs and toads.
   *Order Urodela.* Tailed amphibians, such as salamanders; includes the **mud puppy** *Necturus.*
   *Order Gymnophiona.* Worm-like burrowing forms.
*Class Reptilia.* True terrestrial tetrapods.
  *Subclass Anapsida.*
   *\*Order Cotylosauria.* Primitive stem reptiles.
   *Order Chelonia.* Turtles and tortoises.
  *Subclass Lepidosauria.* Lizards and snakes.

*Subclass Archosauria.* Crocodiles, \*dinosaurs, and \*flying reptiles.
 *\*Subclass Euryapsida.* Ancient marine reptiles.
 *\*Subclass Synapsida.* Ancient mammal-like reptiles.
*Class Aves.* Feathered vertebrates.
 *\*Subclass Archaeornithes.* Ancient birds.
  *Subclass Neornithes.* Modern birds.
*Class Mammalia.* Hairy vertebrates.
  *Subclass Prototheria.* Egg-laying mammals.
  *Subclass Theria.* Give birth to and suckle their young.
   *\*Infraclass Patriotheria.* Ancient, small, primitive forms.
   *Infraclass Metatheria.* Marsupials (pouched mammals).
   *Infraclass Eutheria.* Higher mammals with an efficient placenta; includes the **cat.**

The sequence of classification, when applied to the seven forms studied in this manual, is outlined in Table 1-1.

**TABLE 1-1**
Sequence of classification.

*Protochordates*

|  | **Amphioxus** | **Tunicate** | **Acorn Worm** |
|---|---|---|---|
| Subphylum | Cephalochordata | Urochordata | — |
| Class | — | Ascidiacea | Enteropneusta |
| Genus | *Branchiostoma* | *Ciona* | *Balanoglossus* |
| Species | *B. lanceolatum* | *C. intestinalis* | — |

*Craniates*

| | **Marine Lamprey** | **Spiny Dogfish** | **Spotted Mud Puppy** | **Domestic Cat** |
|---|---|---|---|---|
| Subphylum | Vertebrata | Vertebrata | Vertebrata | Vertebrata |
| Class | ~~Agnatha~~ *Cephalaspidomorphi* Chondrichthyes | | Amphibia | Mammalia |
| Subclass | ~~Cyclostomata~~ | Elasmobranchii | Lissamphibia | Theria |
| Order | Petromyzoniformes | ~~Selachii~~ *Squaliformes* | Caudata | Carnivora |
| Family | *Petromyzontidae* | *Squalidae* | Proteidae | Felidae |
| Genus | *Petromyzon* | *Squalus* | *Necturus* | *Felis* |
| Species | *P. marinus* | *S. acanthias* | *N. maculosus* | *F.* ~~*domestica*~~ *cattus* |

*(handwritten: Superclass Agnatha →)*

# Atlas and Dissection Guide
# for Comparative Anatomy

# ANATOMY OF THE PROTOCHORDATES

# Anatomy of the Protochordates

Although the comparative anatomy laboratory is concerned almost exclusively with the study of vertebrate forms, it is worthwhile first to mention the various existing animal types that lack a backbone but are closely related to the vertebrates. Collectively these are known as protochordates. Almost all these animals, along with the vertebrates, belong to the same phylum, Chordata. All chordates have, at some time during their life, a notochord, a dorsal nerve cord, and gill slits. The phylum Chordata comprises three subphyla: Cephalochordata (amphioxus), Urochordata (tunicates), and Vertebrata. Hemichordates (acorn-worms) are usually considered a separate phylum.

## The Cephalochordates

The cephalochordates have a poorly defined head and a notochord that extends the entire length of the body. The common name, amphioxus, which means "sharp at both ends," refers to any member of the genus *Branchiostoma*. Because of its many primitive and generalized chordate features, *B. lanceolatum* is the form most commonly studied in the laboratory.

Amphioxus is a small animal, only about two to three inches in length, that can be found lying along sandy seashores in temperate zones. It resembles a fish and has an elongated but laterally compressed body. It lacks paired appendages but has a median dorsal fin that is most highly developed at the caudal end. Amphioxus differs from fishes in its lack of a skull, vertebrae, brain, heart, and kidneys. The body surface is semitransparent, since the epithelium covering its body is but a single cell in thickness and the underlying connective tissue is gelatinous. Because of this semitransparency, the angularly shaped muscle masses, the myomeres, can be clearly seen.

During most of the day amphioxus assumes a feeding position by burrowing into the sand, leaving only its head exposed (Fig. 1-1). While in its burrow, amphioxus is primarily engaged in obtaining food from the seawater that flows over its head at regular intervals. To extract this food, which consists of microscopic organisms suspended in the water, amphioxus utilizes a concentrating and water-filtering mechanism located in the pharynx.

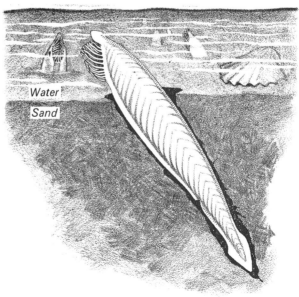

Water

Sand

**FIGURE 1–1**

Amphioxus in normal habitat.

## EXTERNAL MORPHOLOGY (Fig. 1-2)

*Examine a preserved specimen placed in a shallow dish of water, using a hand lens or a dissecting microscope.*

**Median dorsal fin**   Extends the entire length of the body.

**Caudal fin**   An expanded continuation of the median dorsal fin at the posterior end of the body.

**Ventral fin**   An anterior continuation of the caudal fin extending to the atriopore.

**Atriopore**   The opening of an internal cavity, the atrium, located just in front of the ventral fin.

**Rostrum**   The snout-like anterior end of the body that projects over the oral hood (see definition below).

**Buccal cirri**   Twenty-two stiff tentacle-like structures projecting from the free edge of the oral hood. They may keep out large particles from the pharynx and possibly contain taste discriminating elements.

**Myomeres**   The longitudinal series of about 60 angularly shaped muscle segments, known also as myotomes.

**Myosepta**   The connective tissue partitions separating the myomeres.

**Gonads**   The row of rectangular bulging masses located beneath the ventral ends of the myomeres.

## INTERNAL MORPHOLOGY (Figs. 1-3, 1-4, 1-5)

*Examine a cleared, stained whole mount with a disecting microscope or the low power of a compound microscope.*

**Metapleural folds**   A pair of integumentary folds that extend along the ventrolateral surfaces from the anterior pharyngeal region almost to the atriopore. (To see them, focus on the body surface until you note a horizontal line slightly dorsal and parallel to the ventral edge of the body.)

**Oral hood**   A funnel-shaped membrane located at the ventral part of the anterior end of the body. It helps to direct the water vortex, created by the wheel organ (see definition), into the mouth.

**Hatschek's pit**   The ciliated depression that lies in the roof of the funnel-shaped cavity located beneath the rostrum. It aids in generating a ciliary current and secretes mucus that entraps minute particles of food that are in the water.

**Vestibule**   Funnel-shaped cavity enclosed by the oral hood.

**Fin rays**   The internal connective-tissue rods that are present as supporting elements in both the dorsal and ventral fins. They are absent from the caudal fin.

**Notochord**   The supporting rod that extends almost the entire length of the body.

**Nerve cord**   The elongated cylindrical mass of nervous tissue that lies above the notochord. It is almost as long as the notochord.

**Ocelli**   A linear series of pigment cells lying along the ventral wall of the nerve cord. They may possibly serve as photoreceptors.

**Eye spot**   A mass of pigment cells located at the anterior end of the nerve cord. Its sensory function has not been definitely established.

**Wheel organ**   A series of finger-like projections, lined with cilia, that lies on the inner lateral and dorsal walls of the oral hood. The cilia beat in such a way as to suggest a turning wheel, and they help to create a vortex of water that is directed toward the mouth by the oral hood.

**Velum**   The vertical membrane located just behind the base of the wheel organ. The velum may regulate the amount of water entering the pharynx.

**Mouth**   The opening in the velum.

**Velar tentacles**   Short tender projections extending from the margin of the velum that rings the mouth. By bending across the mouth, they may prevent sand or particles from passing into it.

**Pharynx**   The chamber, interrupted by gill slits, that extends from the mouth to the intestine.

**Gill bars**   Slender parallel oblique bars, supported by **branchial rods,** that make up the side wall of the pharynx. Cross pieces (synapticulae) link the bars.

**Gill slits**   Elongated openings between the gill bars.

**Atrium**   The large cavity surrounding the pharynx. Its ventral boundary is visible as a line extending below the pharynx to the atriopore.

**Intestine**   The digestive tube that extends from the pharynx to the anus. In some books the intestine is divided into several segments that are not easily demarcated: the short esophagus soon widens into a stomach, or midgut; behind the point of origin of the hepatic diverticulum, the midgut narrows into the ileocolic ring, beyond which the still narrow hindgut extends to the **anus.** The ileocolic ring is a belt of cilia, which moves the food mass along the gut beyond the point where enzymes enter it.

**Hepatic diverticulum**   The pouch that projects forward as an outpocketing of the intestine (midgut). It extends along the right side of the posterior part of the pharynx. Also known as the intestinal diverticulum, or cecum, this pouch is involved in the production of indigestive enzymes. Some claim that the currents generated by the cilia in the alimentary tract propel food particles into the midgut and also into the hepatic diverticulum. Others dispute this activity because they have not been able to observe it happening.

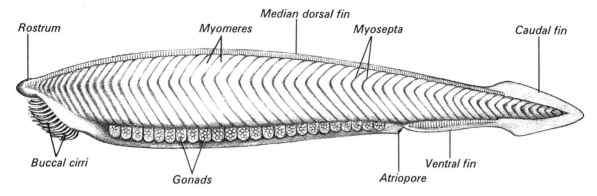

**FIGURE 1–2**
External morphology.

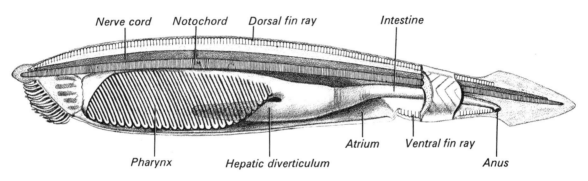

**FIGURE 1–3**
Cleared whole mount.

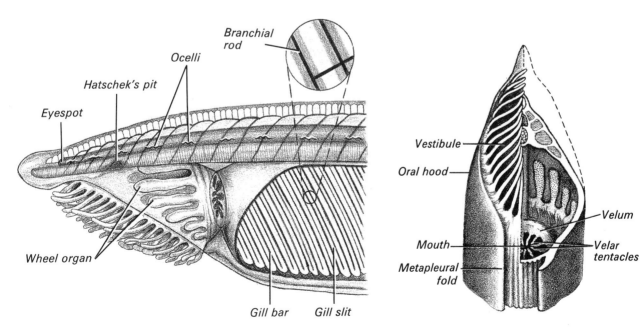

**FIGURE 1–4**
Anterior end (enlarged lateral view).

**FIGURE 1–5**
Anterior end (ventral view; cut away at right).

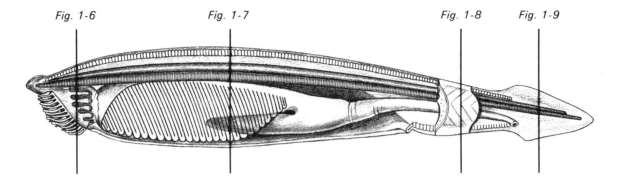

Fig. 1-6    Fig. 1-7    Fig. 1-8    Fig. 1-9

Levels of cross sections.

## CROSS SECTIONS

*Using the low power of a compound microscope, examine cross sections through the buccal cavity, pharynx, intestinal region, and tail. In addition to the structures described in the preceding sections, several other features of the vascular, excretory, and reproductive systems will be seen.*

All the cross sections present the following features in common.

**Epidermis**  The outer-cellular layer of the body, consisting of a single layer of columnar cells. It is covered by an acellular cuticle and rests upon a thin layer of connective tissue.

**Dorsal fin**  A short projection from the dorsal midline region that is supported by the gelatinous **fin rays.**

**Myomeres**  Masses of muscle under the skin, separated by connective-tissue partitions, the **myosepta.**

**Nerve cord**  The somewhat triangular structure located in the midline, slightly ventral to the dorsal fin. It has a very narrow cavity, the **neurocoel.**

**Notochord**  The larger rounded structure that lies beneath the nerve cord. It is surrounded by a connective-tissue sheath.

## Section Through Oral Hood (Fig. 1-6)

**Vestibule**  The central cavity of the **oral hood,** located under the front part of the notochord. The thickened ciliated epithelial masses of the **wheel organ** are found on the inner hood wall.

**Buccal cirri**  A number of these, cut transversely, are present at the lower boundary of the vestibule.

## Section Through Pharynx (Fig. 1-7)

**Metapleural folds**  These project down from the ventrolateral corners of the body.

**Transverse muscle**  This lies in the wrinkled body wall between the metapleural folds. It extends from the myomeres on one side to those on the other. When it contracts, it compresses the atrial cavity dorsal to it from side to side, thus expelling its water.

**Pharynx**  The narrow cavity limited laterally by gill bars that extend between the notochord and transverse muscle. The **gill bars,** which are separated by

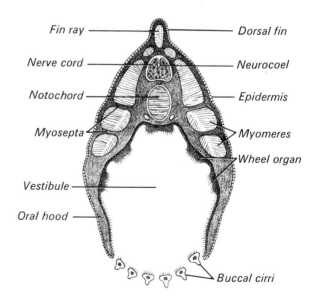

Fin ray — Dorsal fin
Nerve cord — Neurocoel
Notochord — Epidermis
Myosepta — Myomeres
— Wheel organ
Vestibule
Oral hood — Buccal cirri

**FIGURE 1-6**
Cross section through oral hood.

7

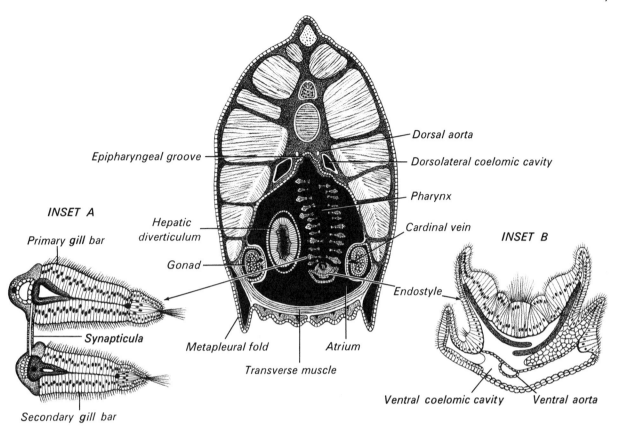

**FIGURE 1–7**
Cross section through pharynx. Inset A: Gill bars (enlarged). Inset B: Endostyle (enlarged).

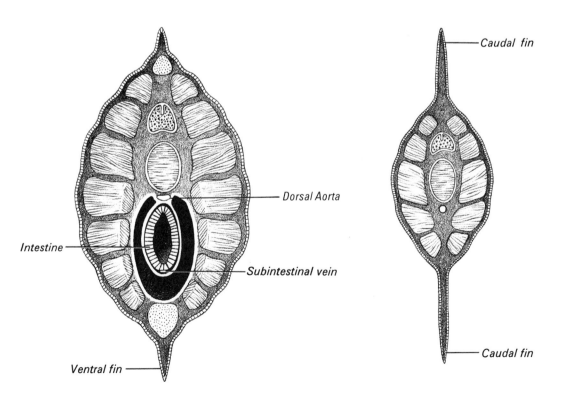

**FIGURE 1–8**
Cross section through intestinal region (posterior to atriopore).

**FIGURE 1–9**
Cross section through tail.

gill slits, consist of two types, primary and secondary. The secondary, or tongue, bars are smaller and do not contain a coelomic cavity near their lateral end (inset A). The gill bars are braced by small bridges called **synapticulae.**

**Endostyle** The deeply grooved structure in the floor of the pharynx. It contains both ciliated and glandular cells (inset B).

**Epipharyngeal groove** Located in the roof of the pharynx opposite the endostyle.

**Atrium** The cavity that surrounds the pharynx laterally and ventrally.

**Gonads** These are located ventrolaterally in most sections. They protrude into the atrium from the body wall, still bounded by the atrial lining. An ovary consists of many large nucleated cells. A testis consists of much smaller germ cells, which appear as *a mass of small dark dots.*

**Hepatic diverticulum** In sections near the posterior end of the pharynx, this hollow oval-shaped structure is seen on the right side of the pharynx. It lies within the atrial cavity.

**Coelom** Present in the cross section as three sacs. One, the **ventral coelomic cavity,** lies beneath the endostyle; the other two, the **dorsolateral coelomic cavities,** are located on both sides of the epipharyngeal groove.

**Blood vessels** The paired **dorsal aortae** may be seen between the notochord and the dorsal end of each of the paired coelomic cavities. The **ventral aorta** is enclosed within the ventral coelomic cavity. **Cardinal veins** are present close to the midpoint of the medial border of the gonads.

### Section Through Intestinal Region (Fig. 1-8)

**Intestine** This is the large, hollow, oval structure lying ventral to the unpaired **dorsal aorta.** Beneath the intestine is the **subintestinal vein.**

**Ventral fin** This fin is seen in cross sections of the intestine at levels posterior to the atriopore. At levels anterior to the atriopore, metapleural folds will be seen in its place.

### Section Through Tail (Fig. 1-9)

**Caudal fin** This is narrower and higher than either the dorsal or ventral fins.

## The Urochordates

There is a general consensus that the urochordates, like the cephalochordates, meet the criteria for inclusion in the phylum Chordata. However, a large gap separates both groups from the hemichordates, which have remained at a low evolutionary level and whose taxonomic status is unclear. The relationship of the urochordates and cephalochordates with each other and with vertebrates is as yet uncertain.

Studies of life cycles of urochordates have shown that some of them are basically annual forms. Their larvae, which have a short free-swimming life period, do not feed but are specialized for rapid selection of a suitable habitat. The larvae have remained since Darwin's time a major focus for discussions of chordate evolution. The affinities of the urochordates with vertebrates are more apparent in the larval stage than in the adult.

The marine animals belonging to this subphylum are widely distributed and occur in a variety of shapes, sizes, and life conditions. There are three classes of urochordates, the most generalized being Ascidiacea, or tunicates, which are also known as sea squirts. A common form studied belong to the genus *Ciona*.

### THE ADULT TUNICATE (Figs. 1-10, 1-11)

*Examine an adult specimen that has been cleared, stained, and embedded in plastic. If available, examine a preserved specimen, even though these are frequently distorted.*

**Tunic** The transparent, tough, outer membrane, whose presence gives these animals their common name, contains a cellulose-like substance (tunicin). The tunic is secreted by the mantle, a thin cellular layer that also contains scattered bundles of muscle fibers and thus is capable of contraction.

**Base** The stalk-like end of the body that in sessile forms is attached to a fixed substrate, such as a rock.

**Siphons** Two spout-like projections, each having an aperture, are located at the opposite end to the base. The terminal one is the **incurrent siphon,** while the subterminal one is the **excurrent siphon.**

**Pharynx** This is the largest body organ, extending from the middle of the incurrent siphon almost to the base. Its thin wall is perforated by numerous tiny openings (gill slits). The pharynx is absent from the excurrent siphon. Along the side of the pharynx opposite the excurrent siphon extends the faintly stained **endostyle,** a mucus-secreting organ.

**Atrium** This is the cavity that lies between the pharynx (and other viscera; see below) and the mantle. It extends into the excurrent siphon. Contraction of the mantle forcefully expels water out of the atrial and pharyngeal cavities through both siphons. This activity is responsible for the other common name by which these animals are known, sea squirts.

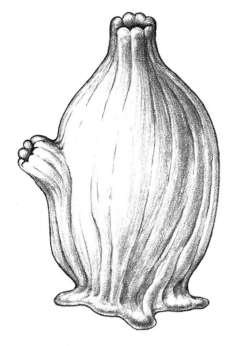

**FIGURE 1–10**
*Ciona:* An adult tunicate.

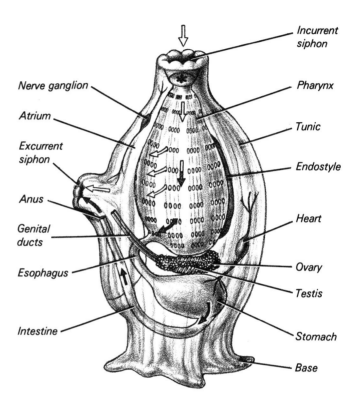

Incurrent siphon

Nerve ganglion

Pharynx

Atrium

Tunic

Excurrent siphon

Endostyle

Anus

Heart

Genital ducts

Ovary

Esophagus

Testis

Intestine

Stomach

Base

**FIGURE 1–11**
*Ciona:* A cleared and stained specimen.

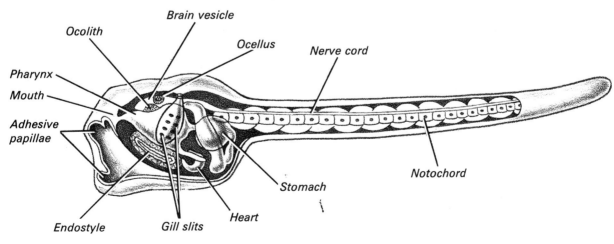

Brain vesicle

Ocolith

Ocellus

Nerve cord

Pharynx

Mouth

Adhesive papillae

Stomach

Notochord

Endostyle

Gill slits

Heart

**FIGURE 1–12**
A tunicate larva.

**Esophagus**    This is a very short organ beginning at the narrow termination of the pharynx. It receives food particles (microorganisms) that have been sucked into the animals through its incurrent siphon by the action of cilia lining the pharynx. The particles become trapped by mucus (secreted by the endostyle) that coats the inside of the pharynx and are conveyed by cilia to the esophagus.

**Stomach**    This more conspicuous organ has a C-shaped configuration and is darkly stained.

**Intestine**    This straight segment of the digestive tract extends from the stomach to the **anus.** Its diameter is similar to that of the stomach and it may be stained only partially or not at all. Undigested material is eliminated through the anus into the atrium and is expelled externally via the excurrent siphon.

**Gonads**    These are represented by the dark mass located within the stomach loop. Tunicates are monoecious (hermaphrodites) but are self-sterile. Thus the ovary and testis has its own **genital duct.** This is the conspicuous slender tube extending from the gonad alongside the intestine to the base of the excurrent siphon. It usually represents the sperm duct since the oviduct, which parallels it, is visible only when it contains eggs.

## THE TUNICATE LARVA (Fig. 1-12)

A tadpole-like larva develops within 18 to 24 hours after fertilization, which occurs in the water following extrusion of germ cells through the excurrent siphon. The larva is free-swimming, being propelled by a muscular tail. After a few hours it will become transformed into an adult.

*Examine a prepared slide containing a tunicate larva.*

**Notochord**    This is the well-defined structure that extends through most of the length of the tail.

**Nerve cord**    This dorsal hollow tube lies above the notochord and terminates in the body as an enlarged **vesicle** (a brain precursor). Eye-like and ear-like structures are associated with the sensory vesicle but frequently are not well defined.

**Gill slits**    A few of these as well as an **endostyle** are present.

**Adhesive papillae**    These serve to attach the larva to an object, after which the organism undergoes retrograde metamorphosis, when it becomes sac-like, losing its notochord and most of its nerve cord.

# The Hemichordates

While hemichordates are definitely related to the chordates, it is very questionable whether they have a true notochord. Thus these marine animals are classified in a separate phylum of their own rather than as a subphylum of Chordata. Those preferring such a classification do not at the same time demand changing the group name. For while these animals seemingly lack a genuine notochord, they nevertheless resemble chordates from other viewpoints.

While the relationship between hemichordates and chordates is unclear, those between hemichordates and echinoderms are firmly established. It is reflected in the remarkable echinoderm characteristics of the tornaria larva (the developmental state of some hemichordates), the details of coelom development, and the probable homology of various structures in the two groups.

There is reason to hypothesize that echinoderms and hemichordates were derived from common sessile or semi-sessile ancestral forms that collected their food externally—as all extant hemichordates do. The drastic metamorphosis of echinoderm larvae, contrasted with the more modest one of the tornaria larva, suggests that the hemichordates may have diverged less from that common ancestor than have the echinoderms.

The hemichordates fall into two classes: the pterobranchs, tiny sessile, stalked, colonial organisms (Fig. 1-13), and the enteropneusts, called acorn-worms or tongue-worms. The pterobranchs are rare deep-sea forms and thus are not routinely studied, but they will be described briefly to identify their major structural features. They are enclosed in a housing that they secrete around themselves, and the body is divided into three characteristic regions: **proboscis, collar,** and **trunk.** The collar bears large, ciliated, tentaculated arms, and only one pair of gill slits is present. Pterobranchs are considered to be more primitive than enteropneusts.

The enteropneusts are worldwide in distribution, live in shallow waters, and burrow into the sand. This group will therefore be studied as exemplified by the genus *Balanoglossus.*

## EXTERNAL FEATURES (Fig. 1-14)

*Examine an enteropneust, preferably one that has been cleared, stained, and embedded in plastic. If a preserved specimen is used, place it into a pan of water and study it with the aid of a hand lens. Note the following features.*

**Proboscis**    The conical anterior segment of the body used in burrowing and feeding.

**Collar**    The short, second body segment. The proboscis is attached to the collar by a narrow **stalk.** The

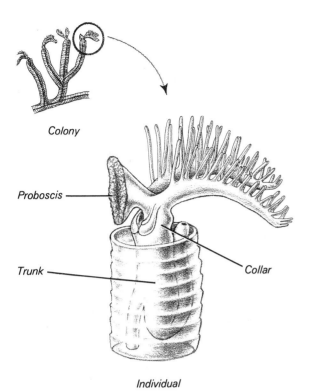

Colony

Proboscis

Trunk

Collar

Individual

**FIGURE 1–13**
*Rhabdopleura:* A pterobranch hemichordate.

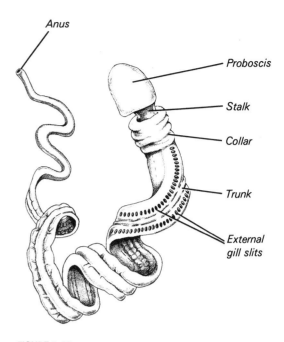

Anus

Proboscis

Stalk

Collar

Trunk

External
gill slits

**FIGURE 1–14**
*Balanoglossus:* An enteropneust hemichordate.

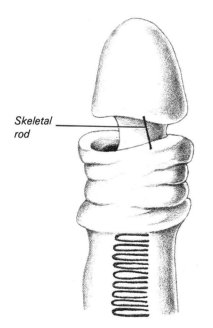

Skeletal
rod

**FIGURE 1–15**
Anterior segment (cleared and stained).

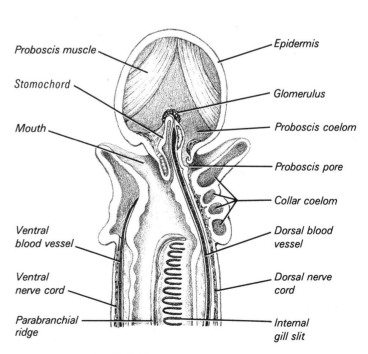

Proboscis muscle

Stomochord

Mouth

Ventral
blood vessel

Ventral
nerve cord

Parabranchial
ridge

Epidermis

Glomerulus

Proboscis coelom

Proboscis pore

Collar coelom

Dorsal blood
vessel

Dorsal nerve
cord

Internal
gill slit

**FIGURE 1–16**
Anterior segment (sagittal section).

collar, which also assists in the burrowing process, is supported by a **skeletal rod** (Fig. 1-15).

**Trunk**   The longest segment of the body. It extends from the end of the collar to the **anus.** Ingested sand and debris pass out through the anus as a core.

**External gill slits**   These are paired, numerous, small and linearly arranged on the anterior part of the trunk.

## INTERNAL ANATOMY (Fig. 1-16)

**Epidermis**   This is uniformly ciliated on the proboscis. The cilia aid feeding by creating a current that carries mucus food particles into the mouth. The mucus is derived from the numerous secretory cells located in the epidermis. Mucus also serves to strengthen the burrow with a firm lining and thereby facilitates the movement of the animal within it.

**Mouth**   Difficult to see in an intact specimen, it is located inside the front of the collar ventral to the proboscis stalk. Organic material serves as the source of food.

**Proboscis coelom**   The cavity inside the proboscis that enables it to inflate and deflate during burrowing. Similarly, there is a **collar coelom** that helps anchor the body as the deflated proboscis is projected forward in the mud by the action of its **proboscis muscle.** Distention of the proboscis occurs by entry of seawater into it through the **proboscis pore.**

**Stomochord**   Projecting forward from the roof of the mouth into the proboscis coelom is a structure that somewhat resembles a notochord, thus giving this group its name, hemichordate. This structure may help stiffen the proboscis. The homology between stomochord and notochord has, however, been strongly challenged. The mass of tissue at the tip of the stomochord is known as the **glomerulus** because it may be an excretory organ.

**Dorsal nerve cord**   This is located along the entire length of the collar and trunk regions. Only in the former is it partially hollow; otherwise it is solid throughout. A **dorsal blood vessel** runs beneath and parallel to the nerve cord.

**Ventral nerve cord**   This is solid and extends within the ventral trunk wall. It is connected with the dorsal nerve cord by a circumpharyngeal nerve ring that runs in the junction between the collar and trunk. Running along with the ventral nerve cord is the **ventral blood vessel.**

**Internal gill slits**   These are V-shaped and lead into gill pouches, which in turn open into the external gill slits. As a result, water taken into the pharynx via the mouth is passed to the outside.

**Parabranchial ridge**   This shelf partially separates the respiratory and digestive segments of the pharynx.

# Evolution

Evolutionary considerations relative to the protochordates focus on two basic issues: the relationship of the chordates to the nonchordates, and the relationship of the protochordates to the chordates.

## CHORDATE–NONCHORDATE RELATIONS

To appreciate this issue, it is first necessary to describe, in general terms, the various characteristics of multicellular organisms that relate chordates to members of other phyla. Chordates originate from embryos that have three basic layers, and all their organ systems develop from these layers: (1) the ectoderm that covers the organism, (2) the middle mesoderm that, among other things, lines the body cavity and forms the internal skeleton, and (3) the endoderm that lines the interior of the gut. The gut has separate openings for the mouth and anus. The anus is derived from the blastopore, which is an opening in the surface of the ball of cells that is formed at a very early embryonic stage.

These characteristics are shared to a substantial, but not complete, extent by chordates and echinoderms (starfishes, sea cucumbers, and so forth) as well as members of two smaller phyla (Chaetognatha, Pogonophora). Taken on the whole, therefore, the features listed separate these four phyla from all others. Moreover, because their mouths are not formed from the blastopore, they are collectively called **deuterostomes** (second mouths).

More specifically, echinoderms and chordates have been considered to be interrelated because of their similarities to the hemichordates. That group contains two markedly distinct types of organisms, pterobranchs and acorn-worms. The pterobranchs are colonial, deep-sea animals that use tentacles for feeding and thus are echinoderm-like. The acorn-worms are elongated, marine forms that bear pharyngeal gill slits and thus are chordate-like.

The linkage between echinoderms and chordates is, apparently, further strengthened by the similarity between the larvae of various echinoderms (e.g., starfishes, sea cucumbers) and those of lower chordates (e.g., acorn-worms) as well as by possible biochemical and immunological connection. However, the closeness of the relationship is not firmly established, since it is possible that the similarity of the larvae may simply reflect convergence of independently evolving organisms, and, furthermore, some doubt has been raised concerning the validity of a biochemical relationship. ~~Thus, it is difficult to link both groups definitively, since, unlike echinoderms, the protochordates have no hard parts, so they have probably~~

*Pikaea : a protochordate fossil from the Burgess Shale, BC.*

13

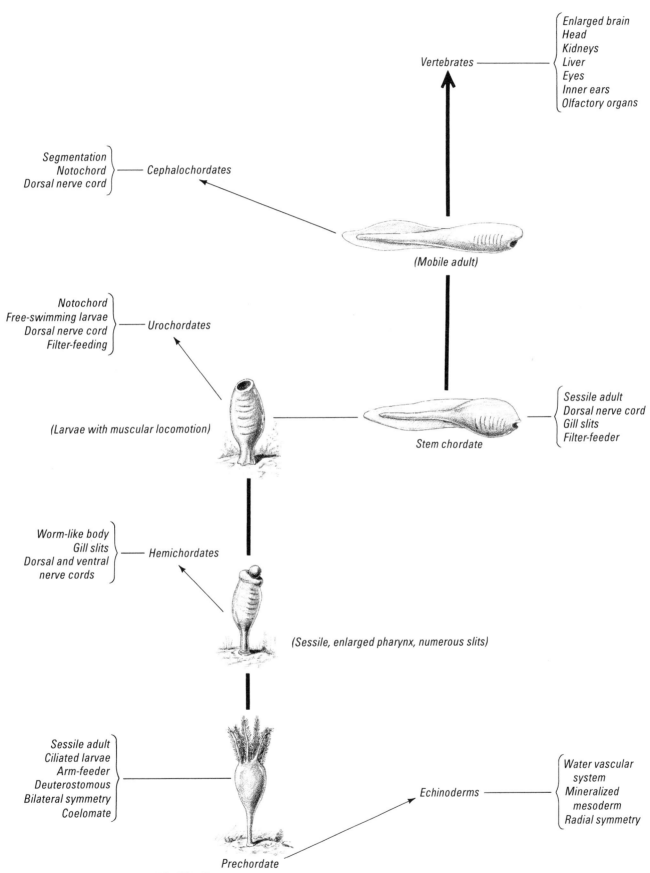

Segmentation
Notochord        } — Cephalochordates
Dorsal nerve cord

Vertebrates — { Enlarged brain
Head
Kidneys
Liver
Eyes
Inner ears
Olfactory organs

(Mobile adult)

Notochord
Free-swimming larvae  } — Urochordates
Dorsal nerve cord
Filter-feeding

(Larvae with muscular locomotion)

Stem chordate

Sessile adult
Dorsal nerve cord
Gill slits
Filter-feeder

Worm-like body
Gill slits              } — Hemichordates
Dorsal and ventral
nerve cords

(Sessile, enlarged pharynx, numerous slits)

Sessile adult
Ciliated larvae
Arm-feeder         }
Deuterostomous
Bilateral symmetry
Coelomate

Echinoderms — { Water vascular
system
Mineralized
mesoderm
Radial symmetry

Prechordate

**FIGURE 1–17**
Hypothetical scheme of the evolution of vertebrates (after Romer).

not left a fossil record. Nevertheless, the embryological evidence at hand is sufficient to establish some sort of a relationship between the chordates (including the hemichordates) and the nonchordate phylum Echinodermata.

## PROTOCHORDATE–CHORDATE RELATIONS

We now turn to the issue of how the various protochordate phyla fit into the scheme of things. To do so, we should bear in mind the principal diagnostic features that characterize chordates, namely, a notochord, a dorsal hollow nerve chord, and pharyngeal gill slits. All these should be present (or be considered lost in the animal's evolutionary history) to define an animal as a chordate.

*Balanoglossus,* which is a hemichordate, possesses gill slits as well as a short rod-like structure in its muscular proboscis. This structure is probably not a true notochord since it does not function as a structure for support; it may not have an endodermal origin; and it has a cavity that opens into the pharynx. Thus this structure is frequently called the **stomochord.** Another feature that adds doubt to the status of the hemichordates is the anal opening at the end of the body, so no postanal tail exists (an additional chordate diagnostic feature). As a result of these nonchordate features, hemichordates are classified in an entirely separate phylum and are not considered a chordate subphylum.

For urochordates, only the free-swimming larva has a notochord, hollow nerve chord, and one pair of gill slits, but it shows no sign of segmentation. As expected, no anus is present in the nonfeeding larva, so a postanal tail is not possible. In the course of being transformed into a sessile adult, two additional pairs of gill slits arise and, in time, they all become converted into a complex basketwork. Thus, tunicates are considered a true chordate subphylum, but they are not as complex as hemichordates or cephalochordates. Furthermore, their adult appearance reflects a loss of features during the course of evolution. Urochordates, however, are linked to cephalochordates by similar feeding methods as well as by the behavior of their larvae.

Amphioxus, as an example of a cephalochordate, possesses five chordate features. The notochord extends to the extreme anterior end of the body; the dorsal nerve chord is slightly expanded anteriorly; numerous gill slits are present; and a conspicuous postanal tail exists along with many segmental muscles. The presence of a full complement of chordate characteristics suggests that cephalochordates are the protochordates most closely related to the vertebrates. Thus, it seems

likely that the cephalochordates represent an offshoot from a chordate line that also produced both urochordates and vertebrates.

In summary, chordates are more closely related to echinoderms and, especially, to hemichordates than they are to other phyla. In addition, the vertebrates are more closely related to cephalochordates than to urochordates. And, finally, it seems reasonable to assume that the echinoderms, the hemichordates, and the three simple chordate groups diverged from a common lineage (Fig. 1-17). Although each group apparently evolved along its own path, they all retained some of their common features, depending on individual needs, while losing or modifying others as well as acquiring new structures.

# Function

*Balanoglossus,* which is a worm-like animal, moves primarily by cilia that cover its body. It appears to be capable of burrowing into the sea bottom by thrusting its proboscis into mud, expanding it by pressing seawater into the diverticulum of the collar, and, with the grip so obtained, drawing the trunk forward. Feeding is facilitated by a current of water that is set in motion by the beating of the cilia, which also cover the long proboscis. Food particles become trapped in mucus, which is secreted by cells on the proboscis, and are swept back into the mouth and then through the pharynx into a simple alimentary tract. The water escapes through the numerous (100 to 150) gill slits. The digestive and absorptive functions are carried out by ciliated epithelium that lines the intestinal canal.

The urochordate larva has no blood vessels, but the adult contains a well-developed heart, which is located at the base of the animal. The heart beats alternately in each direction for about a minute at a time, pumping acellular blood. A singular nerve ganglion is located between the siphons. It receives sensory stimuli via axons that form a nerve network, which extends between the ganglion and receptor cells that are located in the musculature of the atrial wall. These stimuli can induce a motor response, depending on the nature of the stimulation. An endocrine-type organ, the neural gland, is located under the nerve ganglion. It apparently excretes a pituitary hormone that is comparable to oxytocin, which presumably serves to bring about contraction of the smooth muscles in the organ. Although excretory cells may be grouped, they do not form organs. Thus we see that the tunicate has unique characteristics associated with its circulatory, excretory, and nervous systems.

Amphioxus has a persistent notochord and skeletal elements that support its oral structures, pharynx, and fins. Like urochordates, amphioxus is a filter-feeder, but it has a different structural organization. Although it lacks a heart or capillary system, the circulatory system is nevertheless similar to the vertebrate plan in that it has a complex set of blood vessels.

Functionally, amphioxus is particularly interesting because of its mode of moving and feeding. It ejects itself from a (half-) burial site by erratic body-twisting movements. These convulsions may be triggered by light stimulation of the ocelli, which activate signals that pass along the axons of giant nerves and extend to the myomeres, or by chemoreceptors located adjacent to the oral hood, which can, apparently, also induce migration to different sites. Once free, the animal swims to a new burial location, much the same way as a fish, by alternating contractions of the myomeres that bend the body in opposite directions. Since the notochord is stiff, much of the muscular energy of contraction is expended in bending, and its elasticity then produces a rapid flick backward. This action produces the propulsive force toward its next burial site.

Once the amphioxus is positioned in its new temporary location with its mouth projecting upward, feeding is its major activity (see Fig. 1-1). The oral hood spreads open, perhaps with the assistance of the anterior tip of the notochord that is embedded in its roof. Then the cilia on the buccal cirri, wheel organ, and gill bars generate a current of water that is directed inward toward the mouth, down the pharynx, and out the gill slits and atriopore. The buccal cirri may serve as a crude filter mechanism to extract sand particles, while minute food particles are caught in mucus that is secreted by the endostyle. Those food particles are then propelled upward by the force of a current that is generated by the cilia on the inner surface of the gill bars. Then they are directed backward into the midgut by the cilia in the epipharyngeal groove. The food embedded in the mucus stream is digested, probably by enzymes secreted by the hepatic diverticulum. Absorption, for the most part, takes place in the short hind gut.

# ANATOMY OF THE LAMPREY

# Morphology

*Handwritten margin notes:*
*- Hagfish belongs to Class Myxini*
*- Lamprey " " " Cephalaspidomorphi*
*- most primitive fossil vertebrates: conodonts*

# Anatomy of the Lamprey

The vertebrates constitute the largest of the four chordate subphyla. A vertebrate has a cranium (brain-case), a vertebral column (which, in the lamprey, is present as fragments), and a well-defined head. The earliest vertebrates were the jawless fishes that are placed in the class Agnatha. ~~All living agnathans belong to the subclass Cyclostomata, or round-mouthed fishes, which consist of the lampreys and hag-fishes and are the most primitive group of living vertebrates~~ (Fig. 1-1). It has been suggested that cyclostomes are descendants of an extinct subclass of jawless, fish-like creatures, the ostracoderms, which are the oldest and most primitive fossil vertebrates known, having vanished about 350 million years ago. However, although the cyclostomes do exemplify a primitive stage of development, they are not ancestral vertebrates.

All eight genera of lampreys belong to the order Petromyzoniformes. They may be marine or fresh-water and are found throughout the temperate parts of the world: in the oceans, along the coasts, or in large inland lakes. The usual form studied in the laboratory is the marine lamprey *Petromyzon marinus*, whose relatively large size enables detailed anatomical ob-servation of both its primitive and its specialized char-acteristics. Its primitive characteristics include the absence of jaws, true teeth, and paired appendages, and the presence of a continuous notochord, segmental musculature, a straight alimentary tract, and a poorly developed brain, skeleton, and reproductive system.

Its specialized characteristics are its eel-like shape, a head without "bony armor," and its mode of feeding.

The lamprey attaches itself by the mouth cup to other fishes, rasps at its prey's flesh with its horny tongue, and then sucks its blood. The digestive and respiratory systems are modified to conform to this parasitic habit, and the lamprey is able to secrete an anticoagulant that permits continued utilization of the host's blood.

The marine lamprey has become adapted to a fresh-water habitat. Thus, many spend the entire life cycle in the Great Lakes, where they cause intense depletion of the natural fish population.

**FIGURE 1–1**
Cyclostomes.

## EXTERNAL MORPHOLOGY (Fig. 1-2)

*Examine a preserved specimen of an adult lamprey. Note its eel-like shape. Its streamlined appearance is appropriate to the swift movements of this animal. It has a gray, scaleless, slimy skin, which is extremely tough. Its overall appearance matches its feeding habits. Identify the following features.*

**Head**   Cylindrical in shape, it extends posteriorly as far as the last external gill slit on each side.

**Trunk**   Cylindrical, but it becomes laterally compressed near the tail region and extends from the end of the head to the cloacal aperture (see definition below).

**Tail**   Extends posteriorly from the cloacal aperture.

**Fins**   Three median fins, the **anterior dorsal fin,** the **posterior dorsal fin,** and the **caudal fin,** are present. All are supported by cartilaginous fin rays.

**Buccal funnel**   Supported by an internal ring of cartilage, this hood-shaped structure is a sucking disk that attaches securely to the host. Its opening is fringed with numerous finger-like structures, the **buccal papillae,** and its interior is studded with rows of brown **horny teeth** (Fig. 1-3).

**Nostril**   Single median mid-dorsal opening located at the top of the head. It is the external aperture of the olfactory apparatus.

**Pineal organ**   The pineal organ lies just beneath the surface of the skin, and its position is demarcated by the small oval area located just posterior to the nostril. It detects changes in light and initiates diurnal color changes in the pigmentation of the skin.

**Eyes**   Lidless, functional organs for sight. There is one on each side of the head.

**Lateral line system**   Consists of groups of pores that are sense organs that detect vibrations and movements in the water. Groups of such pores may be seen (with the aid of a hand lens) extending caudally from the top of the lateral eye, dorsolaterally beneath the eye, and caudally on the ventral surface from the buccal funnel. This receptor system is probably sensitive to changes in the current and turbulence of the water around the animal.

**External gill slits**   Seven oval apertures behind the eye on each side of the head.

**Myomeres**   Muscle segments whose outline may be seen through the skin of the trunk and tail.

**Cloacal aperture**   The opening of the shallow pit on the underside of the body at the posterior end of the trunk. The urogenital papilla, through which the excretory and genital products are emitted, may protrude from the aperture. The anus, a slit-like opening of the intestine, is just anterior to the papilla.

## SAGITTAL SECTION (Fig. 1-5)

*Examine a sagittal section of the anterior portion of the lamprey, cut close to the median plane, and identify the following features.*

**Myomeres**   Each consists of bundles of longitudinal muscle fibers that attach onto connective tissue **myosepta** that separate the myomeres.

**Brain**   The lobulated structure above the anterior end of the notochord. It extends into the neural canal as the **spinal cord.** The **adenohypophysis** has three parts seen in freshly cut specimens (Fig. 1-4).

**Pineal organ**   Shaped like an inverted pyramid, it lies above the brain.

**Nostril**   This external apeture leads into the narrow canal that opens into the expanded **olfactory sac.** The canal continues caudoventrally as the elongated **nasohypophyseal pouch,** which ends blindly.

**Cranial cartilages**   Two median cartilages, anterior dorsal and posterior dorsal, are present.

**Notochord**   This skeletal axis beneath the spinal cord supports the body during myomere contraction.

**Dorsal aorta**   The thin tube beneath the notochord.

**Buccal funnel**   Lined with rows of horny teeth, it has a supporting **annular cartilage** that can be seen in section at both ends of the buccal funnel.

**Mouth**   The aperture at the back of the buccal funnel dorsal to the **tongue.** The tongue, too, is covered with horny (lingual) teeth.

**Pharynx**   The chamber sloping caudoventrally from the mouth. A cavity, the **hydrosinus,** extends anteriorly from the dorsal pharyngeal wall. The floor of the pharynx is supported by the long median **lingual cartilage** and the well-developed **lingual muscle.**

**Esophagus**   This, the first part of the digestive tract, is a thick-walled tube that is the dorsal continuation of the pharynx. It extends beneath the dorsal aorta until it expands somewhat to form the intestine.

**Branchial tube**   The ventral continuation of the pharynx, this tube lies beneath the esophagus and ends blindly. Its opening is guarded by a curtain-like **velum.** Its walls contain seven **internal gill slits** on each side. This tube is known also as the respiratory tube because its gill slits lead into gill pouches.

**Ventral aorta**   This blood vessel extends from the **ventricle** cranially beneath the branchial tube.

**Inferior jugular vein**   A median blood vessel that lies beneath the pharynx and extends caudally to open up into the **sinus venosus** of the heart.

**Branchial basket cartilages**   These can be seen in section beneath the inferior jugular vein. They actually extend laterally and upward within the body wall, which they support (see Fig. 1-14).

21

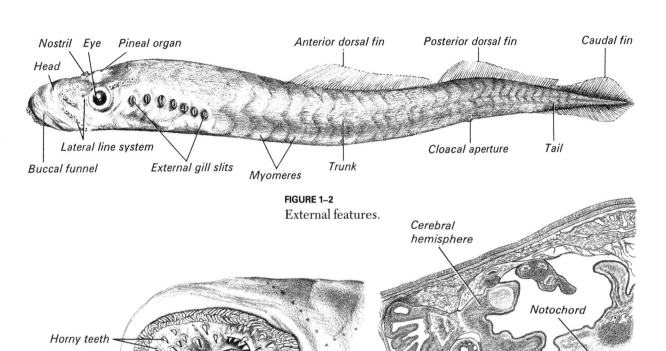

Nostril  Eye  Pineal organ
Head
Lateral line system
Buccal funnel
External gill slits
Myomeres
Trunk
Anterior dorsal fin
Posterior dorsal fin
Caudal fin
Cloacal aperture
Tail

FIGURE 1–2
External features.

Horny teeth
Buccal papillae

FIGURE 1–3
Head (ventral view).

Cerebral hemisphere
Notochord
Nasohypophyseal pouch
Parts of the adenohypophysis

FIGURE 1–4
Head (near mid-sagittal view).

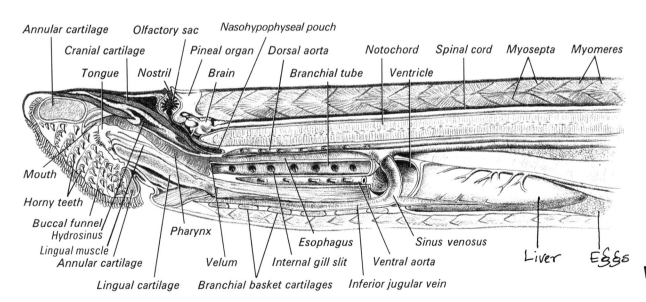

Annular cartilage  Olfactory sac  Nasohypophyseal pouch
Cranial cartilage  Pineal organ  Dorsal aorta
Tongue  Nostril  Brain  Branchial tube
Notochord  Spinal cord  Myosepta  Myomeres
Ventricle
Mouth
Horny teeth
Buccal funnel
Hydrosinus
Lingual muscle
Annular cartilage
Lingual cartilage
Pharynx
Velum
Branchial basket cartilages
Esophagus
Internal gill slit
Ventral aorta
Sinus venosus
Inferior jugular vein
Liver  Eggs

FIGURE 1–5
Sagittal section through anterior end.

## CROSS SECTIONS

*Examine cross sections through the eyes, the pharynx, the trunk, and the tail.*

### Section Through Eyes (Fig. 1-6)

**Pharynx** The central cavity is surrounded by four masses of **pharyngeal muscles.**

**Eyes** Located on either side of the pharynx, each consists of two parts: a large **lens** and a cup-shaped **retina,** made up of neural and pigmented layers. **Ocular muscles** extend medially from the retina.

**Lingual cartilage** The median cartilage beneath the pharynx. On either side of the cartilage are lingual muscles.

**Pharyngeal glands** Embedded in the lingual muscles on the ventrolateral sides of the pharynx, they secrete an anticoagulant that keeps the blood of the host animal from clotting. Their inconspicuous ducts open up on the underside of the tongue.

**Cranial cartilage** One of several such cartilages making up the chondrocranium. This cartilage supports the **nasohypophyseal pouch,** above which are the **olfactory sac** and the **pineal organ.**

**Carotid arteries** Located on both sides of the chondrocranium, they supply the eyes and brain.

**Myomeres** Seen on either side of the brain.

### Section Through Branchial Tube (Fig. 1-7)

**Spinal cord** The dorsalmost midline structure.

**Notochord** The spherical mass directly beneath the spinal cord.

**Anterior cardinal veins** Vessels located on either side of the notochord.

**Dorsal aorta** The blood vessel that lies beneath the notochord.

**Esophagus** This part of the digestive tract is characterized by the numerous folds of its inner lining that extend from its thick walls.

**Branchial tube** It has a thin wall and is located immediately beneath the esophagus.

**Gill pouches** These are respiratory sacs or internal gills, lined with **gill lamellae,** and are located on either side of the larynx.

**Ventral aorta** An unpaired vessel posterior to the fourth gill pouch; anterior to the fourth gill pouch it is paired.

**Lingual muscle** The prominent muscle mass beneath the ventral aorta.

**Inferior jugular vein** The lowermost midline structure. At the anteriormost part of the gill region it divides into two jugular veins.

### Section Through Trunk (Fig. 1-8)

**Posterior cardinal veins** Located on either side of the dorsal aorta.

**Intestine** Located in the mid-ventral part of the body cavity.

**Gonad** This is a large unpaired median organ located above the intestine. In the male the gonad is the **testis,** which consists of numerous leaf-like lobes. In the female it is the **ovary,** which consists of a mass of eggs.

**Kidney** These are located on either side of the gonads. They are drained by large ducts that run along the ventral border of the kidney.

### Section Through Tail (Fig. 1-9)

**Fin rays** These support the fin.

**Caudal artery** Small blood vessel lying beneath the notochord. Beneath the caudal artery is the **caudal vein.**

## THE AMMOCOETES LARVA

The adult marine lampreys migrate from the sea to fresh water as the spawning season approaches. After spawning has taken place, they die. About three weeks after external fertilization, small lamprey larvae, known as **ammocoetes,** hatch. They exist burrowed in the mud of streams and brooks for the next three to seven years, during which time they attain a length of about six inches. They then undergo a profound metamorphosis to the adult stage in the course of a few weeks. Metamorphosis is the reorganization of old structures, such as the pharynx, which becomes divided longitudinally, and the formation of new structures, such as teeth and eyes. All of these changes are essential for a successful life as a semiparasitic adult lamprey. After metamorphosis the new adults return to the sea where they live for a year or two before returning to fresh water to spawn and die.

Study of the larva is of value because it presents many primitive vertebrate characteristics not seen in the adult. Morphologically, the ammocoetes larva superficially resembles amphioxus, and it exists as a filter-feeder, much like amphioxus. But unlike amphioxus and other primitive chordates that filter-feed from currents of water propelled by cilia, the ammocoetes drives its feeding current by muscular activity coordinated by cranial nerves.

23

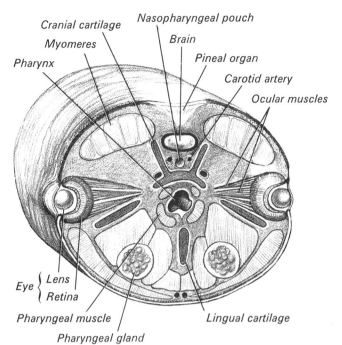

FIGURE 1–6
Cross section through eyes.

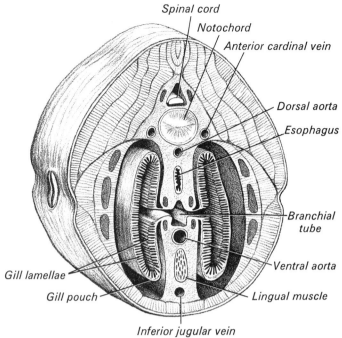

FIGURE 1–7
Cross section through branchial tube.

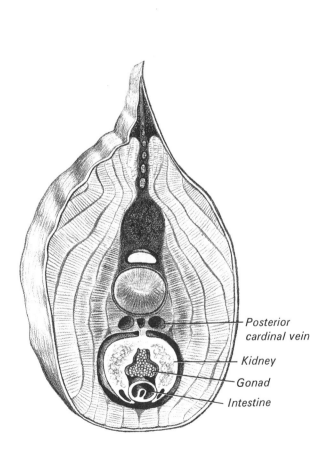

FIGURE 1–8
Cross section through intestine.

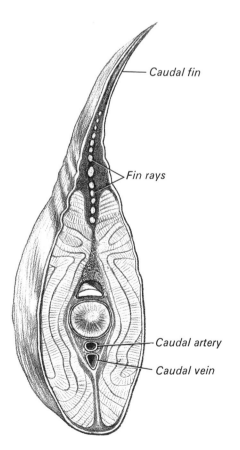

FIGURE 1–9
Cross section through tail.

### Whole Mount (Fig. 1-10)

*Examine a whole mount of a small larva with the low power of a compound microscope.*

**Dorsal fin**    The fin located along the upper surface of the body. It extends around the tail as the **caudal fin.**

**Spinal cord**    The dark rod that extends cranially from the tail and expands anteriorly as the brain. The **brain** may exhibit the **prosencephalon, mesencephalon,** and **rhombencephalon** (depending on slide quality).

**Sense organs**    These consist of a pair of nonfunctional **eyes,** one lying on each side of the midbrain, a median **nostril** extending beneath the forebrain, and oval **auditory vesicles,** located lateral to the hindbrain.

**Notochord**    The light-colored longitudinal band ventral to the spinal cord and to most of the brain.

**Oral hood**    The expanded cup-shaped membrane at the anterior end of the larva that encloses a chamber, the **buccal cavity.**

**Oral papillae**    Finger-like projections attached to the sides of the oral hood that serve as straining and sensory devices at the entrance to the buccal cavity.

**Velar folds**    A pair of muscular flaps, each attached to the roof and sides of the buccal cavity at the posterior end of the cavity. Their rhythmic movements cause a stream of water to be directed backward into the pharynx.

**Pharynx**    The expanded anterior part of the digestive tract. It has seven **gill pouches,** or outpocketings, on the lateral wall. Each pouch is lined with **gill lamellae** and has a small round **external gill slit** (which can be seen if the microscope is focused sharply on the surface of the pouch). During metamorphosis, the pharynx divides into a dorsal digestive tube and a ventral branchial tube.

**Subpharyngeal gland**    The large, dark, elongated mass, also called the endostyle, is located beneath the greater part of the pharynx. In the embryo it secretes mucus, which passes through a pore into the pharynx. Certain of its cells are transformed into the thyroid gland of the adult.

**Esophagus**    The short slender tube extending caudally from the pharynx.

**Intestine**    The elongated portion of the digestive tract that extends from the end of the esophagus to the **anus,** located at the base of the caudal fin.

**Heart**    This organ is located beneath the anterior part of the esophagus. It appears as a clear structure.

**Liver**    This organ is located behind the heart. It contains a large clear vesicle, the **gall bladder.**

### Section Through the Pharynx (Fig. 1-11)

*With the low power of a compound microscope, examine a cross section of the pharynx, taking into consideration that the appearance will depend on the location of the pharyngeal cut.*

**Myomeres**    Muscle segments are more numerous in the dorsal section than in the ventral section.

**Neural canal**    The large dorsal midline cavity containing the **spinal cord.**

**Notochord**    Located beneath the spinal cord, it is surrounded by a thick **perichordal sheath.** At each ventrolateral side of the notochord is a large **anterior cardinal vein.**

**Dorsal aorta**    Small midline structure located beneath the notochord.

**Pharynx**    Occupies most of the cross section. **Gill lamellae** project both medially and laterally into the pharynx.

**Hyperpharyngeal ridge**    A ciliated mid-dorsal ridge. Distinguished from the **hypopharyngeal ridge,** a ciliated mid-ventral ridge.

**Ventral aorta**    This appears single or paired, depending on the level of the section, beneath the hypopharyngeal ridge.

**Subpharyngeal gland**    Lies beneath the ventral aorta. Whether it can be observed depends on the level of the section.

**External jugular veins**    These lie near the ventral body wall, one on each side of the midline.

**Branchial basket cartilages**    These may be seen, cut in section, in a number of places near the external body wall.

### Section Through the Intestine (Fig. 1-12)

*Note the presence of the myomeres, the spinal cord, the notochord, and the dorsal aorta, as in the preceding section.*

**Fin folds**    Both dorsal and ventral fin folds should be noted.

**Posterior cardinal veins**    Located on either side of the dorsal aorta.

**Kidneys**    Paired structures located beneath the posterior cardinal veins.

**Intestine**    Occupies most of the peritoneal cavity. Its ventral side is invaginated as the **spiral fold,** which encloses the **intestinal artery.**

25

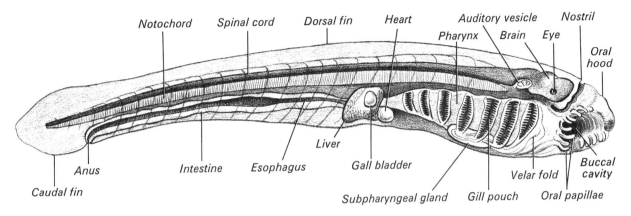

**FIGURE 1–10**
Cleared whole mount.

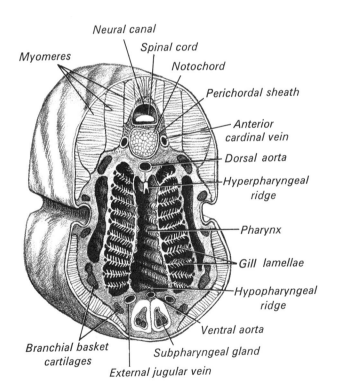

**FIGURE 1–11**
Cross section through pharynx.

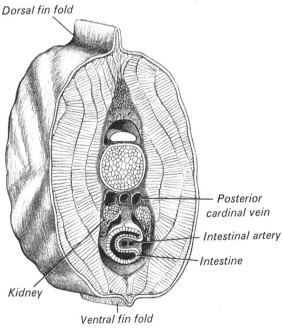

**FIGURE 1–12**
Cross section through intestine.

## EVOLUTION

Cyclostomes belong to a small class of vertebrates, the agnathans, most of whom are extinct. They are characterized by the absence of jaws. The cyclostomes are important because they occupy the significant position in the phylogeny of vertebrates of being the first to have evolved from their chordate source. All other vertebrates trace their ancestry to agnathans.

The jawless vertebrates have advanced beyond the protochordates by their possession of a head (with a cranium, brain, and paired eyes), incomplete vertebrae (represented by cartilaginous elements on the dorsal surface of the notochord), and bone (in the scales and armor of the extinct forms). However, they lack certain features that are present in higher vertebrates, such as jaws, girdles, and typical paired appendages.

The cyclostomes represent the only members of their class to have an eel-like form. Their lack of armor, or scales, essentially precludes the likelihood of finding a significant fossil record of their ancestors. They probably evolved from an extinct jawless fish from which some of the other ancient agnathans belonging to the subclass Ostracodermi originated (Fig. 1-13). Thus, it is unlikely that the first vertebrates were fully armored but, rather, the fossil ostracoderms must have had ancestors that lacked armor and larvae that permitted growth. Conceivably, this occurrence would account for the finding of a fossil that some have interpreted as representing an extinct lamprey-like cyclostome.

The two existing types of cyclostomes, the lamprey and hagfish, may have evolved independently. Although they are modified for a parasitic life, they do provide some information about the soft tissues of their class. However, caution is necessary to avoid misinterpreting characteristics. Thus the adult lamprey is studied to gain an understanding of its common vertebrate features as well as its specialized parasitic characteristics.

The larva of the lamprey, ammocoetes, is even more primitive than the adult. This is especially true in terms of its mouth parts, pharynx, gonads, and some digestive organs. Thus the ammocoetes larva may be considered similar to one group of ancient ostracoderms, and the larva is studied to gain an insight into the organization of the early vertebrates.

## FUNCTION

Agnathous vertebrates do not have true teeth, whereas cyclostomes have epidermal tooth analogs on the buccal funnel and on the tongue. In the lamprey, the horny teeth are yellow, conical, and regularly spaced. When worn, the teeth are replaced by new ones that develop below them. The teeth on the buccal funnel help to hold the prey, while the tongue teeth rasp its side so that it bleeds.

Cyclostomes are the only living vertebrates that are monorhinous; i.e., they have a single nasal opening, which is located in the median line in front of the eyes (see Fig. 1-5). The nasal cavity extends backward from the olfactory sac deep into the skull to the tip of the notochord, where it terminates as the blind nasopharyngeal pouch. This sac apparently pumps water continuously over the olfactory epithelium so that a fresh sample is always present to help detect prey or spawning areas. The sensory data are conveyed to the brain by the pair of olfactory nerves.

Cyclostomes have no dermal skeleton. The specialized mouth parts of the cyclostomes are supported by cartilages. The rest of the head skeleton consists of cartilages of unknown homology, including a branchial basket framework that surrounds the gill slits and heart (Fig. 1-14).

The presence of a horizontal partition between the esophagus and branchial tube, which has a velum at its anterior end, is associated with the parasitic feeding habit of the lamprey. Thus, the animal can remain attached to its prey, with its mouth sealed, while passing water back and forth through its gill slits, thereby maintaining respiration. Blood and fluid from the prey are barred by the velum from entering the branchial tube, which functions only when the animal is not feeding.

The three-chambered heart, in which the conus arteriosus is essentially absent, pumps the blood forward through the ventral aorta. The red blood cells contain hemoglobin, but its molecular weight is 17,000 instead of the usual 68,000, so each hemoglobin molecule takes up one molecule of oxygen. This occurrence ensures efficient transport of oxygen.

The kidney contains many nephrons, which facilitate the excretion of a great deal of water. The marine lamprey has several structural adaptations that permit it to live in seawater without becoming dehydrated.

The brain of cyclostomes is primitive (Fig. 1-15). The anterior part is compressed in response to crowding by the terminal mouth and dorsal nasal chamber. Only a shallow groove separates the large olfactory bulbs from the short cerebrum. The optic lobes are prominent, and the medulla is relatively large. As expected, the cerebellum, which controls motor coordination and maintains equilibrium, is rudimentary in these sluggish, parasitic forms.

**FIGURE 1–13**
Ostracoderms: Agnathan fossils related to cyclostomes (reconstructions).

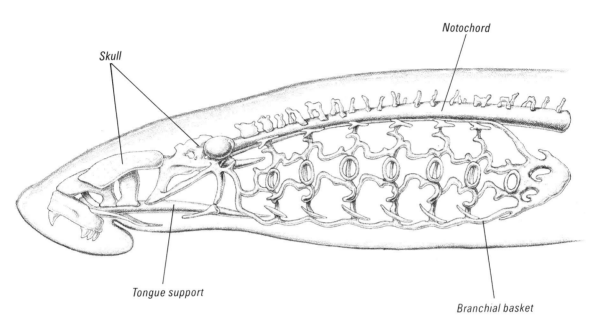

*Skull*

*Notochord*

*Tongue support*

*Branchial basket*

**FIGURE 1–14**
Skeleton of the lamprey.

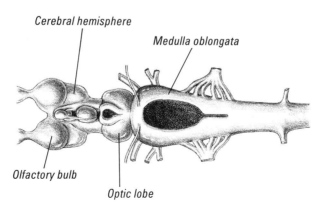

*Cerebral hemisphere*

*Medulla oblongata*

*Olfactory bulb*

*Optic lobe*

**FIGURE 1–15**
Brain of the lamprey.

# ANATOMY OF THE DOGFISH SHARK

# External Morphology

# Anatomy of the Dogfish Shark

The dogfish shark, known also as the spiny dogfish, is an example of a typical vertebrate with a brain encased in a cranium and vertebrae that are associated with the notochord in their development. Because it has scales, gills, and paired appendages in the form of fins, the dogfish shark belongs to the superclass Pisces. Its cartilaginous skeleton makes it a member of the class Chondrichthyes, which includes, in addition to the sharks, skates, rays, and chimerae. Because its gill slits open separately to the outside, it belongs to the subclass Elasmobranchii.

The spiny dogfish is a small shark. Adult males are two to three feet in length, and females are slightly larger. (Some of the large sharks are more than 40 feet in length.) Because they prefer water temperatures of about 10°C, they migrate north in the spring and south in the fall.

Two dogfish shark species are commonly studied. One, *Squalus acanthias*, is found along the North Atlantic coast; the other, *Squalus* ~~acanthias~~ *suckleyi*, abounds along the Pacific coast. Their carnivorous and predacious habits are responsible for the annual destruction of large numbers of commercially valuable fish, and in addition they destroy fishing nets. At present their usefulness to man is limited to laboratory study.

The dogfish shark is dissected for three reasons. First, it exhibits the organizational pattern of the vertebrate body in an almost generalized condition. Thus, a study of this animal provides a basis for a better understanding of the specialized organ systems of higher verte-

brates. Second, it is easily dissected because its skeleton is ~~entirely~~ *mostly* cartilaginous. Finally, because of its manageable size, its structures can be demonstrated with a clarity that would not be possible in a smaller form. This allows such complex structures as the brain, the spinal cord, the cranial nerves, the inner ear, and the eye muscles to be disclosed with relative ease.

The anatomy of both species of spiny dogfish is essentially the same, and both are quite similar to their European counterpart, *Scyllium*, and to the smooth dogfish, *Mustelus*, found along the North Atlantic coast. For ease of handling in the laboratory, many of the animals studied are immature, and this underdeveloped state should be borne in mind when the urogenital system is studied.

## INTEGUMENT

*A small thin piece of skin cut from the body surface can be viewed directly under the compound microscope. A fuller understanding of the structure of the skin may be gained by studying prepared slides. If such slides are not available, scales can be more easily seen after thin pieces of skin are soaked overnight in glycerine solution and then mounted. For examination of individual scales, place a piece of skin for a few hours in a solution of sodium hypochlorite. A coverslip should not be used with these delicate preparations.*

## Placoid Scales (Figs. 1-1, 1-2)

The rough sandpaper quality of the skin surface is due to the presence of placoid scales embedded in it. Each scale consists of a diamond-shaped, bony **basal plate** embedded in the stratum laxum of the dermis (see below), a constricted **neck,** and a **spine** projecting caudally above the epidermal surface. The base is anchored to the stratum compactum (see below) by fibers. A centrally placed **pulp cavity** is present within the spine. This cavity contains blood vessels that nourish the dentine that makes up the bulk of the spine. An enamel-like substance protects the exposed surface of the spine.

## Skin (Fig. 1-2)

The skin consists of two major layers: a thin outer **epidermis** and a thick inner **dermis.** The epidermis is a stratified epithelium, the surface cells of which are flat. The cells of the basal layer, the **stratum germinativum,** are columnar, are mitotically active, and are responsible for the formation of the entire epidermis. Numerous unicellular, mucus-secreting cells are distributed along the surface between the squamous epithelial elements. The dermis is made up of a surface layer of loose connective tissue, the **stratum laxum,** and of a denser connective-tissue layer, the **stratum compactum.** Blood vessels and nerves ramify within the dermis. Close to its epidermal boundary the dermis contains numerous small pigment cells called **melanophores,** which are responsible for the gray color of the dorsal and lateral surfaces of the dogfish shark's body. The melanophores are absent from the lighter ventral surface and also from localized areas along the lateral surfaces. As a result, these localized areas appear as white spots (see Fig. 1-3).

## EXTERNAL FEATURES (Figs. 1-3 to 1-7)

*Examine the intact dogfish specimen to reveal the following features.*

**Head**  Broad and somewhat flattened.

**Trunk**  Fusiform in shape in order to facilitate locomotion. The trunk gradually tapers posteriorly.

**Tail**  Compressed laterally, it bears a caudal fin (see below). It is the principal organ of locomotion.

**External gill slits**  Five pairs of oblique openings, located on the sides of the head, that communicate with the pharynx. The name elasmobranch, which means "gill plate," refers to the plate of skin between the external openings of each of the gill slits.

**Spiracle**  Modified first gill slit located above and behind the eye. It serves for water intake, especially when the jaws are grasping food.

**Eyes**  Located laterally near the top of the head, they cannot be simultaneously focused on a single object. The eyelids are immovable.

**Lateral line canal**  A linearly arranged series of minute openings. On each side the line extends from a point just behind the spiracle to the tail. If a transverse cut is made across the lateral line and the cut edge turned up, the canal can be seen on the skin edge. This special sensory apparatus, innervated by the vagus, or tenth cranial nerve (N. X), is thought to serve as a receptor for low-frequency vibrations and hydrostatic pressure changes. Thus, the lateral line canal presumably helps facilitate locomotion and orientation of the animal in an environment of turbulence and current changes. The lateral line canal and the similar canals located on the head constitute the lateral line system (see Exercise 7, Sense Organs).

**Pectoral fins**  Pair of flexible appendages located ventrolaterally behind the gill slits. They have short bases and are connected by a ventral cartilaginous bar.

**Pelvic fins**  Pair of flexible appendages located ventrolaterally at the caudal end of the trunk. They are smaller than the pectoral fins.

**Dorsal fins**  Two unpaired appendages located along the dorsomedial line. One is somewhat behind the level of the pectoral fin; the other is somewhat behind the level of the pelvic fin.

**Fin spines**  Located at the anterior border of each dorsal fin. They are sharp and posteriorly directed and may be considered modified scales. The presence of fin spines is reflected in the species name *acanthias* (Greek *akantha,* thorn).

**Caudal fin**  Major part of the tail, consisting of a large **dorsal lobe** and a smaller **ventral lobe.** This type of tail, in which the lobes are unequal in size, is called heterocercal.

**External nares**  Openings situated on the ventral surface of the snout. Each naris, or nostril, is divided by a flap of skin into a medial aperture, which is oval and is thought to have an excurrent function, and a lateral aperture, which is round and is presumed to be the incurrent orifice. Rather than communicating with the pharynx, the nares end blindly and thus serve only as olfactory organs (see Exercise 7, Sense Organs).

**Mouth**  Located subterminally, the mouth facilitates the seizure of prey swimming below the shark.

**Labial pouch**  Blind pocket located at each corner of the mouth. Extending forward from each labial pouch is a furrow, the **labial groove.**

33

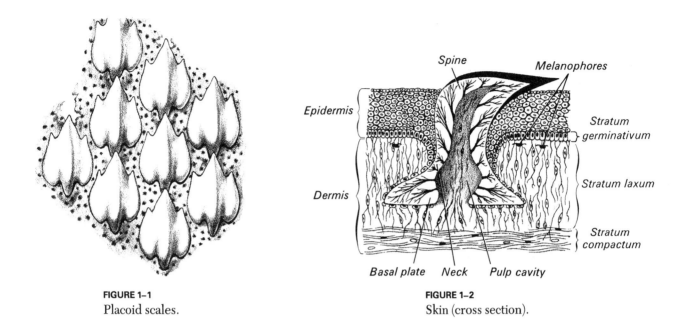

**FIGURE 1–1**
Placoid scales.

**FIGURE 1–2**
Skin (cross section).

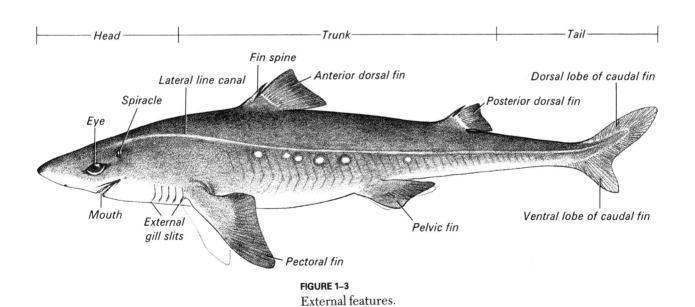

**FIGURE 1–3**
External features.

**Teeth**    These structures, thought to be modified placoid scales, are arranged in a series of rows paralleling the curvature of the jaws. The first two rows of teeth, located on the crest of each jaw, are usually the functional teeth. They are flat and sharp-edged and tend to be three-pointed. After they are worn out, the functional teeth are replaced by a forward movement of teeth from the back rows. The functional teeth serve to hold and tear food rather than to chew it.

**Endolymphatic pores**    Two small openings located on the dorsal surface of the head between the spiracles. They are the external openings of the endolymphatic ducts, which extend up from the internal ear (see Exercise 7, Sense Organs).

**Ampullae of Lorenzini**    Numerous tiny pores in the skin of the top and bottom of the head. Their presence can be observed by pressing the skin surface and noting the sites from which mucus oozes. The ampullae are special sensory endings that serve as thermoreceptors and probably also as mechanoreceptors. These endings are innervated by fibers from the seventh cranial nerve (N. VII).

**Claspers**    Modified copulatory organs located at the median border of each pelvic fin in the adult male dogfish. The claspers are elongated projections indented on one side by a deep groove.

**Cloaca**    A region located posteriorly on the ventral surface of the body that serves as a common exit for terminal products of the digestive, urinary, and reproductive systems.

**Papilla**    A cone-shaped structure, possessing a pore at its apex, that projects from the roof of the cloaca. In the male this structure is known as the **urogenital papilla** since both urine and sperm are emitted through the pore. In the female it is known as the **urinary papilla** since the pore is associated with the elimination of urine only.

**Abdominal pores**    Openings located posteriorly on the lateral edge of the cloaca of both sexes. They may serve as outlets for excess coelomic fluid.

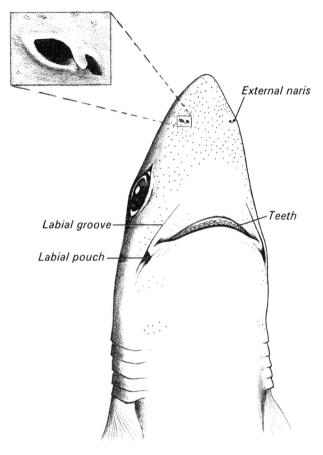

FIGURE 1–4
Head (ventral view).

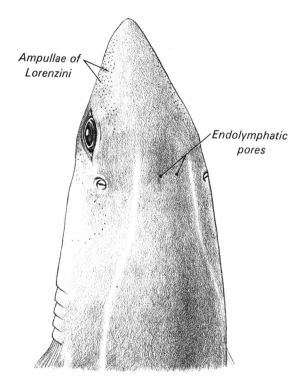

FIGURE 1–5
Head (dorsal view).

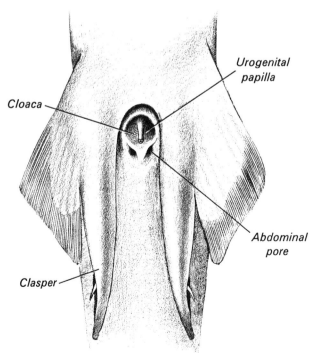

FIGURE 1–6
Male cloacal region (ventral view).

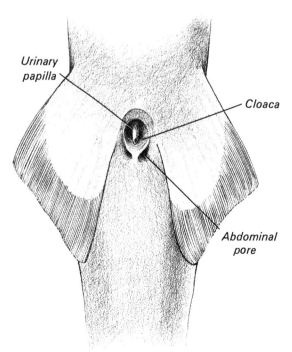

FIGURE 1–7
Female cloacal region (ventral view).

# Anatomy of the Dogfish Shark

The skeleton of the dogfish (Fig. 2-1) is entirely carti-laginous. It can be subdivided into axial and appen-dicular parts. The axial skeleton is made up of the skull and vertebral column. The skull consists of the **chondrocranium,** which encloses the brain and sense organs, and the **splanchnocranium,** which is the frame-work of the jaws and gill arches. The appendicular skeleton consists of the cartilaginous supports of the median (dorsal and caudal) fins, the pectoral and pelvic girdles, and their fins.

The skeleton of the dogfish, like that of other verte-brate forms, supports or protects the softer parts of the body, such as the brain, spinal cord, sense organs, and gills. The skeleton also facilitates locomotion—both indirectly, by providing the basis for the streamlined external body shape, and directly, by providing the structural framework for the fins. Thus in a broad sense the skeleton defines the mode of life of the dogfish.

*Prepared specimens of the skeleton should be studied. Because these skeletons are very delicate and easily damaged by careless handling, they are usually stored in glass jars that are filled with 5% or 10% formalin. It is preferable to study the specimens in their glass con-tainers. If they are removed from the containers, place them in shallow trays with enough preservative to cover them. When moving the specimens with instru-ments, take care not to perforate or disarticulate the cartilaginous skeleton.*

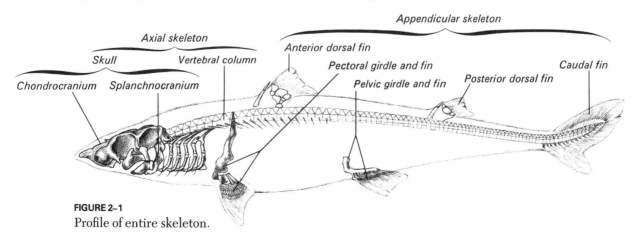

**FIGURE 2–1**
Profile of entire skeleton.

## AXIAL SKELETON

### Skull: Chondrocranium

This part of the skull consists of a single irregularly shaped cartilaginous mass, the chondrocranium, which encloses the brain. The olfactory and otic capsules are fused with the chondrocranium.

On the *dorsal* side of the chondrocranium are the following features (Fig. 2-2).

**Rostrum**  Scoop-like anterior projection of the chondrocranium enclosing the **precerebral cavity.** This cavity, which opens dorsally, is in life filled with a gelatinous material. The two large apertures located at the base of the rostrum are the **rostral fenestrae.**
**Olfactory capsules**  Rounded swellings located on either side of the base of the rostrum. Their thin walls are readily damaged during preparation and thus may be found to be incomplete.
**Orbits**  Located on both sides of the chondrocranium. Their anterior limit is demarcated by an **antorbital process** that continues backward as the **supraorbital crest** and terminates as the **postorbital process.**
**Otic capsules**  Irregularly shaped masses protecting the inner ear, the capsules are fused posteriorly to the chondrocranium. Each is bounded dorsolaterally by the **supraotic crest,** which is a backward continuation of the postorbital process. Each crest terminates in a very small posterolateral projection of the cranium.
**Epiphyseal foramen**  Located in the median line just behind the precerebral cavity. Through it passes the stalk-like epiphysis (pineal body).
**Endolymphatic fossa**  A large deep depression located in the median line in the region of the otic capsules. It contains a pair of small anterior **endolymphatic foramina** and a pair of larger **posterior perilymphatic foramina.**

On the *ventral* side of the chondrocranium are the following features (Fig. 2-3).

**Rostral carina**  Keel-shaped structure located on the mid-ventral surface of the rostrum.
**Antorbital shelf**  Ridge that demarcates the anterior boundary of the orbit.
**Infraorbital shelf**  That portion of the floor of the chondrocranium that extends back as a median ridge from the antorbital shelves.
**Basal plate**  Expanded posterior part of the chondrocranium extending back from the infraorbital shelf.
**Carotid foramen**  Located in the median line at the junction of the infraorbital shelf and basal plate.

**Postotic process**  The ring-shaped process located at each posterolateral corner of the ventral surface of the chondrocranium. The openings in these processes are known as the **postotic fenestrae.**

On the *lateral* side of the chondrocranium is one definite structure, the **optic pedicel,** and a number of foramina (Fig. 2-4).

**Optic formen**  A large opening, located anteroventrally in the orbit, that provides passage for the optic nerve (N. II).
**Oculomotor foramen**  Located just above the root of the optic pedicel, it is the opening through which the oculomotor nerve (N. III) gains entrance into the orbit.
**Trochlear foramen**  A small opening located anterodorsally and lying a short distance above the optic foramen. The trochlear nerve (N. IV) emerges into the orbit through it.
**Superficial ophthalmic foramina**  A series of small openings located parallel to and a little below the supraorbital crest. They can also be seen on the dorsal side of the chondrocranium (see Fig. 2-3). Through them pass the branches of the trigeminal nerve (N. V) and facial nerve (N. VII).
**Deep ophthalmic foramen**  Located just in front of the most anterior superficial ophthalmic foramen. A branch of the trigeminal nerve passes into the orbit through this foramen.
**Trigeminofacial foramen**  A large foramen located at the posterior border of the orbit behind the optic pedicel. Through it passes the main branches of the trigeminal and facial nerves. This opening is also known as the orbital fissure.
**Hyomandibular foramen**  Located in the otic capsule slightly behind the trigeminofacial foramen. Through it passes a branch of the facial nerve known as the hyomandibular nerve.

On the *posterior* side of the chondrocranium are the following features (Fig. 2-5).

**Foramen magnum**  Large medial opening through which the spinal cord passes.
**Occipital condyles**  Processes located ventrolaterally on either side of the foramen magnum that articulate with the first trunk vertebra.
**Vagus foramina**  A pair of openings located laterally on each side of the occipital condyles. The vagus nerve (N. X) passes through each.
**Glossopharyngeal foramina**  Another pair of openings, each of which is located near the posterolateral corners of the chondrocranium. The glossopharyngeal nerves (N. IX) pass through them.

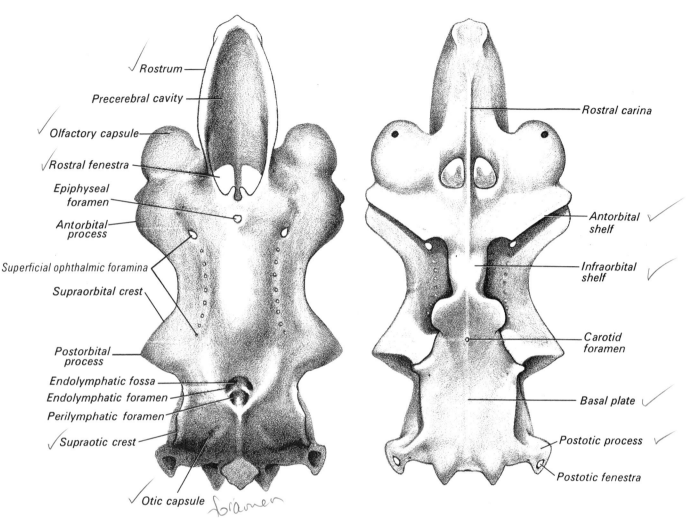

Rostrum

Precerebral cavity

Olfactory capsule

Rostral fenestra

Epiphyseal foramen

Antorbital process

Superficial ophthalmic foramina

Supraorbital crest

Postorbital process

Endolymphatic fossa
Endolymphatic foramen
Perilymphatic foramen

Supraotic crest

Otic capsule   foramen

Rostral carina

Antorbital shelf

Infraorbital shelf

Carotid foramen

Basal plate

Postotic process

Postotic fenestra

**FIGURE 2–2**
Chondrocranium (dorsal view).

**FIGURE 2–3**
Chondrocranium (ventral view).

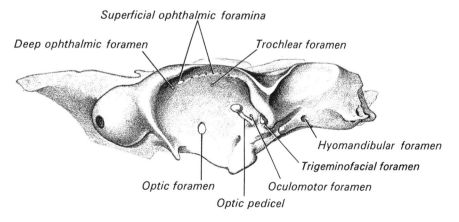

Superficial ophthalmic foramina

Deep ophthalmic foramen

Trochlear foramen

Optic foramen

Optic pedicel

Oculomotor foramen

Trigeminofacial foramen

Hyomandibular foramen

**FIGURE 2–4**
Chondrocranium (lateral view).

## Skull: Splanchnocranium (Figs. 2-6, 2-7, 2-8)

The splanchnocranium is the visceral skeleton of the skull. It consists of seven cartilaginous **visceral arches.**

**Mandibular arch**   This is the first, most anterior, and the largest of the visceral arches. It is modified into two teeth-bearing parts. The dorsal part consists of the two **palatopterygoquadrate cartilages,** which have fused in the anterior midline and have formed an arch that constitutes the upper jaw. Each of the palatopterygoquadrate cartilages bears an **orbital process,** which projects up into the orbit, and a **quadrate process,** which projects up from the lateral end of this cartilage. The ventral part of the mandibular arch consists of the two **Meckel's cartilages** fused at the midline, which together constitute the lower jaw. Arched spine-like **labial cartilages** project forward on each side at regions where the two jaws meet.

**Hyoid arch**   The second visceral arch consists of five cartilages. Laterally, a pair of short **hyomandibular cartilages** articulate with the posterior end of the chondrocranium. These articulate at their ventral ends with a pair of J-shaped **ceratohyal cartilages,** whose long arms lie beneath the chondrocranium. They are connected across the midline to a short cartilaginous bar, the **basihyal cartilage.**

**Visceral arches three to seven**   Each of these gill or branchial arches consists dorsally of **pharyngobranchial cartilages** that articulate laterally with short **epibranchial cartilages.** The general pattern for these arches is for epibranchial cartilages to connect with the following series of ventral cartilages: **ceratobranchials, hypobranchials,** and finally a single median **basibranchial.** Individual differences in the arches are these: the third arch has a single basibranchial; arches four through seven have a common basibranchial cartilage; the sixth and seventh arches also lack hypobranchials and their pharyngobranchials are fused dorsally. Cartilaginous spine-like processes known as **gill rakers** and **gill rays** project from opposite sides of the epibranchial and ceratobranchial cartilages of the gill arches.

## Vertebral Column (Figs. 2-9, 2-10)

The vertebral column consists of two distinctive types of vertebrae.

**Trunk vertebrae**   Each typically consists of a biconcave **centrum.** The small canal running longitudinally through each centrum and the diamondshaped spaces between two articulating centra are filled with the gelatinous notochord. The fused cartilaginous plates above the centrum form a **neural arch** from which projects a **neural spine.** Between adjacent spines are triangular **intercalary plates** of cartilage. The roof and sides of the vertebral or **neural canal** are formed by neural arches, neural spines, and intercalary plates, and the floor is formed by the tops of centra. The spinal cord lies within the neural canal. Dorsal roots of each spinal nerve emerge through foramina in the intercalary plates, and ventral roots emerge through foramina in the neural arches. From the ventrolateral border of each centrum project **transverse processes** to which the rib cartilages are attached.

**Caudal vertebrae**   Each consists of a centrum, neural arches, and a neural canal. Attached to the ventral surface of each centrum is an additional pair of fused cartilaginous plates that form the **hemal arch.** These arches form the **hemal canal,** which contains the caudal artery and vein. Projecting ventrally from each hemal arch is a **hemal spine.**

## APPENDICULAR SKELETON

The appendicular skeleton in the dogfish consists of the median fin cartilages and the pectoral and pelvic girdles and their fins. The girdles in the dogfish do not articulate with the vertebral column.

## Median Fin Cartilages (Figs. 2-11, 2-12)

These consist of the skeletal supports for the two dorsal fins and a caudal fin.

**Dorsal fins**   Each fin has as its skeletal framework a **basal cartilage,** to which are attached **radial cartilages** that bear **ceratotrichia** (fibrous dermal fin rays). A **fin spine** is located at the anterior end of the basal cartilage.

**Caudal fin**   This fin consists of the many caudal vertebrae whose hemal arches are elongated. It is also supported by ceratotrichia.

## Pectoral Girdle and Fins (Fig. 2-13)

The anterior paired fins are supported by a U-shaped cartilage located just posterior to the splanchnocranium. The ventral part of the girdle consists of the **coracoid bar,** from which a **scapular cartilage** extends dorsally on each side. The latter articulates with the skeleton of the pectoral fin at the **glenoid surface;** a

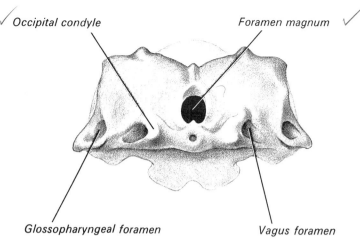

Occipital condyle

Foramen magnum

Glossopharyngeal foramen

Vagus foramen

**FIGURE 2–5**
Chondrocranium (posterior view).

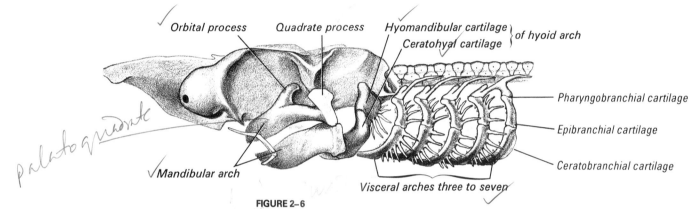

Orbital process

Quadrate process

Hyomandibular cartilage
Ceratohyal cartilage
} of hyoid arch

Pharyngobranchial cartilage

Epibranchial cartilage

Ceratobranchial cartilage

Palatoquadrate

Mandibular arch

Visceral arches three to seven

**FIGURE 2–6**
Splanchnocranium (lateral view).

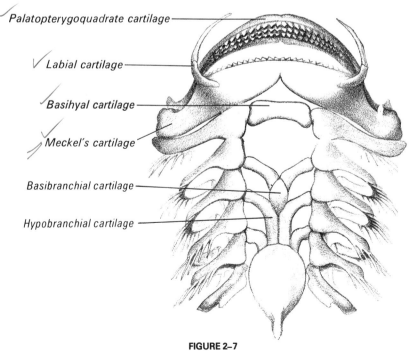

Palatopterygoquadrate cartilage

Labial cartilage

Basihyal cartilage

Meckel's cartilage

Basibranchial cartilage

Hypobranchial cartilage

**FIGURE 2–7**
Splanchnocranium (ventral view).

medially projecting **suprascapular cartilage** is attached to its free end.

The skeleton of the pectoral fin consists of three **basal cartilages:** the smallest and outermost cartilage is the **propterygium;** adjacent to it is a wide triangular cartilage, the **mesopterygium;** the innermost cartilage, the **metapterygium,** is long and narrow. Small **radial cartilages,** arranged in rows, articulate with the basals. Distal to the radials are **ceratotrichia** that provide additional support to the fin. The basal and radial cartilages are collectively known as **pterygiophores.**

## Pelvic Girdle and Fins (Figs. 2-14, 2-15)

This girdle consists of a transverse **ischiopubic bar,** from the end of which a short **iliac process** extends dorsally. The pelvic fin has two basal cartilages. One, the long curved **metapterygium,** articulates at the **acetabular surface** with the bar. The other, the small **propterygium,** lies anterior to the metapterygium. The metapterygium articulates with numerous radial cartilages, which in turn bear the ceratotrichia. Modified radials form the **clasper cartilages** in the male.

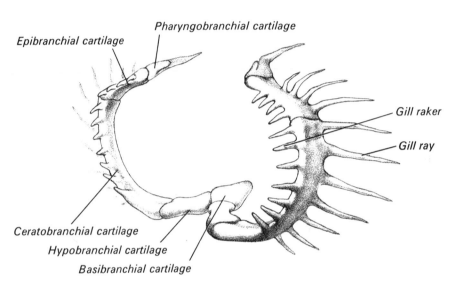

**FIGURE 2–8**
Third visceral arch.

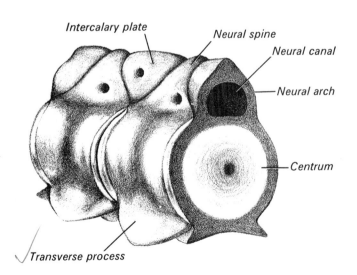

**FIGURE 2–9**
Trunk vertebrae.

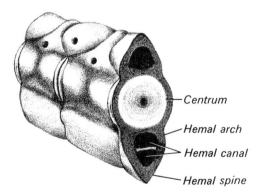

**FIGURE 2–10**
Caudal vertebrae.

43

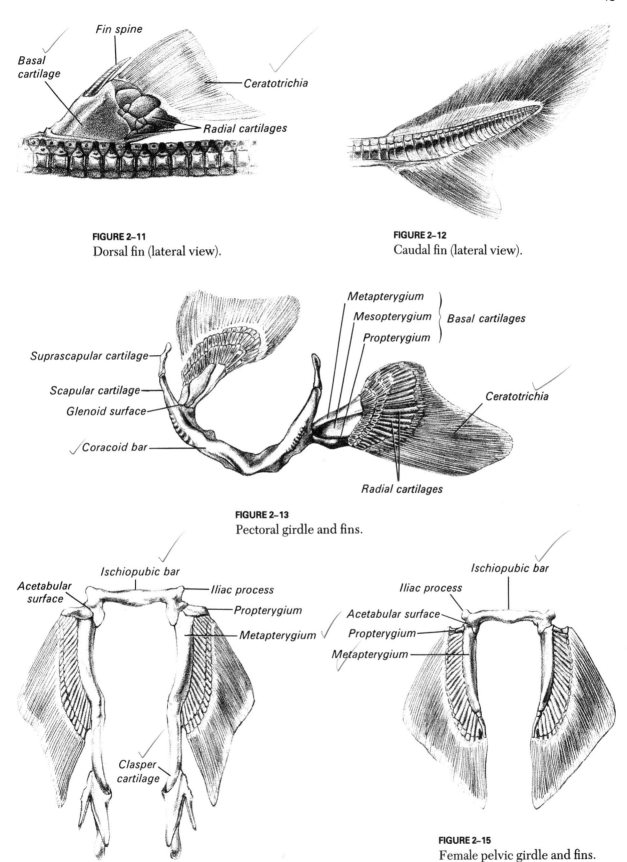

**FIGURE 2-11**
Dorsal fin (lateral view).

**FIGURE 2-12**
Caudal fin (lateral view).

**FIGURE 2-13**
Pectoral girdle and fins.

**FIGURE 2-14**
Male pelvic girdle and fins.

**FIGURE 2-15**
Female pelvic girdle and fins.

# Anatomy of the Dogfish Shark

The muscular system of the dogfish shark is simple in comparison to that of higher vertebrates. For convenience, the somatic muscles will be divided into four regional groups. (A fifth group, the eye muscles, is discussed in Exercise 7, Sense Organs.) A description of the locations of the muscles will follow. Their origins, insertions, and actions are summarized in Table 3-1 at the end of the exercise.

## TRUNK AND TAIL MUSCULATURE (Fig. 3-1)

*Remove a two-inch segment of tail skin from the mid-dorsal to the mid-ventral line, make a clean section across the tail, and identify the following muscle bundles and tissues.*

**Myomeres**  These metamerically arranged muscle bundles exhibit a zigzag pattern.

**Myosepta**  The white connective tissue septa between the myomeres.

**Horizontal septum**  A horizontal connective-tissue partition extending laterally from both sides of the centrum. This septum divides the axial musculature into dorsal and ventral components.

**Epaxial muscles**  The muscle bundles lying above the horizontal septum.

**Hypaxial muscles**  The muscle bundles lying below the septum.

**Vertical septa**  Connective-tissue partitions extending vertically from the top of the vertebrae to the mid-dorsal line of the body and from the bottom of the vertebrae to the mid-ventral line. These septa, together with the vertebrae, divide the axial musculature into right and left halves.

**Linea alba**  A white line of connective tissue present on the mid-ventral body surface, which is the site of attachment of the ventral vertical septum (see Fig. 3-3).

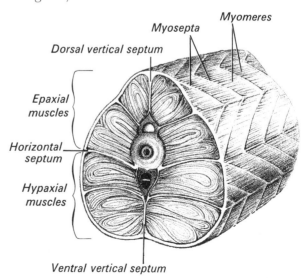

**FIGURE 3–1**

Tail musculature.

## APPENDICULAR MUSCULATURE (Fig. 3-2)

*Remove the skin of a pectoral and/or pelvic fin and identify the following.*

**Levator muscles**    Elevating muscles that are present on the dorsal surface of the fin.
**Depressor muscles**    Lowering muscles located on the ventral surface of the fin.
**Radial muscles**    Located on the sides of the immovable (but flexible) dorsal fins.

## BRANCHIOMERIC MUSCULATURE

*From one side of the head, remove all of the skin between the spiracle and pectoral girdle. Take care to avoid stripping off the superficial muscles.*

These superficial muscles are derived from splanchnic mesoderm rather than myomeres. They are paired and lie anterior to the pectoral fin, where they control the movement of the gill arches and jaws. They may be divided into two groups, constrictors and levators.

### Constrictor Muscles (Figs. 3-2, 3-3)

These are six sets of superficial muscles lying between the eye and the pectoral fin. They compress the gill pouches, forcing water out, and also close the mouth. The constrictors are divided into dorsal and ventral components.

The dorsal constrictors are as follows.

**First dorsal constrictor**    This is made up of two muscles: the **spiracular** (craniomaxillary), a slender muscle located just anterior to the spiracle, and the **mandibular adductor** (quadratomandibular), a thick muscle located below the spiracle and behind the eye.
**Second dorsal constrictor** (or dorsal hyoid constrictor)    A broad muscle located behind the spiracle. The cranial part of this muscle, which lies dorsal to the mandibular adductor, is sometimes considered a separate muscle, the **epihyoid**, or hyomandibular levator.
**Third to sixth dorsal constrictors**    These are sequentially arranged, and each is separated from the next by a vertical raphe.

The ventral constrictors are as follows.

**First ventral constrictor** (or intermandibular)    This is a flat, ventrally located muscle that lies behind the mandibular adductor.
**Second ventral constrictor** (or ventral hyoid constrictor)    This muscle is continuous with its dorsal counterpart. It passes around to the undersurface of the head where it has a cranial extension, the **interhyoid**, which is sometimes considered a separate muscle. (The interhyoid may be exposed by cutting across the intermandibular muscle and separating the two. They may be distinguished by the direction of the fibers, the intermandibular being oblique, the interhyoid being more nearly transverse.)
**Third to sixth ventral constrictors**    These are arranged sequentially and are aligned with their dorsal counterparts.

### Levator Muscles (Fig. 3-2)

These muscles elevate the mandibular, hyoid, and gill arches.

**First levator** (or palatoquadrate levator)    A small muscle located in front of and slightly deeper than the spiracular muscle.
**Second levator** (or hyoid levator)    This muscle lies directly under and fuses with the cranial portion of the second dorsal constrictor (epihyoid). This muscle is therefore not readily discernible.
**Third to sixth levators** (or cucullaris)    These form a single triangular muscle mass between the third to sixth dorsal constrictors and epaxial musculature.

## HYPOBRANCHIAL MUSCULATURE (Fig. 3-3)

These muscles are derived from the hypaxial parts of the anterior trunk myomeres. They are located between the lower jaw (Meckel's cartilages) and the pectoral girdle and anchor the ventral parts of the splanchnocranium.

*On the ventral surface a common coracoarcual muscle is evident on each side of the midline. Cut across and reflect the intermandibular and interhyoid muscles of one side to disclose the median coracomandibular muscle. Lateral and just dorsal to it is the coracohyoid muscle. If this muscle is transected and the cut ends deflected to one side, the coracobranchials, a muscle group, will be seen.*

**Common coracoarcuals**    Short muscles that together form a triangle whose base is at the pectoral girdle.
**Coracomandibular**    A median longitudinal muscle

hyomandibular
levator/hyoid
levator/2nd
levator

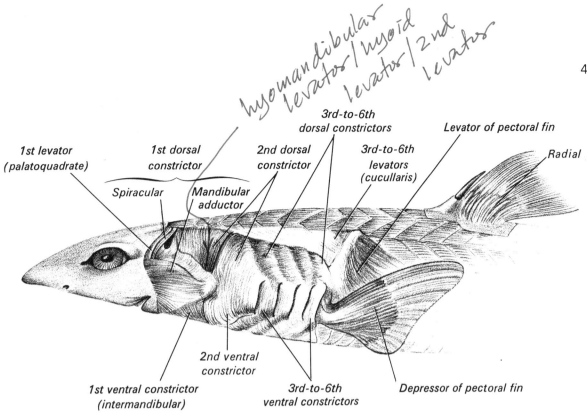

1st levator
(palatoquadrate)

1st dorsal
constrictor

2nd dorsal
constrictor

3rd-to-6th
dorsal constrictors

3rd-to-6th
levators
(cucullaris)

Levator of pectoral fin

Radial

Spiracular

Mandibular
adductor

2nd ventral
constrictor

1st ventral constrictor
(intermandibular)

3rd-to-6th
ventral constrictors

Depressor of pectoral fin

**FIGURE 3–2**
Appendicular and branchiomeric musculature.

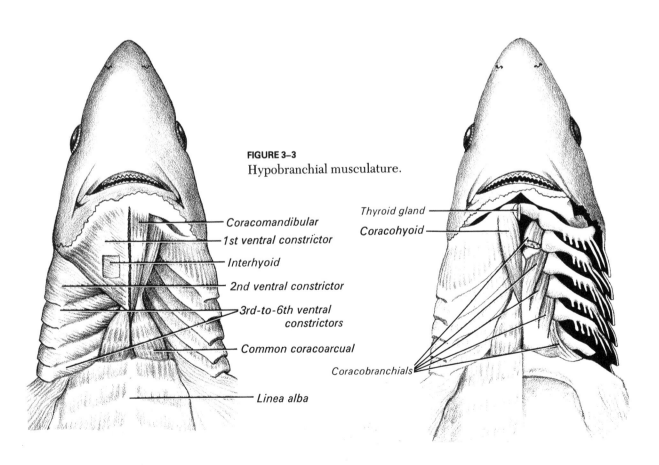

**FIGURE 3–3**
Hypobranchial musculature.

Coracomandibular

1st ventral constrictor

Interhyoid

2nd ventral constrictor

3rd-to-6th ventral
constrictors

Common coracoarcual

Linea alba

Thyroid gland

Coracohyoid

Coracobranchials

lying dorsal to the first ventral constrictor and common coracoarcual. Although it appears to be single, this muscle is probably paired. The dark mass located under its anterior end is the **thyroid gland.**

**Coracohyoids**    A pair of strong, obliquely running slender muscles located dorsal and lateral to the coracomandibular.

**Coracobranchials**    A muscle group extending obliquely laterally and inserting on the ventral ends of the visceral arches by a series of individual slips.

**TABLE 3-1**

Summary of musculature of the dogfish shark.

| Muscle | Origin | Insertion | Action |
|---|---|---|---|
| *Appendicular Muscles* | | | |
| Pectoral levator | Scapular process | Fin pterygiophores | Elevates pectoral fin |
| Pectoral depressor | Coracoid bar | Fin pterygiophores | Depresses pectoral fin |
| Pelvic levator | Iliac process | Fin pterygiophores | Elevates pelvic fin |
| Pelvic depressor | Ischiopubic bar | Fin pterygiophores | Depresses pelvic fin |
| Radial | Basal cartilage | Ceratotrichia | Abducts dorsal fin |
| *Branchiomeric Muscles* | | | |
| **Dorsal Constrictors:** | | | |
| First | | | |
|   Spiracular | Otic capsule | Palatoquadrate cartilage | Elevates the cartilage |
|   Mandibular adductor | Palatoquadrate cartilage | Mandibular arch | Closes mouth |
| Second | Otic capsule | Hyomandibular cartilage | Compresses gill chambers |
| Third to Sixth | Vertical raphe | Vertical raphe | Compress gill chambers |
| **Ventral Constrictors:** | | | |
| First | Mid-ventral raphe | Mandibular arch | Elevates floor of mouth |
| Second | Hyomandibular cartilage | Ceratohyal cartilage | Compresses gill chambers |
| Third to Sixth | Vertical raphe | Vertical raphe | Compress gill chambers |
| **Levators:** | | | |
| First (Palatoquadrate) | Otic capsule | Palatoquadrate cartilage | Raises upper jaw |
| Second (Hyoid) | Fused to epihyoid muscle | Fused to epihyoid muscle | Elevates hyoid arch |
| Third to Sixth (Cucullaris) | Fascia of epaxial muscle | Scapular cartilage | Elevate scapular cartilage |
| *Hypobranchial Muscles* | | | |
| Common coracoarcual | Coracoid bar | Mandibular adductor | Helps to open mouth |
| Coracomandibular | Coracoid bar | Mandibular arch | Opens mouth |
| Coracohyoid | Coracoid bar | Basihyal cartilage | Helps in feeding |
| Coracobranchial | Coracoid bar | Visceral cartilage | Helps in swallowing |

# Digestive and Respiratory Systems

# Anatomy of the Dogfish Shark

In the dogfish shark the coelom, or body cavity, consists of the pericardial cavity (see Exercise 5, Circulatory System), which lies anterior to the pectoral girdle, and the pleuroperitoneal cavity posterior to it. The two cavities are separated by a partition, the transverse septum.

The visceral organs that collectively make up the digestive system are suspended in the pleuroperitoneal cavity. The smooth and shiny membrane lining this cavity and its organs, as well as the surface of the membranes suspending the organs, is known as the **peritoneum.** The lining of the inner surface of the body wall is known as the **parietal peritoneum;** that enveloping the external surface of an organ (viscus) is called the **visceral peritoneum.** Double layers of peritoneum suspending a digestive organ, i.e., extending between the visceral and parietal peritoneum are called a **mesentery.** Such a duplication connecting two organs is an **omentum.** Narrow mesenteries, or omenta, are **ligaments.**

The basic pattern of distribution of the peritoneum is shown in Fig. 4-1. The primitive gut is suspended between dorsal and ventral mesenteries. In a more advanced embryonic stage (Fig. 4-2) the liver and spleen are interposed in the mesenteries and, as a result, form mesenteric ligaments and omenta. In the adult most of the lower two-thirds of the ventral mesentery disappears and its remnant is evident as the **falciform ligament** and the **lesser omentum.**

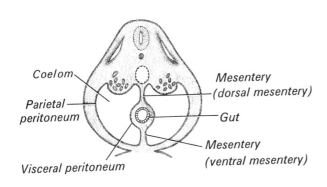

**FIGURE 4–1**
Arrangement of peritoneum in primitive stage.

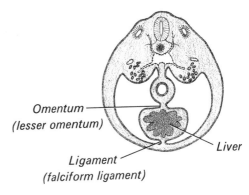

**FIGURE 4–2**
Differentiation of mesenteries.

## DIGESTIVE SYSTEM (Fig. 4-3)

After food has been grasped by the teeth of the dogfish shark, it passes into the buccal cavity of the mouth, then through the pharynx, which is primarily a respiratory center, and into the esophagus.

*Make a longitudinal incision just to the left of the mid-ventral line from the left side of the cloaca forward as far as the pectoral girdle. Make transverse cuts from the anterior and posterior ends of the longitudinal incision. Fasten the resultant flaps down with hooks and chains to provide good exposure of the peritoneal cavity. If desired, cut open the digestive tract in order to observe the characteristics of the internal surface.*

### Digestive Organs

**Esophagus** This is the upper part of the J-shaped organ that extends down from the transverse septum. On the internal surface are numerous finger-like projections, or **papillae.**

**Stomach** This is the lower part of the J-shaped organ and is made up of two parts. The larger **cardiac portion,** which is continuous with the esophagus, is characterized internally by longitudinal folds, or **rugae.** The smaller segment of the stomach is the **pyloric portion,** which ends in the constricted **pylorus,** where the pyloric sphincter is located.

**Small intestine** This continuation of the digestive tract from the pylorus is made up of two parts. The first segment, which is short and narrow, is the **duodenum.** The second, longer part of the small intestine is the **ileum.** It is thicker and encloses the **spiral valve,** an internal structure that increases the intestinal surface area for absorption.

**Colon** The short portion of the digestive tract extends from the ileum to the point of entrance of the **rectal gland.** The rectal gland is a larger finger-like structure that projects cranially and is thought to remove excess salt from the circulation.

**Rectum** The last portion of the digestive tract extends from the colon to the anus. The anus projects into the cloaca.

### Digestive Glands

**Liver** Consists of long **right** and **left lobes** and a short **median lobe.** Along the right edge of the latter is the thin-walled, green **gall bladder.** The **common bile duct** extends from the anterior end of the gall bladder down to the dorsal side of the duodenum, which it perforates. The bile is conducted to the gall bladder from its site of formation in the liver through **hepatic ducts** too minute to be seen.

**Pancreas** Consists of two distinct portions: a flat round **ventral lobe** lying in the curve of the duodenum, and a long narrow **dorsal lobe** located posterodorsally to the duodenum and stomach. The narrow **pancreatic duct,** which can be found by carefully picking away the pancreas bit by bit, extends from the posterior margin of the ventral pancreatic lobe to the duodenum.

**Spleen** This dark triangular organ, which is molded against the outer curvature of the stomach, is not a digestive gland but belongs to the circulatory system. However, since it is such a conspicuous part of the abdominal viscera, it is mentioned at this point.

### Mesenteries

The parietal peritoneum can be seen lining the pleuroperitoneal cavity. The visceral peritoneum adheres to the outer surface of the digestive organs. The following mesenteries should be identified.

**Falciform ligament** This short double layer of peritoneum extends between the anterior end of the liver and the mid-ventral body wall. It is also known as the suspensory ligament of the liver, or as the mesohepar.

**Lesser omentum** This part of the ventral mesentery extends between the liver and neighboring border of the stomach and duodenal region. It consists of two parts, the **hepatogastric ligament,** bridging the space between the stomach and liver, and the **hepatoduodenal ligament,** extending between the liver and duodenum. Along the free border of the latter ligament pass the common bile duct and two major blood vessels.

**Mesogaster** Extends between the stomach and mid-dorsal body wall.

**Gastrosplenic ligament** Mesenteric connection between the stomach and spleen.

**Mesentery** The mesentery proper, or mesointestine, is that part of the dorsal mesentery extending between the greater part of the small intestine and the mid-dorsal body wall.

**Mesorectum** This is the part of the mesentery that extends from the rectum and rectal gland to the mid-dorsal line and suspends these organs.

## RESPIRATORY SYSTEM

Water taken into the mouth and pharynx bathes the gill filaments as it passes to the outside. The respira-

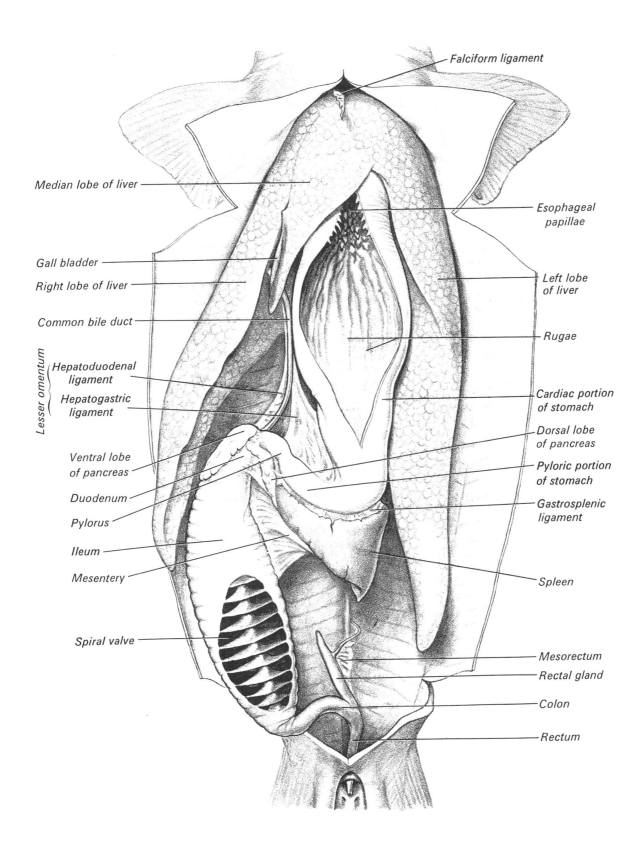

**FIGURE 4–3**
Digestive system (blood vessels omitted).

tory exchange of carbon dioxide and oxygen takes place at the surface of the gill filaments, which are parallel folds of richly vascular tissue suspended from each side of a gill septum.

*Insert one blade of a pair of scissors into the left corner of the mouth and cut the body wall along a line that runs posteriorly under the gill slits as far as the pectoral girdle. To expose a gill completely, cut dorsally and ventrally from the ends of two adjacent gill slits.*

## Mouth (Fig. 4-4)

**Buccal cavity**   The chamber of the mouth is characterized by transverse mucosal folds.

**Teeth**   Present as a series of similar structures (homodont teeth) arranged in rows along the border of each jaw. Other rows of replacement teeth lie behind them.

**Tongue**   This small organ forms the floor of the mouth and in the shark is immovable.

## Pharynx (Figs. 4-4 to 4-7)

**Pharyngeal cavity**   The chamber of the pharynx is characterized by longitudinal mucosal folds.

**Spiracle**   This structure is homologous with the first (hyoid) gill slit and is located on each side in the anterolateral wall of the pharynx. It is rounded rather than elongated.

**Internal gill slit**   Five pairs of elongated apertures located further back along the pharyngeal wall each open into a gill chamber.

**Gill chamber**   The cavity lying between an internal and an external gill slit. It is also known as a branchial chamber, or gill pouch.

**Holobranch**   An entire gill. This is an arch-shaped, respiratory-tissue partition located between each pair of adjacent gill chambers. A holobranch consists of three elements: the gill, or **interbranchial septum,** which is the central framework that extends outward to the body wall, and a pair of vascular folds, the half-gills, or **demibranchs,** on either side of the septum. In the dogfish shark there are nine demibranchs on each side. These are arranged as four holobranchs and one demibranch, the demibranch being located anterior to the first gill chamber. There is no demibranch on the posterior wall of the last gill chamber. The spiracle has only a minute demibranch, or **pseudobranch,** on its anterior wall.

**Gill filaments**   These thin folds, or plates, known also as gill or primary lamellae, are arranged in a row and give to the demibranch its corrugated appearance. Secondary lamellae extend perpendicularly along the surface of the primary lamellae. These secondary lamellae are small, closely packed folds that can be seen with a hand lens (see Fig. 10-8).

**Gill rays**   Cartilaginous spines that support the interbranchial septa.

**Gill arch**   This is the main skeletal framework of each gill. Gill rays and gill rakers extend from opposite sides of the epibranchial and ceratobranchial cartilages of these visceral arches.

**Gill rakers**   Cartilage-supported projections from the gill arches that face the pharynx. They protect the gills by screening them from food particles in the water. The particles are diverted backward toward the esophagus.

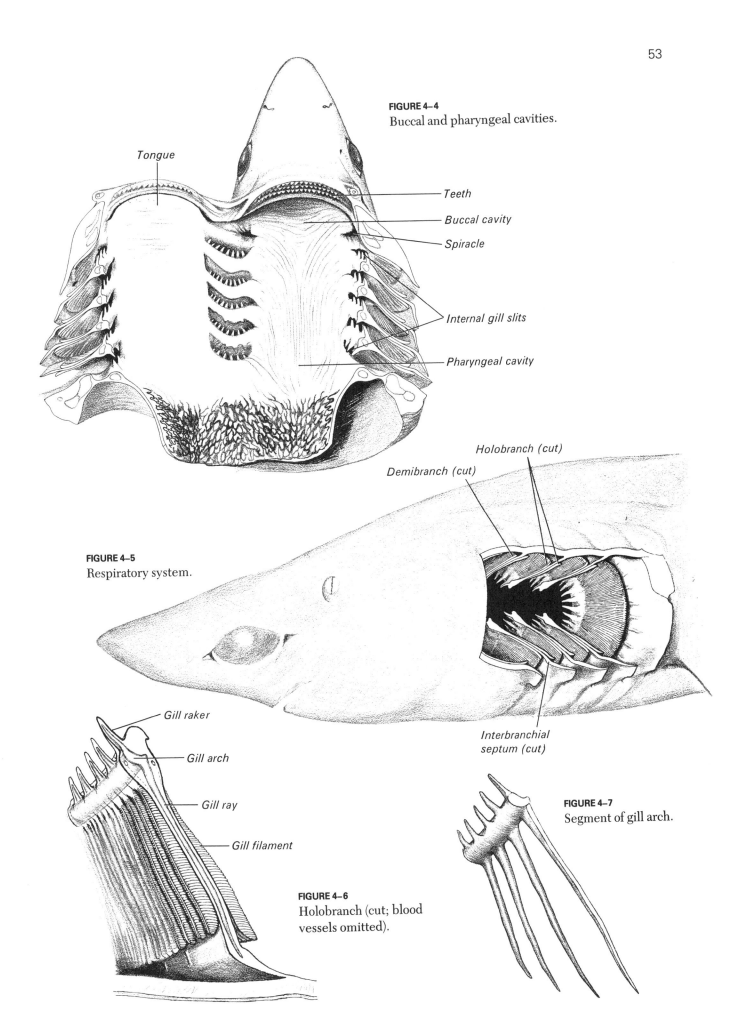

53

**FIGURE 4–4**
Buccal and pharyngeal cavities.

Tongue

Teeth

Buccal cavity

Spiracle

Internal gill slits

Pharyngeal cavity

Holobranch (cut)

Demibranch (cut)

**FIGURE 4–5**
Respiratory system.

Interbranchial septum (cut)

Gill raker

Gill arch

Gill ray

Gill filament

**FIGURE 4–7**
Segment of gill arch.

**FIGURE 4–6**
Holobranch (cut; blood vessels omitted).

# Anatomy of the Dogfish Shark

## HEART

### Exterior (Fig. 5-1)

*Remove the remaining skin from the ventral surface of the head between the mouth and pectoral girdle. Expose the pectoral girdle by teasing away some of the overlying muscle tissue and continue the mid-ventral incision cranially from the abdomen through the center of the pectoral girdle. Remove the muscles that lie just in front of the pectoral girdle until a membrane is reached. Cut through the membrane to expose the cavity in which the heart lies.*

**Pericardium**  This is the glistening smooth membrane that lines both the heart and its cavity. As the **parietal pericardium**, it lines the pericardial cavity that surrounds the heart. It is continuous with the **visceral pericardium** that forms the intimate covering of the heart musculature.

**Pericardial cavity**  The very narrow space between the heart and parietal pericardium. It is, in life, filled with serous fluid secreted by the parietal pericardium.

**Atrium**  Large, thin-walled, single-chambered sac that lies anterior to the sinus venosus and dorsal to the ventricle.

**Ventricle**  Thick-walled, single-chambered sac, on whose ventral surface are paired **coronary arteries.** (Veins corresponding to these arteries empty into the coronary sinus, but they cannot be seen.) The ventricle extends anteriorly as a muscular tube, the **conus arteriosus.**

**Sinus venosus**  Thin-walled triangular sac, the base of which is applied to the transverse septum. This sac may be readily seen by lifting up the posterior end of the ventricle.

### Interior (Fig. 5-2)

*Since exposure of the interior of the heart causes a certain amount of damage to the sinus venosus, it is advisable to defer this part of the exercise until after the systemic veins have been studied. The valves of the heart can best be seen after the intact organ has been excised from the pericardial cavity. To do this, first cut across the ventral aorta behind its posterior most pair of branches and then cut completely through the sinus venosus just posterior to the atrium. Expose the valves of the heart in the excised specimen by cutting through the wall of the conus arteriosus with a longitudinal incision and then making a similar cut through the dorsal atrial wall. Then wash out the atrial chamber.*

**Sinoatrial aperture**  Opening between the sinus venosus and atrium. It is guarded by a **sinoatrial valve.**

**Atrioventricular aperture**  Opening between the atrium and ventricle. It is guarded by an **atrioventricular valve.**

**Semilunar valves**  These are pocket-like valves located along the wall of the conus arteriosus.

## ARTERIAL SYSTEM

*Pick away the muscle and connective tissue around the ventral aorta to expose it and its branches more clearly. These blood vessels are present as bilateral pairs unless otherwise indicated.*

### Afferent Branchial Arteries (Figs. 5-1, 5-3)

**Ventral aorta** This unpaired, thin-walled uninjected vessel is a forward continuation of the conus arteriosus.

**Afferent branchial arteries** Three primary pairs arise from the ventral aorta. The posterior pair arises near the conus arteriosus and each divides near its origin. The next pair arises a little further forward and remains undivided. The anterior pair represents the terminal branches of the ventral aorta and arises as a result of its bifurcation. Each of the anterior branchial arteries divides upon reaching the gill region. Thus, five pairs of afferent branchial arteries, numbered from 1 to 5 in an anterior-to-posterior direction, are formed. These vessels carry unoxygenated blood from the heart to the gills.

### Efferent Branchial Arteries (Fig. 5-3)

*If the pharynx is unopened (see Exercise 4, Digestive and Respiratory Systems), then cut from the right angle of the mouth along a line that passes through the ventral parts of the gill clefts as far back as the pectoral girdle. Carefully strip away the mucous membrane that covers the roof of the mouth and pharynx and expose the blood vessels beneath it.*

**Efferent branchial arteries** These are four pairs of arteries, numbered 1 to 4, that extend posteromedially, carrying blood away from the gills. Each arises from the union of a **pretrematic artery** from the anterior side of a gill slit with a **posttrematic artery** from the posterior side. The trematic arteries unite at the dorsal and ventral margins of the gill slit and together form an **arterial collector loop** that encircles the gill slit. Several cross-connecting branches extend between the pre- and posttrematic arteries of each gill at the base of its interbranchial septum. The efferent branchial arteries all empty into a midline vessel, the **dorsal aorta.**

**External carotid artery** This small vessel arises on each side near the ventral end of the first collector loop and runs cranially along the lower jaw.

**Hyoidean epibranchial artery** This vessel, known also as the efferent hyoidean artery, arises from the

dorsal end of the first collector loop. It runs cranially across the roof of the mouth, turns toward the midline, and terminates by dividing into two branches. One branch, the **stapedial artery,** passes directly into the orbital region of the cartilaginous skull to supply the recti muscles and adjacent structures. The other, the **internal carotid artery,** veers toward the midline along the ventral surface of the skull to meet and unite briefly with the internal carotid artery from the other side. The united internal carotid arteries enter the cranial cavity, where they separate almost immediately and turn laterally and forward to supply the brain. Shortly after entering the chondrocranium, the internal carotid joins the efferent spiracular artery (see below).

**Afferent spiracular artery** Originates at about the middle of the pretrematic artery of the first collector loop. This vessel runs cranially for a short distance and then turns dorsally to disappear in the spiracle, where it branches to supply the lamellae of the pseudobranch. (The afferent branchial artery can frequently be found by removing the skin posterior to the spiracle and then carefully dissecting away some of the underlying tissue.)

**Efferent spiracular artery** Formed by reunion of the lamellar branches of the afferent spiracular artery. This vessel runs anteromedially into the chondrocranium, where it joins the internal carotid. It may also be considered a branch of the internal carotid. In its course the efferent spiracular artery passes ventral to the stapedial artery, where it gives rise to the large **ophthalmic artery.** The ophthalmic vessel passes into the orbit and supplies the eye. (The efferent spiracular may be seen more clearly by cutting away the cartilage midway between the hyoidean epibranchial artery and the upper jaw.)

**Hypobranchial artery** This vessel arises from the base of the second collector loop and may receive contributions from the bases of the other collector loops. The hypobranchial courses posteriorly beneath the pharynx to the level of the anterior end of the conus arteriosus, where it divides into two branches. The dorsal branch, the **pericardial artery,** passes back to supply the roof of the pericardial cavity. The proximal parts of the pericardial arteries of each hypobranchial artery are united across the midline, dorsal to the conus arteriosus, by a transverse vessel. The ventral branch of each hypobranchial artery is the **coronary artery.** Together, the paired coronary arteries supply the musculature of the ventricle of the heart.

**Esophageal artery** A vessel that arises from the second efferent branchial artery near its point of origin from the dorsal end of the arterial collector loop. The esophageal artery runs posteriorly, giving off

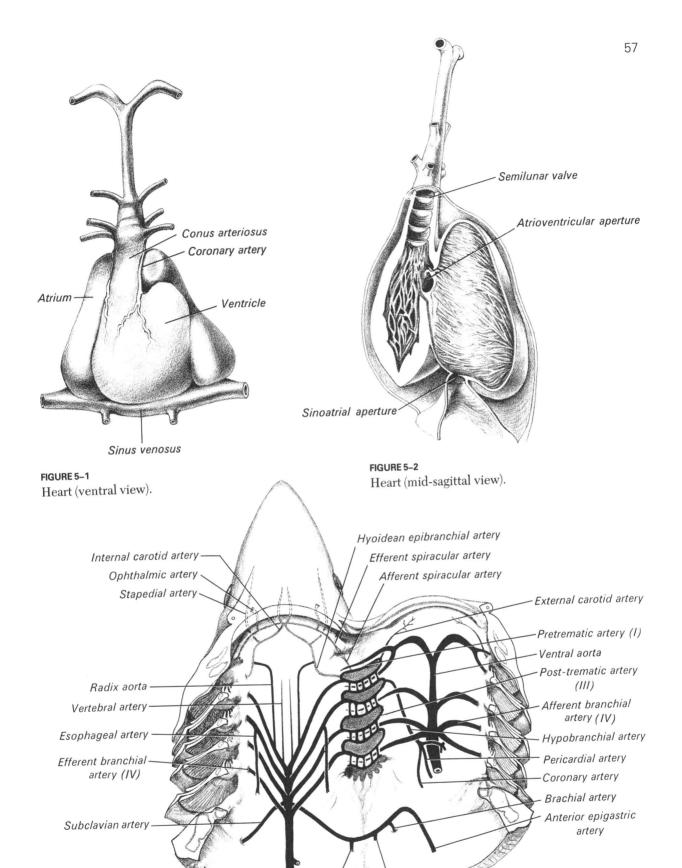

57

**FIGURE 5–1**
Heart (ventral view).

Conus arteriosus
Coronary artery
Atrium
Ventricle
Sinus venosus

**FIGURE 5–2**
Heart (mid-sagittal view).

Semilunar valve
Atrioventricular aperture
Sinoatrial aperture

**FIGURE 5–3**
Efferent branchial and associated arteries.

Internal carotid artery
Ophthalmic artery
Stapedial artery
Radix aorta
Vertebral artery
Esophageal artery
Efferent branchial artery (IV)
Subclavian artery
Dorsal aorta
Lateral artery
Ventrolateral artery

Hyoidean epibranchial artery
Efferent spiracular artery
Afferent spiracular artery
External carotid artery
Pretrematic artery (I)
Ventral aorta
Post-trematic artery (III)
Afferent branchial artery (IV)
Hypobranchial artery
Pericardial artery
Coronary artery
Brachial artery
Anterior epigastric artery

branches to the dorsal surface of the pharynx, and further caudally to supply the esophagus.

**Radix aorta**   Commences at the first efferent branchial, near the site where the latter joins its corresponding vessel. It extends forward to join the hyoidean epibranchial artery. The radices aortae are also known as the paired dorsal aortae since they are the anterior extensions of the unpaired dorsal aorta.

**Vertebral artery**   This slender vessel, which arises from the first efferent branchial artery, is only occasionally evident. When seen, it is located on each side just medial to the radix aorta.

**Subclavian artery**   A vessel arising from each side of the dorsal aorta between the third and fourth efferent branchial arteries. Each subclavian artery curves laterally toward the pectoral fin, giving off several branches in the following sequence: a small lateral artery, a ventrolateral artery, and an anterior epigastric artery. The first two branches are not always clearly evident. The **lateral artery** proceeds caudally in the body wall parallel to the lateral line. The **ventrolateral artery** also courses posteriorly to supply its adjacent muscles. The prominent **anterior epigastric artery** runs along the inside of the body wall approximately midway between the lateral and mid-ventral lines, giving off segmental arteries to the myomeres. The anterior epigastric arteries anastomose with corresponding branches from the iliac arteries in the posterior part of the pleuroperitoneal cavity. Beyond the point of origin of the anterior epigastric artery, the subclavian artery continues into the pectoral fin as the **brachial artery.**

## Dorsal Aorta and Its Branches (Fig. 5-4)

*Identify the arteries of the peritoneal cavity, taking care not to destroy any of the associated veins. Expose again the peritoneal cavity, which was opened to study the digestive system (see Exercise 4, Digestive and Respiratory Systems).*

**Dorsal aorta**   This large arterial trunk is formed beneath the heart by the four pairs of efferent branchial arteries. It descends to penetrate the transverse septum and to enter the pleuroperitoneal cavity, where it gives off a series of unpaired and paired abdominal branches.

The unpaired branches are these:

**Celiac artery**   This vessel arises just past the site at which the aorta passes into the pleuroperitoneal cavity. The celiac artery gives off small branches to the gonads, the esophagus, and the cardiac end of the stomach, and then passes into the lesser omentum. There it gives rise to the gastrohepatic artery and continues on as the pancreaticomesenteric artery (see below).

**Gastrohepatic artery**   This vessel divides almost immediately into the **gastric artery,** which ramifies and passes laterally to supply the muscular wall of the stomach. The thinner **hepatic artery** curves upward sharply and, along with the bile duct, enters the substance of the liver, which it supplies.

**Pancreaticomesenteric artery**   This vessel, known also as the intestinopyloric trunk, runs caudally behind the pylorus, where it gives off three branches, and continues on as the anterior intestinal artery (see below). Its three branches are the **duodenal artery,** a short vessel supplying the duodenum; the **pyloric artery,** which nourishes the pylorus and adjacent parts of the stomach; and the **intraintestinal artery,** which penetrates the ventral wall of the intestine and passes down the core of the spiral valve.

**Anterior intestinal artery**   A continuation of the pancreaticomesenteric artery, this runs down the right side of the valvular intestine. Along its course it gives off **annular arteries** at the points of attachment of the spiral valve to the intestinal wall.

**Posterior intestinal artery**   This vessel arises from the dorsal aorta near the posterior end of the mesentery and supplies the left side of the intestine by annular arteries, which fuse with those from the other side.

**Gastrosplenic artery**   This artery, known also as the lienogastric, arises from the dorsal aorta a little below the posterior intestinal artery. It passes to the spleen, from where it sends branches to the stomach. In some specimens the posterior intestinal and gastrosplenic vessels arise from a common channel, the **anterior mesenteric artery.**

**Posterior mesenteric artery**   This vessel arises from the dorsal aorta at the level of the rectal gland and passes along the anterior margin of the mesorectum to supply this organ.

**Caudal artery**   This is the continuation of the dorsal aorta into the tail region. It can be seen, in cross sections through the tail, lying within the hemal arch.

The paired abdominal branches are these:

**Subclavian artery**   (See the previous section on efferent branchial arteries.)

**Genital arteries**   These arise from the celiac artery or dorsal aorta, or both, and enter the gonads. They are known as spermatic arteries in the male and as ovarian arteries in the female.

**Segmental arteries**   A series of segmentally arranged arteries arising along the entire course of the dorsal aorta. Some of these vessels can be seen by carefully

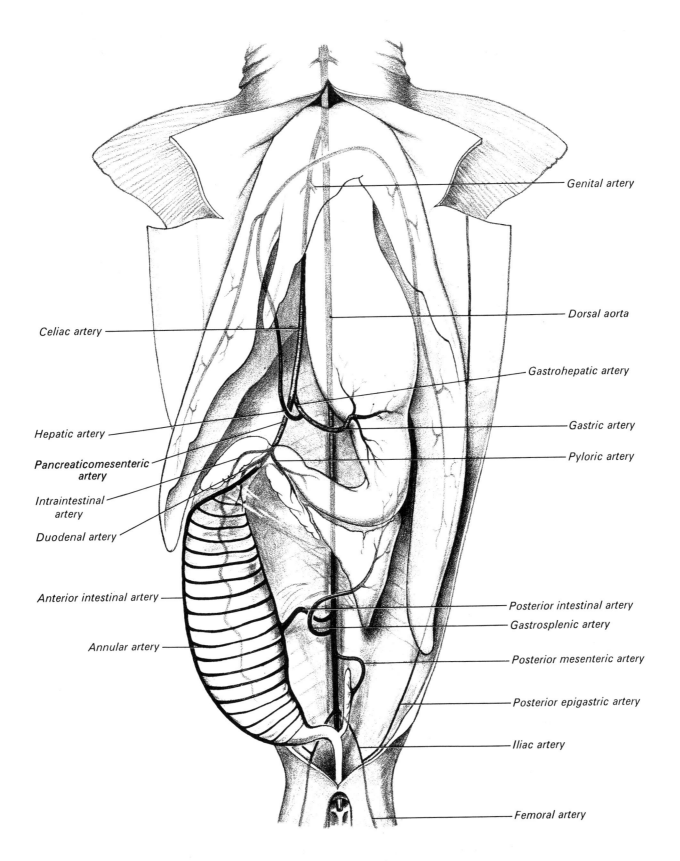

Genital artery

Dorsal aorta

Gastrohepatic artery

Gastric artery

Pyloric artery

Celiac artery

Hepatic artery

**Pancreaticomesenteric artery**

Intraintestinal artery

Duodenal artery

Anterior intestinal artery

Annular artery

Posterior intestinal artery

Gastrosplenic artery

Posterior mesenteric artery

Posterior epigastric artery

Iliac artery

Femoral artery

**FIGURE 5–4**
Dorsal aorta and its abdominal branches.

turning the lateral margin of one of the kidneys toward the midline. Typically, each vessel divides into three branches, the parietal, the renal, and the vertebromuscular arteries, but this division may prove difficult to demonstrate. The **parietal artery** is the lateral continuation of the segmental artery. Embedded within the musculature of the body wall, it encircles the pleuroperitoneal cavity as far as the mid-ventral line. The **renal artery** is best seen at the site at which it is dorsal to the kidneys. In the female, several renal arteries are modified as **oviducal arteries** that supply the oviduct. The **vertebromuscular artery,** a very small branch of the segmental artery, supplies the vertebral column, the spinal cord, and the epaxial muscles adjacent to the neural arches.

**Iliac arteries**   This pair arises from the aorta just anterior to the cloaca and emerges from behind the kidneys. Each gives off a **posterior epigastric artery,** which ascends along the lateral body wall to anastomose with the anterior epigastric artery from the subclavian. The iliac artery continues into the pelvic fin as the **femoral artery.**

## VENOUS SYSTEM

The venous circulation consists of ~~three~~ systems—~~two~~ _four_ ... _three_
sets of portal vessels and a single set of systemic veins—all of which drain into the sinus venosus of the heart via the common cardinal veins. The portal systems are venous pathways that begin in capillaries in one area of the body and end as capillaries in another. The systemic veins begin as capillaries and terminate by emptying into a common chamber, the sinus venosus.

### Hepatic Portal System (Fig. 5-5)

The hepatic portal system begins in capillaries in the organs of the digestive tract. These capillaries unite to form progressively larger veins, which finally come to form the hepatic portal vein, a large channel that passes into the liver. In the liver this vessel breaks up into smaller and smaller vessels and ultimately terminates as capillary-like vessels. The blood from these vessels drains into large spaces, the hepatic sinuses, which in turn empty into the sinus venosus of the heart by a pair of hepatic veins (see below).

*The branches that unite to form the hepatic portal vein can best be located near the abdominal organs that they drain. After these branches have been located, trace them to the region in the lesser omentum where they join to form the hepatic portal vein.*

**Hepatic portal vein**   This broad channel is located in the lesser omentum alongside and somewhat dorsal to the bile duct. It extends into the right lobe of the liver. The hepatic portal vein receives three tributaries, the gastric, pancreaticomesenteric, and lienomesenteric veins, all in approximately the same region.

**Gastric vein**   This is the most anterior of the tributaries, and its branches may be seen on the walls of the ventral portion of the stomach. The gastric vein passes toward the hepatic portal vein posterior to the gastric artery.

**Pancreaticomesenteric vein**   This tributary is a continuation of the **anterior intestinal vein,** a vessel that ascends along the right side of the intestine, collecting blood from **annular veins** along its course. The pancreaticomesenteric vein passes across the ventral surface of the intestine, just beneath the caudal border of the ventral lobe of the pancreas, and enters the lesser omentum. Before it enters the lesser omentum, it receives the **anterior splenic vein** from the spleen (and nearby parts of the stomach), the **intraintestinal vein,** a vessel that leaves the intestine adjacent to the point at which the corresponding artery enters, and the **pyloric vein,** a vessel that drains the pylorus.

**Lienomesenteric vein**   This middle tributary of the hepatic portal vein begins as the **posterior intestinal vein,** which, after receiving branches from the rectal gland, ascends along the left side of the intestine, collecting in its course annular tributaries. Near the middle of the intestine the posterior intestinal vein passes to the left across the body cavity to the posterior end of the dorsal lobe of the pancreas, where it receives the short **posterior splenic vein** from the spleen. Beyond this point it is known as the lienomesenteric vein. The lienomesenteric vein passes anteriorly along the elongated dorsal lobe of the pancreas and joins the other tributaries that form the hepatic portal vein, just anterior to the tip of the dorsal lobe of the pancreas.

### Renal Portal System (Figs. 5-6, 5-7)

This system of veins begins as capillaries in the tail and terminates as capillaries in the kidneys. Systemic veins (which are not part of the renal portal system) then carry blood from the kidneys toward the heart.

*Make a transverse cut through the tail to form a fresh cross-sectional surface for examinations. Identify the structures visible in this section, especially the caudal vein. Progressing cranially, make a series of deep transverse incisions through the tail parallel to each other and spaced about half an inch apart. The incisions*

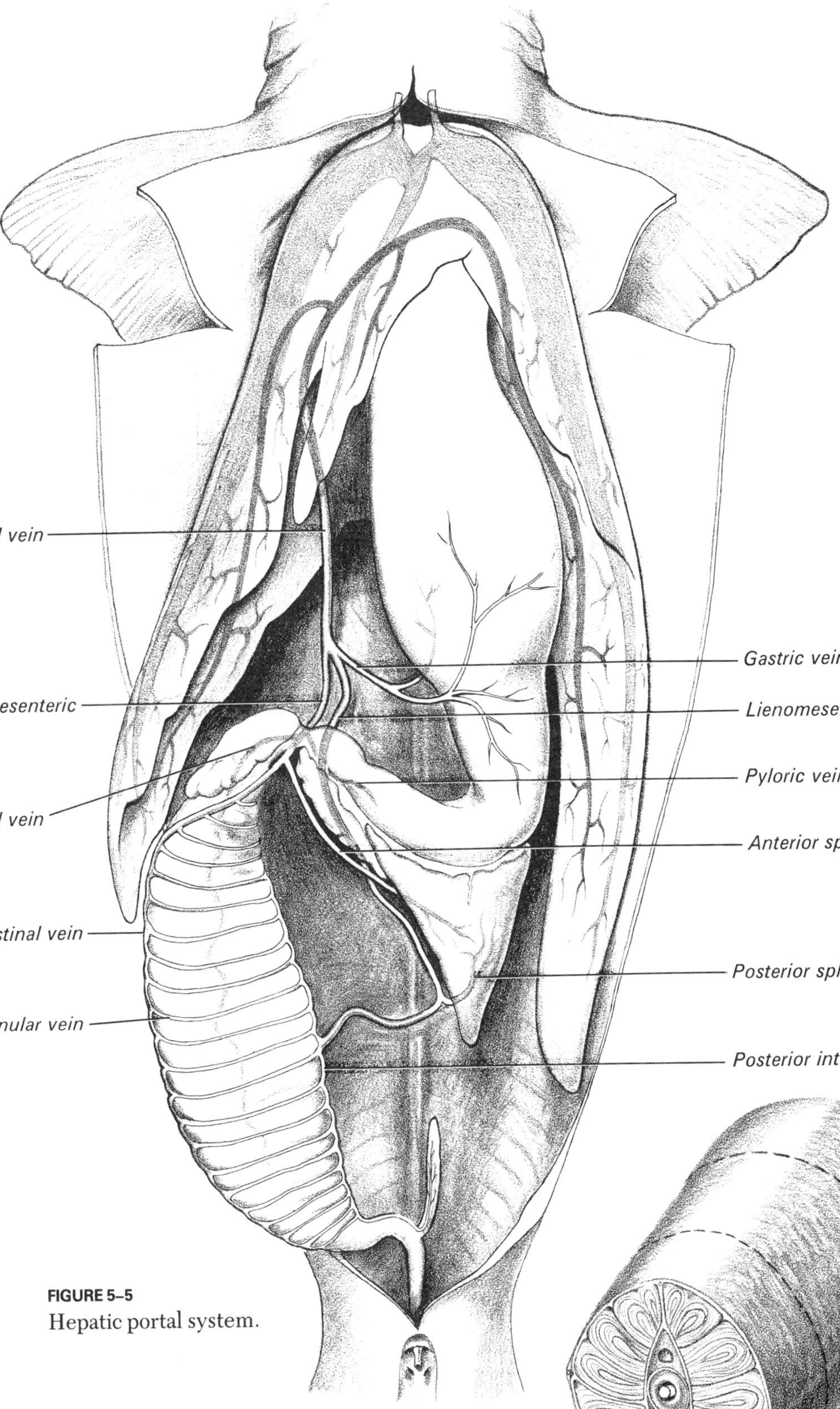

**FIGURE 5–5**
Hepatic portal system.

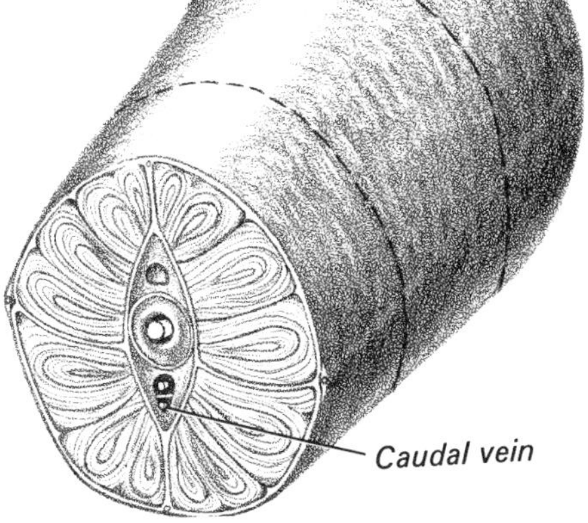

Hepatic portal vein

Pancreaticomesenteric vein

Intraintestinal vein

Anterior intestinal vein

Annular vein

Gastric vein

Lienomesenteric vein

Pyloric vein

Anterior splenic vein

Posterior splenic vein

Posterior intestinal vein

Caudal vein

**FIGURE 5–6**
Cross section of tail.

*should not go through to the ventral surface. By this means it will be possible to follow the course of the caudal vein as it ascends to bifurcate at about the level of the cloacal opening.*

**Renal portal veins**  These veins result from the division of the **caudal vein,** which has coursed cranially in the hemal canal from the tail into the trunk. The renal portal veins lie dorsal and parallel to the lateral margins of the kidneys and give off numerous side branches that empty into the kidney sinuses. These side branches are known as **afferent renal veins.**

## Systemic Veins (Fig. 5-7)

All of the large veins returning blood from the body to the heart open into the sinus venosus. The blood then passes into the atrium, from there into the ventricle, and finally out through the conus arteriosus and into the ventral aorta.

Most of the systemic veins are large tissue spaces without definite walls. They are therefore more properly called venous sinuses. Because of their relative lack of structure their identification is more difficult. These veins, unless otherwise indicated, are present as bilateral pairs.

*Elevate the ventricle and cut transversely across the middle of the sinus venosus, making certain that a sizeable posterior portion is retained for study. Flush out the sinus venosus, pressing on the liver to prevent fluid from reentering this chamber.*

**Hepatic vein**  The opening of this vein into the posterior surface of the stretched sinus may be seen as a slit present on each side of the sinus, just lateral to the midline. A probe inserted into either of these slits will lead into one of the hepatic sinuses of the liver. Venous blood is brought into this sinus by means of the hepatic portal vein.

**Inferior jugular vein**  This drains the floor of the pharynx. It opens on each side into the anterior wall of the sinus venosus near the extreme lateral angle.

**Common cardinal vein**  Known also as the duct of Cuvier, this vessel enters the lateral corner of the sinus venosus. It collects the venous blood on each side from three major tributaries: the anterior cardinal veins, the posterior cardinal veins, and the subclavian veins.

**Posterior cardinal vein**  On each side, this vessel lies immediately lateral to the dorsal aorta and medial to each of the kidneys. The right posterior cardinal vein extends further caudally than does the left. Each of these vessels leads cranially into a very large **posterior cardinal sinus.** The posterior cardinal sinuses appear as bluish thin-walled chambers at the anterior end of the pleuroperitoneal cavity and are confluent across the midline. These sinuses lead into the common cardinal veins via large openings located somewhat lateral to the openings of the hepatic veins and can be probed. The tributaries of each posterior cardinal vein are the **efferent renal veins,** which drain sinuses in the kidneys, and the **parietal veins,** which are segmentally arranged and remove blood from the body wall. **Genital sinuses,** known as spermatic sinuses in the male and as ovarian sinuses in the female, surround each gonad. These, however, may be difficult to see if they are not filled with blood or injection medium. The genital sinuses drain into the posterior cardinal sinuses by broad connections.

**Anterior cardinal vein**  This vessel drains the sides of the head region from a series of sinuses. Each vein empties into the common cardinal vein on its side through a small opening located just anterior to the site of entrance of the posterior cardinal vein. The course of the anterior cardinal vein can be followed by probing anterodorsally while at the same time palpating the surface of the head dorsal to the gills.

**Subclavian vein**  This very short vessel enters the ventrolateral margin on either side of a common cardinal vein. It is formed by two tributaries, the lateral abdominal vein and the brachial vein. The **lateral abdominal vein** is located beneath the parietal peritoneum along the lateral abdominal wall. It commences posteriorly with the confluence of the **cloacal vein** from the wall of the cloaca and the **iliac vein,** which is a continuation of the **femoral vein** from the pelvic fin. The lateral abdominal vein continues anteriorly until it reaches the coracoid cartilage of the pectoral girdle, where it is joined by the **brachial vein,** which drains the pectoral fin. The newly formed subclavian vein then curves abruptly dorsomedially and soon joins the common cardinal vein.

63

these 5 vessels may not be visible due to modification in the dissection

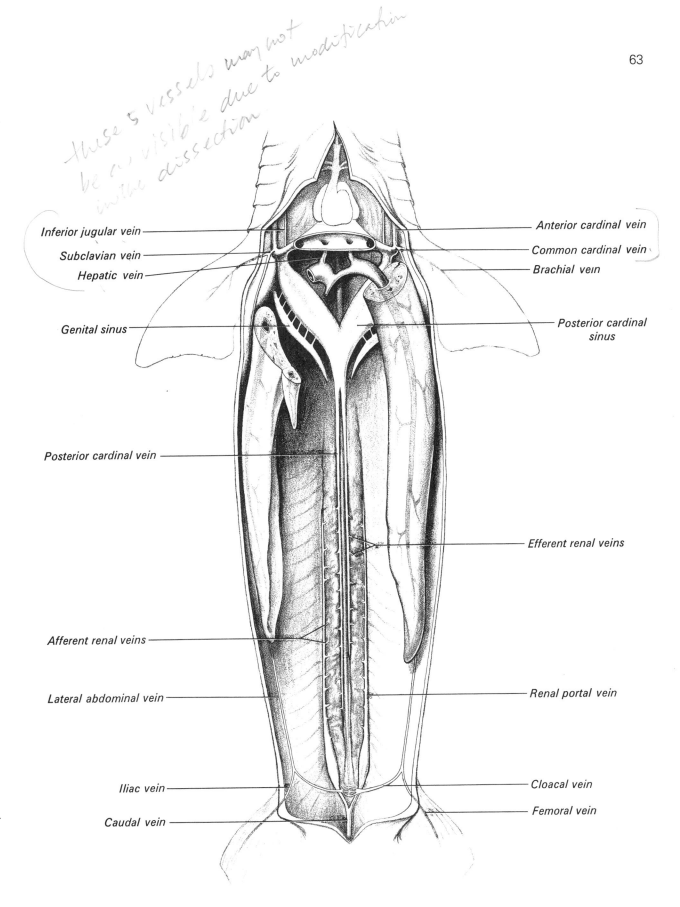

Inferior jugular vein

Subclavian vein

Hepatic vein

Genital sinus

Posterior cardinal vein

Afferent renal veins

Lateral abdominal vein

Iliac vein

Caudal vein

Anterior cardinal vein

Common cardinal vein

Brachial vein

Posterior cardinal sinus

Efferent renal veins

Renal portal vein

Cloacal vein

Femoral vein

**FIGURE 5–7**
Systemic veins and their tributaries.

# Urogenital System

# Anatomy of the Dogfish Shark

Elasmobranchs, such as the dogfish shark and holocephalians, as well as some teleosts, are unique among living fishes in that their act of copulation involves internal fertilization. This process is facilitated by means of **claspers,** or modified medial portions of the two pelvic fins in the male (Fig. 6-1). The medial border of each clasper is grooved and clusters of spermatozoa, or **spermatophores,** are transported to the cloaca of the female along this groove.

The movement of the spermatophores along the clasper is aided by means of muscular **siphon sacs** that lie beneath the skin just in front of the cloaca. The cranial end of each of these sacs ends blindly, and the caudal end communicates with the cranial opening of the clasper groove.

It appears that one clasper at a time is inserted into the female cloaca during copulation. It was formerly thought that prior to copulation the siphon sac was filled with seawater and that upon its contraction water was forced down the clasper groove. It was assumed that spermatophores were forcefully injected into the cloaca by this means and thus into the uterus. But recent evidence suggests that the siphon sac produces a secretion that may lubricate the clasper groove and thereby facilitate fertilization.

The eggs that have ripened in the ovary become freed from their follicles, pass out of the ruptured ovarian wall, and enter the body cavity. They pass to the common opening of the oviducts, which they then enter. Spermatozoa reach the upper portions of the oviduct and fertilize the eggs as they pass through. Groups of fertilized eggs become enclosed temporarily by a thin membranous shell that forms during their downward course. After several months the shell breaks down and the embryos develop within the uterus. They obtain their nourishment from an external **yolk sac** that is suspended from the underside of the embryo (see Fig. 6-3). The yolk is moved by ciliary action into the embryo's intestine where it is digested and absorbed. Vascular villi that line the internal uterine wall are in contact with the surface of the yolk sac. These villi probably allow for fluid exchange between the mother and embryo.

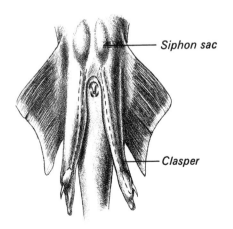

**FIGURE 6–1**
Claspers.

## MALE UROGENITAL SYSTEM (Fig. 6-2)

The structures of the excretory and reproductive systems are closely related because of their common embryological origins.

*The kidneys lie against the back wall of the pleuroperitoneal cavity. Make a slit through the peritoneum about one and one-half inches in length along the lateral border of either kidney and strip it toward the midline to facilitate its study.*

**Testes**   The gonads are a pair of somewhat elongated organs located at the extreme anterior end of the pleuroperitoneal cavity dorsal to the liver.

**Mesorchium**   This is the mesentery that suspends each testis from the mid-dorsal body wall.

**Efferent ductules**   These are several small, delicate ducts, passing through the mesorchium, that connect the tubules of the testis with some of the tubules in the cranial part of the kidney. This cranial part of the kidney is also known as the epididymis.

**Kidneys**   The kidneys (opisthonephroi) are the long, dark, paired retroperitoneal organs lying on either side of the dorsal aorta. They extend from the anterior end of the liver to the level of the cloaca. The cranial part is slender; the caudal part, which is the functional excretory organ proper, is slightly thicker. The caudal part consists of numerous renal corpuscles and nephric tubules. The tubules unite to form the collecting tubules that empty into the urinary ducts.

**Mesonephric ducts**   These are coiled ducts, known also as the archinephric ducts, that lie on the ventral surface of the kidneys. In the male they serve for the passage of both urinary wastes and spermatozoa. The spermatozoa pass into the mesonephric ducts from tubules of the cranial parts of the kidneys.

**Accessory mesonephric ducts**   These ducts, apparently present only in males, are small and difficult to find. Each drains the caudal part of the kidney and empties into the sperm sac (see below).

**Seminal vesicles**   The expanded and straight caudal parts of the mesonephric ducts. Sperm pass through the seminal vesicles on their way to the sperm sacs.

**Sperm sacs**   These can be considered as outpocketings from the ventral side of the seminal vesicles at their posterior ends.

**Urogenital sinus**   This cavity is formed by the medial union of the cavities of the two sperm sacs.

**Cloaca**   A common receiving chamber for urinary and digestive wastes and discharged spermatozoa. It conveys these materials to the outside. It is divided by a transverse fold, the **urorectal shelf,** into a ventral region, the **urodeum,** and a dorsal region, the **coprodeum.**

**Urogenital papilla**   This is a small conical projection from the mid-dorsal wall of the cloaca. It contains a small cavity, the urogenital sinus, that opens at the tip of the papilla as the **urogenital pore.**

## FEMALE UROGENITAL SYSTEM (Fig. 6-3)

**Ovaries**   These oblong lobulated bodies are located at the anterior end of the body cavity dorsal to the liver. Swellings that may be seen on their dorsal surface represent the presence of underlying ova in various stages of development.

**Mesovarium**   The peritoneal duplication that attaches the ovaries to the mid-dorsal body wall.

**Oviducts**   These are slender tubes that extend the length of the body cavity. They begin as a common duct located in the falciform ligament. The opening of the oviduct into the pleuroperitoneal cavity is a vertical slit known as the **ostium tubae.** Each duct loops dorsolaterally around the anterior end of the liver and then begins its downward course. As the duct passes dorsal to the ovary it exhibits an expanded portion, the **shell gland.** In the shell gland the membranous shells are secreted around clusters of ova passing through. The caudal part of each oviduct is enlarged to form the **uterus,** which terminates in an opening leading into the cloaca. The development of the embryos takes place in the uterus.

**Mesotubarium**   The mesentery suspending the oviduct (and uterus) from the mid-dorsal body wall.

**Kidneys**   These structures are located retroperitoneally, as they are in the male. The cranial parts are nonfunctional, and the thicker caudal parts serve for excretion.

**Mesonephric ducts**   These are much smaller than those of the male. They are uncoiled and recessed into the substance of the mesonephros and are thus hard to find, especially in young females. The mesonephric ducts terminate in the urinary sinus. In females the accessory mesonephric ducts are exceedingly difficult to find and may be absent.

**Urinary sinus**   The cavity within the base of the urinary papilla that is formed by union of the ends of the mesonephric ducts.

**Cloaca**   As in the male, the cloaca receives the digestive and urinary waste products. Since the oviduct also opens into the cloaca, newborn young also pass through this chamber. Like that of the male, the female cloaca is divided into a ventral urodeum and dorsal coprodeum by the transverse urorectal shelf.

**Urinary papilla**   The small conical projection enclosing the urinary sinus. Its tip opens into the cloaca through the **urinary pore.**

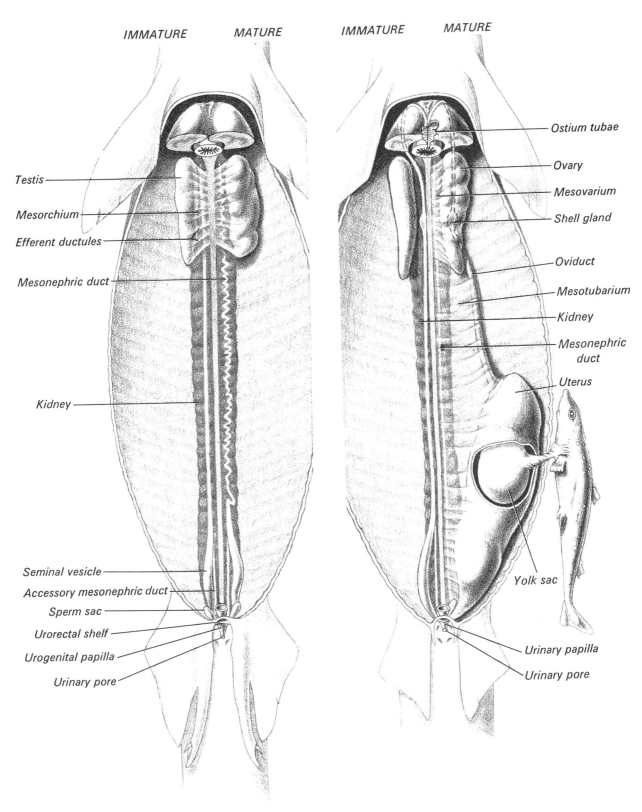

IMMATURE    MATURE    IMMATURE    MATURE

Testis

Mesorchium

Efferent ductules

Mesonephric duct

Kidney

Seminal vesicle

Accessory mesonephric duct

Sperm sac

Urorectal shelf

Urogenital papilla

Urinary pore

Ostium tubae

Ovary

Mesovarium

Shell gland

Oviduct

Mesotubarium

Kidney

Mesonephric duct

Uterus

Yolk sac

Urinary papilla

Urinary pore

**FIGURE 6–2**
Male urogenital system.

**FIGURE 6–3**
Female urogenital system.

# Anatomy of the Dogfish Shark

Since the sense organs and the nervous system are intimately connected, it is advisable to study them in sequence. The sense organs will be dissected first, in order to facilitate exposure of the brain and cranial nerves (see Exercise 8, Nervous System).

In addition to the usual major sense organs, such as the olfactory organ, eye, and ear, fishes also possess special sense organs that are apparently associated with their aquatic mode of life. These special sense organs include the ampullae of Lorenzini and the lateral line system. ~~The ampullae are found only in elasmobranchs.~~

*Carefully remove the skin from the roof and sides of the head to aid in identification of the pores of the lateral line system and the ampullae of Lorenzini.*

## AMPULLAE OF LORENZINI

The ampullae are distributed over the surface of the head. Each consists of a small pore leading into a short canal that terminates in a bulb-like expansion. These expansions contain sensory cells. The ampullae are innervated by fibers from the facial nerve (N. VII) and ~~serve as thermoreceptors and probably as mechano-receptors~~ detect weak electrical fields.

## LATERAL LINE SYSTEM (Fig. 7-1)

The lateral line system is an extension of the lateral line canal onto the head. It is a series of canals that lie beneath the skin and that open to the surface by means of numerous pores. Specialized clusters of epithelial cells, known as **neuromasts,** are located in the canals and are supplied by rami from all three branches of the facial nerve (N. VII). Fine filaments from the sensory cells of the neuromasts, when stimulated, produce an excitation that is transmitted to the cranial nerves. By this means, information concerning low-frequency vibrations and hydrostatic pressure changes are transmitted to the brain. The following individual canals of the lateral line system on the head should be identified.

**Supratemporal canal** Extending across the top of the head at the level of the spiracles, this channel connects the two lateral line canals. It then runs forward to divide into the supra- and infraorbital canals.

**Hyomandibular canal** Beginning near the anterior end of the infraorbital canal (see below), this canal runs back to terminate above and a little behind the angle of the jaws.

**Mandibular canal** This canal is not connected with any of the other canals of the lateral line system and lies just behind the lower jaw.

**Supraorbital canal** Extends forward above the eye almost to the tip of the rostrum and then turns back caudally to join the infraorbital canal.

**Infraorbital canal** Passes ventrally between the eye and the spiracle, zigzags beneath the eye, and then passes forward, medial to the external naris, to the tip of the snout. Along its course it unites with the hyomandibular canal.

## OLFACTORY ORGAN

*Remove the skin from around one of the external nares and cut away the underlying tissues enclosing its associated olfactory sac. The connection of the sac with the olfactory nerve will be seen when the cranial nerves are dissected (Exercise 8, Nervous System). To observe the internal structure of the olfactory sac, make a thin transverse slice through the sac, parallel to the skin surface.*

**Olfactory sacs**  These paired bulbous structures are the organs of smell. They are lined internally by parallel folds, or **lamellae,** whose surfaces are covered by olfactory epithelium. Water entering the naris conveys odors to the sac, stimulating the cilia-like endings of the neurosensory cells of the epithelium. Processes from these cells form a synaptic connection with secondary neurons located in the olfactory bulb. Fibers from the latter then give rise to the olfactory tract, which ends in the olfactory lobe of the brain.

## EYE*

*Remove the skin that borders one of the eyes and makes up the fixed eyelids. Also remove some of the cartilage from above the eye. Carefully free the eyeball from the loose connective tissue surrounding it on all sides. This will permit identification of the six extrinsic muscles of the eyeball. The inferior oblique and inferior rectus muscles can be seen best from the ventral aspect, and the other four from the dorsal view. After identifying all the muscles and their nerves, cut the muscles near their insertions on the eyeball. Then cut the optic nerve and remove the eyeball from the orbit.*

### Ocular Muscles (Fig. 7-2)

These are modified epaxial muscles derived from the first, second, and third cranial myomeres. They rotate the eyeball. The musculature of the eye consists of two groups: a pair of oblique muscles that originate from the anteromedial orbital wall, and four recti muscles originating from the posteromedial wall. The individual muscles are the following.

**Superior oblique**  Inserted on the mid-dorsal part of the eyeball and innervated by the trochlear nerve (N. IV).

**Inferior oblique**  Inserted on the anteroventral side of the eyeball and innervated by the oculomotor nerve (N. III).

*Upon completing this section you can store the shark eye in preservative and compare it later with the sheep (ox) eye (see p. 234).

**Superior rectus**  Inserted on the dorsal surface of the eyeball and innervated by the oculomotor nerve.

**Inferior rectus**  Inserted on the ventral surface of the eyeball just behind the inferior oblique and innervated by the oculomotor nerve.

**Medial rectus**  Passes between the superior and inferior oblique muscles and is inserted on the anterior surface of the eyeball. It too is supplied by the oculomotor nerve. This muscle is also called the internal, or anterior, rectus.

**Lateral rectus**  Inserted on the posterior surface of the eyeball and innervated by the abducens nerve (N. VI). This muscle is also known as the external, or posterior, rectus.

### Other Orbital Structures (Fig. 7-2)

**Optic pedicel**  This cartilaginous structure projects from the site of origin of the recti muscles. It supports the eyeball and aids its rotation.

**Optic nerve** (N. II)  This nerve emerges from the eyeball just ventral to the site of insertion of the medial rectus muscles. At this point it appears as a stout white stump, leaving the orbit via the optic foramen, an opening that is located somewhat anterior to the site of origin of the recti muscles.

**Oculomotor nerve** (N. III)  This nerve enters the orbit via the oculomotor foramen, an opening located above the site of attachment of the optic pedicel. It then divides into four branches.

**Trochlear nerve** (N. IV)  This small nerve enters the orbit through the trochlear foramen located in the anterodorsal part of the orbit.

**Abducens nerve** (N. VI)  This nerve passes into the orbit via the abducens foramen, which is located beneath the site of attachment of the optic pedicel.

**Superficial ophthalmic nerve**  This is a large branch of the combined fifth and seventh cranial nerves. It enters the orbit through the trigeminofacial foramen, runs along the lateral orbital wall, and leaves the orbit via the superficial ophthalmic foramina in its roof.

**Deep ophthalmic nerve**  This structure enters the orbit via the trigeminofacial foramen as a branch of the fifth cranial nerve. It runs beneath the superior rectus muscle and leaves the orbit via the deep ophthalmic foramen, which is located just anterior to the foramina for the superficial ophthalmic nerve.

**Infraorbital nerve**  This is a branch from the combined fifth and seventh cranial nerves. It enters the orbit via the trigeminofacial foramen and runs along the floor of the orbit to emerge beneath the anterior orbital wall. Before leaving the orbit, the infraorbital nerve divides into a medial **maxillary nerve** and a laterally placed **buccal nerve.**

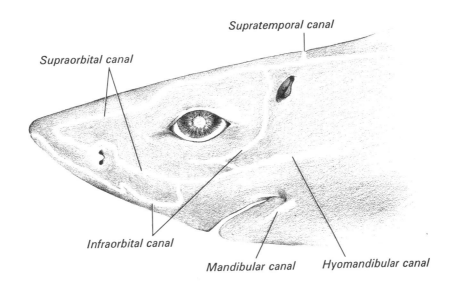

**FIGURE 7–1**
Lateral line system.

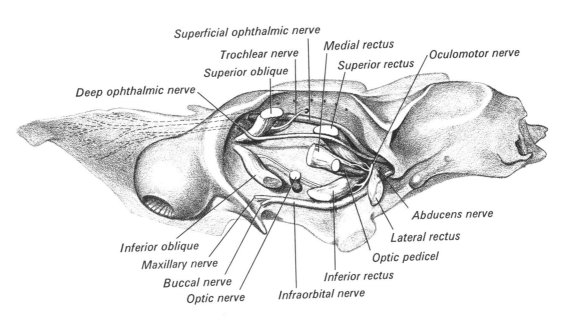

**FIGURE 7–2**
Intraorbital muscles and nerves.

## Eyeball (Fig. 7-3)

**Conjunctiva**   This is the thin transparent layer covering the front of the eyeball. It is continuous with the lining of the immovable eyelids.

**Sclera**   This layer encloses the normally unexposed parts of the eyeball. In the dogfish shark it is largely cartilaginous, providing support for the eyeball.

**Cornea**   This is the continuation of the sclera over the exposed part of the eyeball. At its outer edge the conjunctiva meets this transparent layer.

**Choroid**   The vascular, black-pigmented layer, internal to the sclera, that helps nourish the retina. It facilitates sharp image formation by preventing internal light scattering.

**Suprachoroid**   This thickened vascular layer lies between the sclera and choroid in the back (medial) part of the eyeball.

**Iris**   This curtain-like structure is an extension of the choroid coat that hangs behind the cornea. It limits light entering the eye to that passing through the pupil and lens.

**Pupil**   The central opening in the iris.

**Ciliary body**   A series of folds, forming a thickened ring within the eyeball, that is located at the iridiochoroidal junction. It is formed by both the choroid and the nonsensory part of the retina.

**Retina**   The innermost coat of eyeball made up of a sensory part against the choroid and a nonsensory part over the ciliary body.

**Lens**   The large spherical body behind the pupil. It is transparent and elastic in life but opaque and hard in preserved specimens.

**Suspensory ligaments**   The fibrous strands that extend between the lens and the ciliary body.

**Anterior chamber**   The cavity between the iris and cornea. In life it is filled with a clear liquid called aqueous humor.

**Posterior chamber**   The cavity, filled with aqueous humor, between the iris and suspensory ligament.

**Vitreous chamber**   The large cavity behind the lens and suspensory ligaments that is filled with the gelatinous vitreous humor.

## EAR (Figs. 7-4, 7-5, 7-6)

The dogfish has only an inner ear, which is embedded in the cartilaginous otic capsule of the chondrocranium. The inner ear is the definitive organ of equilibration and is responsible for the maintenance of balance. It consists of a triangular membranous sac, the **vestibule,** and three **semicircular ducts,** collectively known as the **membranous labyrinth.** This labyrinth is filled with a fluid called ~~endolymph.~~ *perilymph* Turning movements

in the three planes of space cause displacement of the endolymph with resulting stimulation of sensory cells, thus providing information with respect to body position. Additional information comes through stimulation of sensory cells in the vestibule.

*Remove the skin and underlying muscles on the left side of the head in the region of the otic capsule. Uncover and examine the endolymphatic fossa and note the presence of a small pair of ducts within it.*

**Endolymphatic ducts**   A pair of ducts, each extending from the endolymphatic pore to the vestibule on its side of the body. The expanded part of the duct within the endolymphatic fossa is the **endolymphatic sac.**

*To expose the membranous labyrinth of the inner ear, carefully shave away the cartilage of the otic capsule, beginning on the dorsal surface and gradually working laterally. The canals that house the semicircular ducts will be seen just before the ducts are reached. Expose the entire length of each of these ducts.*

**Anterior semicircular duct**   This is a vertical membranous tube that extends obliquely anterior from the region of the endolymphatic duct.

**Posterior semicircular duct**   This vertical membranous tube extends obliquely posterior from the region of the endolymphatic duct. It lies in a plane that is at right angles to the other ducts.

**Horizontal semicircular duct**   Oriented horizontally and ventrolaterally to the other two.

**Ampulla**   A swelling in each of the semicircular tubes that is located near the end farthest away from the endolymphatic duct. White masses, known as **cristae,** may be visible inside each ampulla. They are patches of sensory epithelium.

*Continue to remove carefully the cartilage medial to the horizontal duct until the chamber housing the vestibule is reached. Expose this part of the membranous labyrinth and try to identify its components.*

**Utriculus**   The dorsolateral part of the vestibule. It consists of two chambers: the **anterior utriculus,** which receives the anterior and horizontal semicircular duct, and the **posterior utriculus,** which receives the posterior semicircular duct.

**Sacculus**   The median ventral part of the vestibule. Extending ventrally from the posterior corner of the sacculus is an outpocketing, the **lagena.** White patches of sensory epithelium, known as **maculae,** lie within the sacculus and lagena (and utriculus). These patches are overlaid by a mass of calcareous

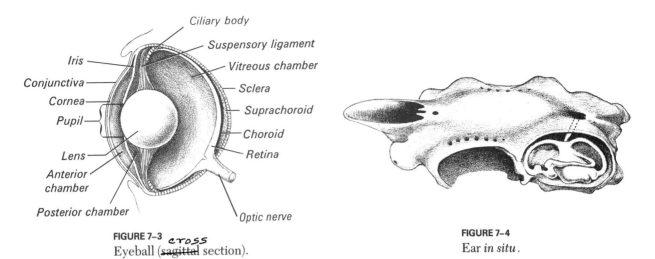

**FIGURE 7–3** *cross*
Eyeball (~~sagittal~~ section).

↗

**FIGURE 7–4**
Ear *in situ*.

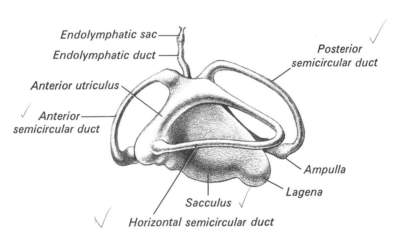

**FIGURE 7–5**
Membranous labyrinth of left ear (lateral view).

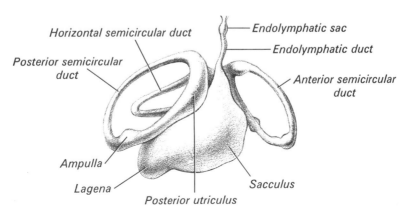

**FIGURE 7–6**
Membranous labyrinth of left ear (medial view).

concretions and sand grains that form **otoliths.** Changes in the position of the head or in linear velocity during movement presumably result in alteration of the position of the otoliths and thus in stimulation of the sensory epithelium. By this means additional information on the position of the body is attained to supplement that derived from the semicircular ducts.

# Anatomy of the Dogfish Shark

The nervous system consists of a cerebrospinal division (brain and spinal cord, Fig. 8-1) and a peripheral division (cranial and spinal nerves). The autonomic division in the dogfish is poorly developed and consists of only a few isolated ganglia and associated fibers.

*Expose the brain by first removing the skin that remains on the dorsal surface of the head and then carefully cut away the cartilage that makes up the roof of the skull. In the removal of the cartilage, take care to* *avoid damaging the roots of the cranial nerves. Study the cranial nerves immediately after the examination of the dorsal brain surface has been completed. Only after the study of these nerves has been completed should the ventral surface be exposed and sagittal sections made and studied. For textual continuity, however, all of the surfaces of the brain are described in the first section. The cranial nerves are described in the second section, and the occipital and spinal nerves are discussed in the third section.*

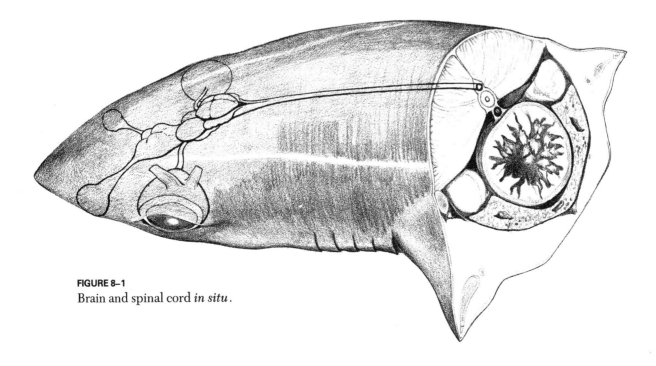

**FIGURE 8–1**
Brain and spinal cord *in situ*.

## BRAIN (Figs. 8-2, 8-3, 8-4)

In the mature dogfish the brain consists of five parts, all of which are enclosed by the protective cartilaginous chondrocranium. The brain tissue itself is invested by a tough outer membrane, the **dura mater,** and a more delicate inner membrane, the **pia mater.**

On the *dorsal* side of the brain (Fig. 8-2) are the following features.

### Telencephalon

**Olfactory bulbs**    The rounded masses that lie just behind and in contact with the olfactory sacs.

**Olfactory tract**    The stalk-like caudal extension that connects each olfactory bulb with the anterior end of the brain.

**Olfactory lobes**    The somewhat rounded masses in which the olfactory tracts terminate at the anterior end of the brain.

**Cerebral hemispheres**    These eminences lie just in back of the olfactory lobes and may be demarcated from them by faint indentations. The hemispheres together constitute the **cerebrum.**

### Diencephalon

**Epiphysis**    A slender stalk that projects anterodorsally from the back part of the diencephalon up through the roof of the skull via the epiphyseal foramen. It is known also as the pineal body.

**Tela choroidea**    Vascular membranous roof that covers the diencephalon.

### Mesencephalon

**Optic lobes**    These rounded bodies, known also as the corpora bigemina, are the visual centers of the brain. They are located behind the diencephalon.

### Metencephalon

**Cerebellum**    The large oval structure partly overlapping the mesencephalon and myelencephalon. Its dorsal surface is demarcated by shallow grooves into four quadrants.

### Myelencephalon

**Auricles**    Ear-like masses that project forward at the anterolateral corners of the myelencephalon. These structures are also known as restiform bodies.

**Medulla oblongata**    The greater part of the myelencephalon. It is continuous with the spinal cord.

*After studying the cranial nerves (see the next section) remove the brain, avoiding any damage to the hypophysis (pituitary gland). To do this, first expose the lateral surface of the brain by carefully picking away small pieces of the left side of the skull. The hypophysis will be seen on the ventral surface just behind the optic nerves. Remove the cartilage that fixes its posterior part until it is entirely free. Elevate the rostral end of the brain and begin cutting the cranial nerves as they come into view. Continue freeing the brain in this manner until all of the nerves have been transected, then cut the spinal cord just caudal to its junction with the brain.*

On the *ventral* surface of the brain (Fig. 8-3) are the following structures.

**Optic chiasma**    The X-shaped structure formed by the crossing of the two optic nerves as they enter the diencephalon.

**Infundibulum**    This structural complex is located just behind the optic chiasma. It consists of a pair of rounded **inferior lobes** and a thin-walled **vascular sac.** The sac is thought to have some secretory function or to be possibly a pressure receptor.

**Hypophysis**    The pituitary gland is usually lost during removal of the brain. If it is intact, its four parts may be identified with the aid of a hand lens. They consist of an **anterior median lobe** located between the inferior lobes of the infundibulum; a **posterior median lobe** located between a pair of **lateral lobes;** and a pair of **ventral lobes** that extend as wing-shaped structures caudally from the posterior median lobe. It is not known if the hypophysis of the dogfish functions as an endocrine gland, as it does in mammals.

*Make a median sagittal section of the brain (using a long slender knife or sharp spatula) to reveal a series of large interconnecting cavities or ventricles. These result from expansion of the original neurocoel and are numbered sequentially in a cranial-caudal direction. Submerging a half-brain in water may aid in recognizing the ventricles.*

In the *mid-sagittal* section (Fig. 8-4) the following brain structures can be seen.

**Lateral ventricles**    Paired cavities, known also as the first and second ventricles, located in the telencephalon. They extend forward as narrow canals through

77

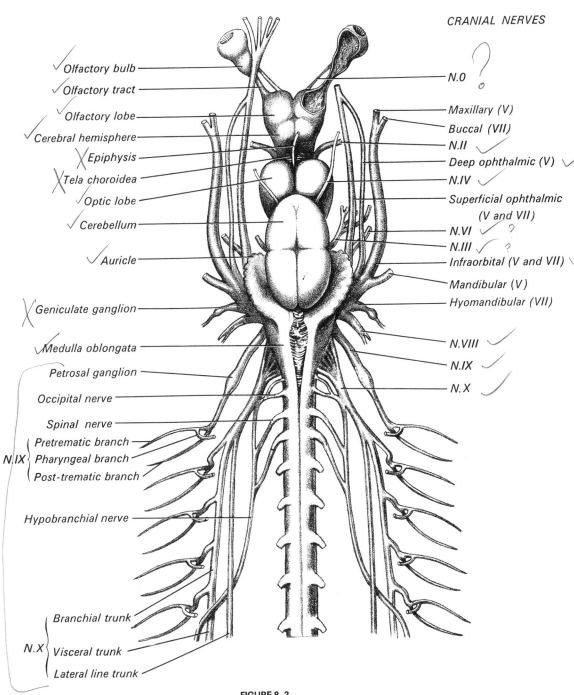

CRANIAL NERVES

Olfactory bulb
Olfactory tract
Olfactory lobe
Cerebral hemisphere
Epiphysis
Tela choroidea
Optic lobe
Cerebellum
Auricle
Geniculate ganglion
Medulla oblongata
Petrosal ganglion
Occipital nerve
Spinal nerve
N.IX { Pretrematic branch
Pharyngeal branch
Post-trematic branch
Hypobranchial nerve
N.X { Branchial trunk
Visceral trunk
Lateral line trunk

N.O
Maxillary (V)
Buccal (VII)
N.II
Deep ophthalmic (V)
N.IV
Superficial ophthalmic (V and VII)
N.VI
N.III
Infraorbital (V and VII)
Mandibular (V)
Hyomandibular (VII)
N.VIII
N.IX
N.X

**FIGURE 8–2**
Brain (dorsal view).

the olfactory tracts into the olfactory bulbs.

**Third ventricle**   This cavity is located mainly in the diencephalon and communicates with the two lateral ventricles through the interventricular foramina. Projecting down into the third ventricle from its roof (tela choroidea) are folds that constitute the **anterior choroid plexus.**

**Hypothalamus**   This forms the floor of the diencephalon and consists of the infundibulum and the hypophysis.

**Epithalamus**   This forms the roof of the diencephalon and consists of the tela choroidea and epiphysis.

**Paraphysis**   The finger-like projection, just anterior to the tela choroidea, that rests against and is considered part of the telencephalon.

**Velum transversum**   This is a deep fold hanging down from the roof of the diencephalon into the third ventricle, just behind the paraphysis.

**Cerebral aqueduct**   The small canal that links the third and fourth ventricles. It also communicates dorsally with the cavity of the cerebellum, the **metacele,** and laterally with those of the optic lobes, the **mesocele.**

**Fourth ventricle**   The cavity of the myelencephalon is continuous with the minute central canal of the spinal cord. Projecting down from its roof (tela choroidea) are folds that constitute the **posterior choroid plexus.**

## CRANIAL NERVES (Fig. 8-2)

The dogfish shark has the ~~10~~ 11 pairs of cranial nerves ~~characteristic of all vertebrates~~. It has an additional pair, the first, which is characteristically well-developed in elasmobranchs. Some nerves contain only sensory fibers, others only motor fibers, and some both.

*Find all the cranial nerves and dissect them carefully, tracing them as far distally as possible.*

**Terminal nerve** (N. O)   This slender sensory nerve runs along the length of the medial side of the olfactory tract.

**Olfactory nerve** (N. I)   This is made up of many fine sensory fibers that arise from the epithelium of the olfactory sac and terminate in the olfactory bulb.

**Optic nerve** (N. II)   This sensory nerve arises in the retina of the eye and extends to the diencephalon, where the fibers cross at the chiasma and continue on to the optic lobes.

**Oculomotor nerve** (N. III)   This motor nerve originates from the ventral surface of the mesencephalon. It innervates four eye muscles: inferior oblique, superior, inferior, and medial recti.

**Trochlear nerve** (N. IV)   A thin motor nerve that emerges from the dorsal surface of the mesencephalon between the optic lobes and cerebellum. It enters the orbit via the trochlear foramen and innervates the superior oblique muscle.

**Trigeminal nerve** (N. V)   A large mixed nerve that arises from the anterior part of the medulla and passes into the orbit through the trigeminofacial foramen. At its origin some of its roots are united with those of the facial nerve, forming a common trigeminofacial root. In the orbit the trigeminal nerve divides immediately into its three characteristic branches, and since its ophthalmic branch is subdivided, the trigeminal nerve may be considered to have four branches:

**Superficial ophthalmic nerve**   This sensory nerve passes across the orbit, which it leaves via the superficial ophthalmic foramina. It runs to the rostrum, where its trigeminal fibers have a general cutaneous sensory function. It also contains facial fibers arising from the common trigeminofacial root. Their function is described below.

**Deep ophthalmic nerve**   Passing along the medial surface of the eyeball, this sensory nerve leaves the orbit via the deep ophthalmic foramen. It joins the superficial ophthalmic nerve above the olfactory nerve.

**Infraorbital nerve**   A thick sensory nerve that runs obliquely across the floor of the orbit. At the anterior orbital margin it divides into a **buccal nerve** (containing N. VII fibers) and a **maxillary nerve.** The maxillary nerve (containing N. V fibers) carries impulses back from the ventral aspect of the snout.

**Mandibular nerve**   This mixed nerve runs ventrally to innervate the branchiomeric muscles and the skin of the lower jaw.

**Abducens nerve** (N. VI)   This motor nerve arises from the ventral surface of the medulla. It runs anterolaterally to enter the orbit, where it innervates the lateral rectus muscle of the eye.

**Facial nerve** (N. VII)   This mixed nerve arises from the medulla and is divided into three branches:

**Superficial ophthalmic nerve**   This nerve has already been described under the trigeminal. Its facial fibers supply both the ampullae of Lorenzini and the lateral line system.

**Buccal nerve**   This is the branch of the infraorbital nerve (see above) containing the N. VII fibers. It carries sensory impulses from the lateral line system on the head.

**Hyomandibular nerve**   This nerve contains only fibers of the facial. Near the brain stem it exhibits a swelling, the **geniculate ganglion.** From this ganglion arises a small palatine nerve that passes

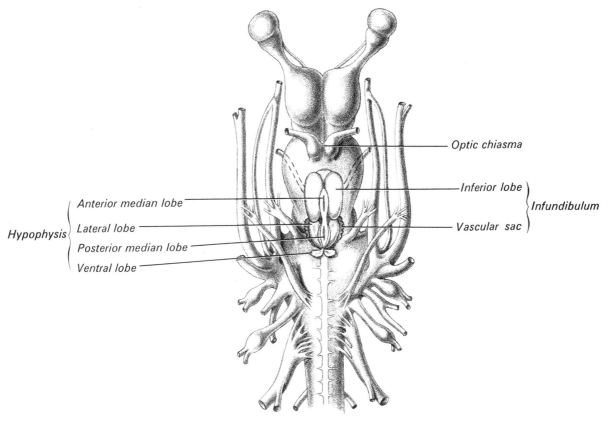

**FIGURE 8–3**
Brain (ventral view).

*Optic chiasma*

*Inferior lobe* } *Infundibulum*

*Vascular sac*

*Hypophysis* {
*Anterior median lobe*
*Lateral lobe*
*Posterior median lobe*
*Ventral lobe*
}

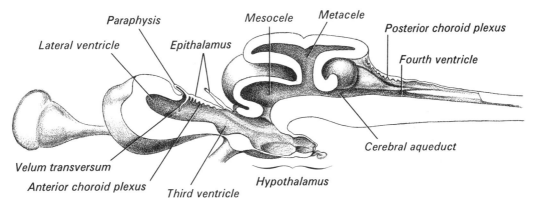

*Paraphysis*     *Mesocele*     *Metacele*

*Lateral ventricle*     *Epithalamus*     *Posterior choroid plexus*

*Fourth ventricle*

*Velum transversum*

*Anterior choroid plexus*     *Third ventricle*     *Hypothalamus*     *Cerebral aqueduct*

**FIGURE 8–4**
Brain (sagittal view).

to the roof of the mouth, where it supplies the taste buds. The main trunk of the hyomandibular nerve passes through the otic capsule, where it ramifies to supply the lateral line system, the skin organs, the muscles of the hyoid arch, the tongue, and the floor of the mouth.

**Statoacoustic nerve** (N. VIII)   This short sensory nerve arises from the anterior end of the medulla, together with nerves V and VII. It innervates the inner ear by means of a vestibular branch to the anterior part of the ear and a saccular branch to the posterior part.

**Glossopharyngeal nerve** (N. IX)   This mixed nerve arises from the medulla just behind nerve VIII. Near its point of origin it exhibits a **petrosal ganglion.** Beyond the ganglion the nerve divides into three branches:

**Pretrematic branch**   This sensory nerve innervates the first demibranch.

**Posttrematic branch**   This mixed nerve supplies the second demibranch with sensory fibers and the muscles of the third visceral arch with motor fibers.

**Pharyngeal nerve**   This innervates the lining of the pharynx.

**Vagus nerve** (N. X)   This large mixed nerve arises from several roots at the posterior end of the medulla. It gives off a **lateral line trunk** that extends beneath the lateral line canal to the caudal end of the body. The vagus nerve then continues as the **branchial trunk,** which gives off four branchial rami to each of the remaining gills. Each of these branches, like those of the glossopharyngeal, ramifies into pretrematic, posttrematic, and pharyngeal nerves. The vagus nerve, posterior to the gill region, is known as the **visceral trunk,** since it provides branches to the heart and abdominal organs.

## OCCIPITAL AND SPINAL NERVES (Fig. 8-2)

*The spinal cord lies in the vertebral canal. To expose it, as well as the occipital nerve and some of the dorsal roots of spinal nerves, remove the musculature and neural arches for a two-inch segment beginning at the site of junction of the chondrocranium and vertebral column. The nerve arising at the junction of the two segments of the axial skeleton is the first spinal nerve. Between it and the vagus arise the occipital nerve roots. Trace the occipital nerve roots to the site at which they unite to form the occipital nerve adjacent to the vagus and then to their continuation as part of the hypobranchial nerve. Another means of exposing the occipital nerve is to find the hypobranchial as it crosses the visceral trunk of the vagus after emerging from the musculature and to trace it medially. To find the ventral roots of the spinal nerves, cut away the lateral walls of the neural arches.*

**Occipital nerve**   Lies posterior to the vagus on each side. Its two roots of origin correspond to the fibers of the hypoglossal nerve of higher vertebrates. By convention, the occipital nerve is not included with the cranial nerves. The roots of the occipital nerve emerge from the ventrolateral surface of the spinal cord (in line with the ventral spinal nerve roots) and then unite. The nerve appears to join the vagus and emerges with it from the chondrocranium. After receiving some filaments from the first few spinal nerves it continues on as the **hypobranchial nerve,** which innervates the skin ventral to the gills and the hypobranchial musculature.

**Spinal nerve**   Each body segment caudal to the brain is provided with a pair of spinal nerves arising from the spinal cord. Each nerve arises from the union of its **dorsal** and **ventral roots.** These roots emerge from the vertebrae through two different sets of foramina. The dorsal root bears a **spinal ganglion.** Shortly after union of the roots, each spinal nerve divides into a **dorsal** and **ventral ramus.** In the region of the bilateral fins several of the ventral rami on each side unite to form a network, or plexus. The **brachial plexus** is formed from the ventral rami of the spinal nerves in the region of the pectoral fins, and the **pelvic plexus** is formed from the ventral rami of the spinal nerves in the region of the pelvic fins.

## Sectional Morphology

# Anatomy of the Dogfish Shark

The optimal procedure in studying morphology is to uncover and identify structures with the minimum amount of disturbance to their natural position. In reality this goal is difficult to attain since in the process of exposing more deeply positioned structures extensive dissection or even transection may be necessary. Such unavoidable dislocation of the natural state makes it difficult to fully appreciate the interrelationship of organs belonging to different systems. This problem probably has become evident in the course of work on the preceding exercises. To help overcome this difficulty and obtain a good three-dimensional understanding of the anatomy of the dogfish shark, the relation of structures in life can be seen at various cross-sectional levels (Fig. 9-1) as well as in sagittal and frontal sections. Since this exercise also serves as a review of material in other exercises in the series Anatomy of the Dogfish Shark, frequent reference will be made to figures in those exercises.

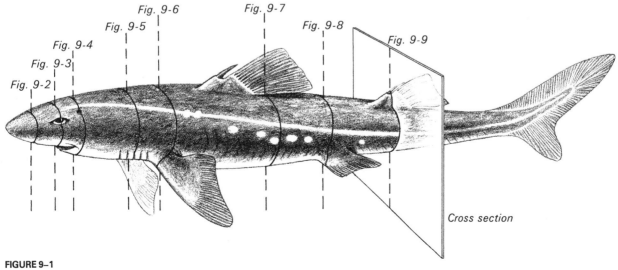

**FIGURE 9–1**

Levels of cross section.

## CROSS SECTIONS

*If suitable specimens are not prepared in advance by the laboratory instructor, use a sharp, long-blade knife or microtome blade to transect a deep-frozen dogfish at the levels shown in Figure 9-1. Then proceed to identify the structures that characterize each segment (with the aid of a hand lens when necessary).*

### Section Through the External Nares (Fig. 9-2)

**Chondrocranium**  This is represented by the scoop-like rostrum that in section has a V-shaped appearance. The **rostral carina** projects downward from the **rostrum** in the mid-ventral plane. Its **precerebral cavity** is filled with a gelatinous material.

**External nares**  The openings located at the ventro-lateral corners. Each naris is partially subdivided by a nasal flap into medial **excurrent** and lateral **incurrent apertures.** The roof of the naris contains numberous septa-like folds, the **olfactory lamellae.** The nares are blinds sacs, since they do not open into the pharynx (as in higher forms), and they serve as olfactory organs.

**Lateral line system**  This special sense organ may be present as the paired **supraorbital canals** located at each mid-lateral edge and an **infraorbital canal** near the mid-ventral line. These canals may be traced into the next segment. They serve to transmit information concerning hydrostatic pressure changes and low-frequency vibrations to the brain.

**Ampullae of Lorenzini**  These special sense organs appear as a cluster of short canals above the external naris. They may also be identified in the subsequent two head sections. They serve as thermoreceptors and possibly also as mechanoreceptors.

**Trigeminal nerve**  The fifth cranial nerve may be represented by the **superficial ophthalmic nerve** located at the apex of the cluster of ampullae of Lorenzini and the **maxillary nerve** located a little above each infraorbital canal. Both of these nerves have a general sensory function over the areas they innervate.

### Section Through the Eyes (Fig. 9-3)

**Chondrocranium**  The section of the chondrocranium encompassing the **orbits** forms the skeletal framework of this segment. The brain is transected at the caudal end of the telencephalon and is enclosed in an oval cavity within the chondrocranium. It reveals two protuberances, the **optic lobes;** a central cavity, the **third ventricle** in the center; and the **optic nerves**

(N. II), which are entering the chiasma (see also Fig. 8-3).

**Eyeball**  The **lens** occupies the bulk of the eyeball. The **cornea** covers the exposed part of the eye and is continued back with the **sclera** over the unexposed part. The vascular **choroid** lies internal to the sclera, while the **retina** is the innermost coat (see also Fig. 7-3).

**Ocular muscles**  Evidence of the presence of three ocular muscles may be found. These include the **superior oblique,** which extends from the antero-medial wall to the mid-dorsal part of the eyeball; the **inferior oblique,** which extends from the anterome-dial wall to the mid-ventral part of the eyeball; and **medial rectus,** which passes forward between the aforementioned two muscles (see also Fig. 7-2).

**Nerves**  Three nerves may be identified in this segment. The **superficial ophthalmic nerve** is located in the dorsomedial corner of the orbit above the superior rectus; the **deep ophthalmic nerve** passes forward beneath the superior rectus; the **infraorbital nerve** runs along the floor of the orbit to emerge as the maxillary and buccal nerves.

### Section Through the Spiracles (Fig. 9-4)

**Chondrocranium**  At this level it is present primarily as the **optic capsules,** which may show evidence of the **ampullae of the horizontal and anterior semi-circular canals** and the **endolymphatic ducts,** which open into the **endolymphatic fossa.**

**Buccal cavity**  The chamber of the mouth lies directly beneath the chondrocranium and at each of its lateral angles is an oblique canal that opens at the **spiracle.** The spiracle is the modified first gill slit of fish. Respiration takes place by water passing through the mouth into the pharynx and then passing out via the gill slits. When a spiracle is present, water also enters through it.

**Splanchnocranium**  This lies beneath the buccal cavity. At this level the mandibular arch is represented by the thick **palatoquadrate cartilage** at each lateral angle the flattened **Meckel's cartilage** extending toward the mid-ventral line. Above the latter lies the **ceratohyal cartilage** of the hyoid arch, which is in contact with the smaller **ceratobranchial cartilage** of the third visceral arch (see also Figs. 2-6, 2-7, 2-8).

**Muscles**  Various muscles belonging to both the branchiomeric and hypobranchial musculature are present. They include the **first levator,** lateral to each otic capsule; **mandibular adductor,** beneath the mandibular arch components; **coracomandibular,** in the mid-ventral plane; and **coracohyoid,** medial

83

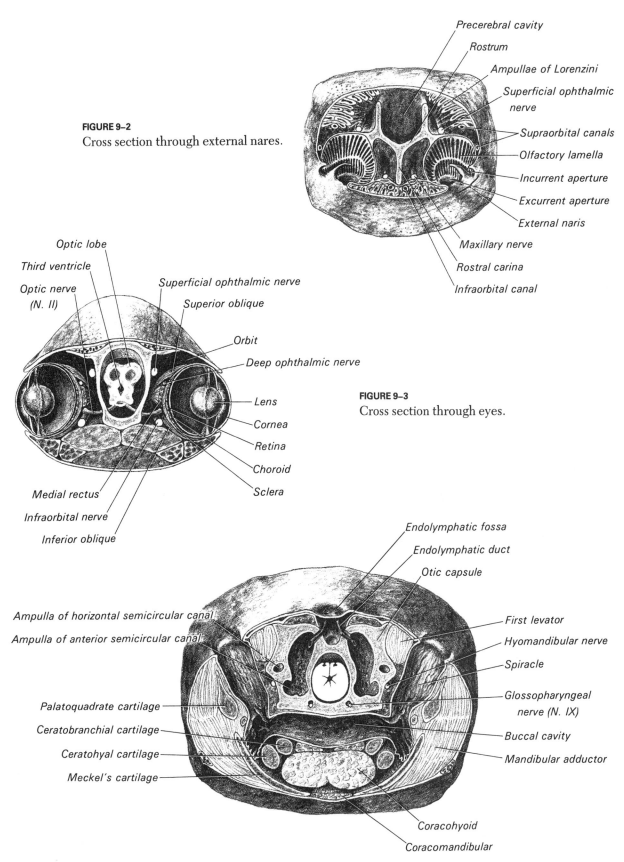

FIGURE 9–2
Cross section through external nares.

Precerebral cavity
Rostrum
Ampullae of Lorenzini
Superficial ophthalmic nerve
Supraorbital canals
Olfactory lamella
Incurrent aperture
Excurrent aperture
External naris
Maxillary nerve
Rostral carina
Infraorbital canal

Optic lobe
Third ventricle
Optic nerve (N. II)
Superficial ophthalmic nerve
Superior oblique
Orbit
Deep ophthalmic nerve
Lens
Cornea
Retina
Choroid
Sclera
Medial rectus
Infraorbital nerve
Inferior oblique

FIGURE 9–3
Cross section through eyes.

Endolymphatic fossa
Endolymphatic duct
Otic capsule
Ampulla of horizontal semicircular canal
Ampulla of anterior semicircular canal
First levator
Hyomandibular nerve
Spiracle
Glossopharyngeal nerve (N. IX)
Buccal cavity
Mandibular adductor
Palatoquadrate cartilage
Ceratobranchial cartilage
Ceratohyal cartilage
Meckel's cartilage
Coracohyoid
Coracomandibular

FIGURE 9–4
Cross section through spiracles.

to the visceral arch (see also Figs. 3-2, 3-3, and 3-4).

**Nerves**  Two sectioned cranial nerves may be identified. They are the **glossopharyngeal nerve** (N. IX), located in the chondrocranium just lateral to the medulla oblongata, and the **hyomandibular nerve,** located between the roof of the spiracle and chondrocranium (see also Fig. 8-2).

## Section Through the Gill Region (Fig. 9-5)

**Vertebral column**  Since this body segment lies beyond the skull, the vertebral column forms the structural frame. Clearly evident is the **centrum** and the **neural arch** that encloses the **spinal cord** (see also Fig. 2-9). A dark mass representing a remnant of the notochord may be within the centrum. The **radices aortae** lie beneath the centrum.

**Pharynx**  This is a dorsoventrally flattened cavity located in the center of the segment. Its roof is supported by several **pharyngobranchial cartilages.** The floor of the pharynx is supported by a mid-ventral **basibranchial cartilage,** lateral to which are **ceratobranchial cartilages.** Laterally the pharynx leads via the **internal gill slit** into the gill chamber (see also Fig. 4-4). The internal gill slit is guarded by papilla-like **gill rakers** that act as strainers. **Gill filaments** appear as thin folds arranged in rows and are responsible for the corrugated appearance of the gill surface. The gill chamber opens to the outside through the **external gill slits.**

**Heart**  At this level the heart is transected because it is located in fishes far forward, dorsal to the pectoral girdle and hypobranchial musculature. The level of the section through the pericardial cavity will probably be near the anterior third of the **atrium.** This part of the heart lies almost directly beneath the middle of the pharynx. Lying ventral to the atrium may be seen the **conus arteriosus** and the apex of the **ventricle** (see also Fig. 5-1). The components of the heart all lie within the pericardial cavity, which is lined by **parietal pericardium.**

**Muscles**  The axial musculature is partially interrupted in the head by the gill region. Thus it extends forward dorsally as the **epibranchial musculature,** laterally as the branchiomeric musculature, and ventrally as the hypobranchial musculature (see also Figs. 3-2 and 3-3). Belonging to the latter group are the **coracobranchials** and the **common coracoarcuals,** which lie lateral and ventral to the heart, respectively. They help in opening the mouth and in swallowing. In addition, small **interbranchial** muscles lie in the roof of the pharynx between the **pharyngobranchial cartilages.**

**Nerves**  At each ventrolateral corner of the branchial

musculature may be seen the **vagus** and **hypobranchial nerves** (see also Fig. 8-2). In its passage caudally from the brain the vagus nerve innervates the lateral line system by means of one branch, and the gills, heart, and abdominal organs by means of another. The hypobranchial nerve, a continuation of the occipital nerve, innervates skin ventral to the gills as well as the hypobranchial musculature.

## Section Through the Pectoral Fins (Fig. 9-6)

**Esophagus**  This organ extends caudally from the pharynx. It is round in cross-section and its inner surface contains numerous finger-like projections, or **papillae.** These provide additional surface area to help facilitate digestion (see also Fig. 4-3).

**Liver**  This gland fills the bulk of the abdominal cavity at this level. It encloses the esophagus both laterally and ventrally. The dogfish shark liver functions in the same manner as higher vertebrates by being a site for the metabolism and storage of food products as glycogen. However, much food is stored as oil, which has a low specific gravity and thus helps make the fish more buoyant.

**Gonads**  These organs lie dorsolateral to the esophagus just beneath the roof of the abdominal cavity and the liver. Identify whether they are **testes** or **ovaries.**

**Musculature**  The pleuroperitoneal cavity is enclosed dorsally by the **epaxial musculature** and ventrally by the **hypaxial musculature.** These two muscle masses represent the dorsal and ventral portions of the myomeres that develop from the embryonic myotomes.

**Scapular cartilage**  This round element belonging to the pectoral girdle is evident mid-laterally between the epaxial and hypaxial muscles (see also Fig. 2-13).

**Pectoral fin**  The elements of the pectoral fin seen at this level are the **pectoral levator** and **pectoral depressor muscles,** which are separated by the **metapterygium, mesopterygium,** and **radial cartilages.**

**Dorsal aorta**  This major blood vessel lies directly beneath the vertebral column. It is formed by the convergence diagonally and mediocaudally of the four pair of efferent branchial arteries from the dorsal angles of the internal gill slits along the posterior part of the pharyngeal roof (see also Fig. 5-3).

## Section Through the Mid-Body (Fig. 9-7)

**Liver**  The right and left **lobes** are seen at the dorsal part of the abdominal cavity. The lobes are separated by the **mesentery.**

**Pancreas**  The pancreas lies in the middle of the abdominal cavity above the stomach. This level reveals

85

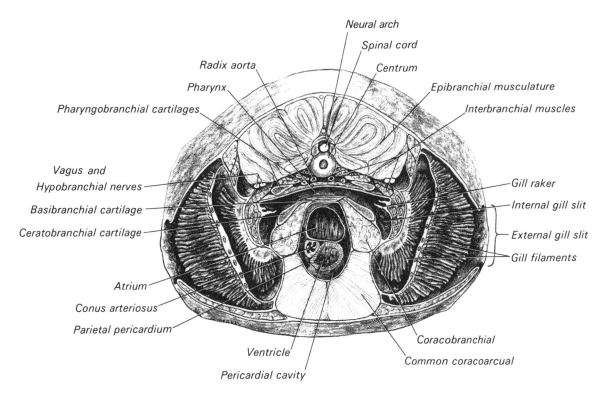

**FIGURE 9–5**
Cross section through gill region.

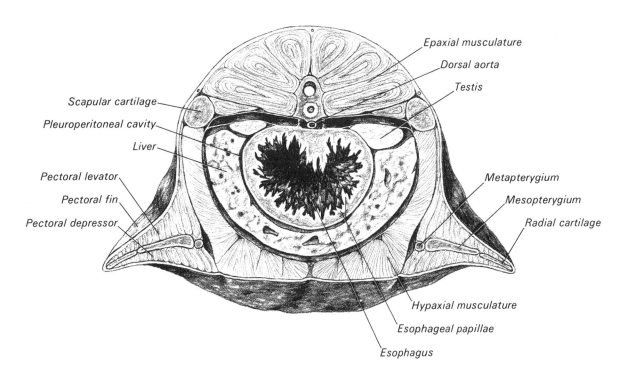

**FIGURE 9–6**
Cross section through pectoral fins.

a section through the dorsal lobe of the pancreas.

**Stomach** This organ is sectioned at the level of the J-shaped end of the pyloric portion.

**Spleen** The spleen may be seen lateral, ventral, and medial to the stomach. It is an organ that functionally belongs to the circulatory system.

**Small intestine** This section through the ileum reveals the **spiral valve**.

**Kidneys** These organs lie on either side of the dorsal aorta retroperitoneally. This is outside of the parietal peritoneum (see also Fig. 4-1). They are flattened bodies that extend nearly the length of the pleuroperitoneal cavity (see also Figs. 6-2, 6-3) and are therefore classified as opisthonephric kidneys.

**Posterior cardinal veins** These vessels lie just lateral to the dorsal aorta and medial to kidneys.

**Mesonephric ducts** These lie on the ventral surface of the kidneys. In the male they serve for the passage of both the urinary wastes and spermatozoa. In the female they are recessed into the substance of the mesonephros and thus may be hard to find.

## Section Through the Cloaca (Fig. 9-8)

**Renal portal veins** These lie dorsal and parallel to the lateral margins of the kidneys.

**Right posterior cardinal vein** This vessel lies between the kidneys in the midline. Only the right vein is present because it extends further caudally than does the left (see also Fig. 5-7).

**Accessory mesonephric ducts** These appear at the ventromedial edges of the kidneys, whose caudal third region they drain. Since little if any urine is excreted by the cranial two-thirds of the kidney, this duct, despite its name, carries most of the urine in the mature male animal.

**Cloaca** This is the common chamber that relieves both the digestive and urinary wastes (and in the male also discharges sperm) and leads them externally.

**Urorectal shelf** A fold of tissue that partitions the cloaca into a dorsal region, the **coprodeum**, and ventral region, the **urodeum**.

**Urinary papilla** This sectioned, cone-shaped projection extends from the dorsal wall of the cloaca. In the male this structure is known as the **urogenital papilla**.

**Pelvic fin** The elements of the pelvic fin seen are the **pelvic levator** and **pelvic depressor muscles**, which are separated by the **metapterygium** and **radial cartilages**.

## Section Through the Tail (Fig. 9-9)

**Basal cartilage** The major cartilage supporting the posterior dorsal fin (see also Fig. 2-11).

**Horizontal septum** A horizontal connecting-tissue partition extending laterally from both sides of the centrum. This septum, together with the vertebra, divides the axial musculature into dorsal and ventral segments.

**Ventral vertical septum** This partition extends vertically from the **hemal spine** of the vertebra to the mid-ventral body wall.

**Caudal vein** This blood vessel lies in the ventral part of the hemal canal that is formed by the **hemal arch**. In the dorsal part of this canal lies the terminal end of the dorsal aorta, or **caudal artery**.

## Sagittal Section Through Head (Fig. 9-10)

*With the aid of a fine hacksaw blade and a long sharp knife, make a mid-sagittal section completely through a severed dogfish head.*

**Skull** The sectioned chondrocranium should reveal its **rostrum** and **rostral carina** (roof and floor). The two components of the mandibular arch of the splanchnocranium that may be identifiable are the median parts of the **palatopterygoquadrate** and **Meckel's cartilages**, both of which have teeth. Just behind the latter is the **basihyal**.

**Brain** The three major segments of the brain that should be noted are the telencephalon, mesencephalon, and rhombencephalon. More specifically, the following brain components within the **cranial cavity** may be identifiable: **cerebral hemisphere**, **optic lobe** (and its ventricle), **cerebellum**, **fourth ventricle** (covered by the posterior choroid plexus), **optic nerve** (entering the optic chiasma), and **hypophysis**.

**Vertebral column** The **neural arches** and **centra** are the major elements that can be seen. They enclose the **spinal cord**. Severed dorsal and ventral nerve roots may be evident.

**Epaxial musculature** The broad mass of epaxial muscles located above the vertebral column.

**Dorsal aorta** This blood vessel lies just beneath the vertebral column.

**Posterior cardinal sinus** The enlarged cavity located just beneath the dorsal aorta starting from beyond the coeliac artery. It represents an expanded posterior cardinal vein.

87

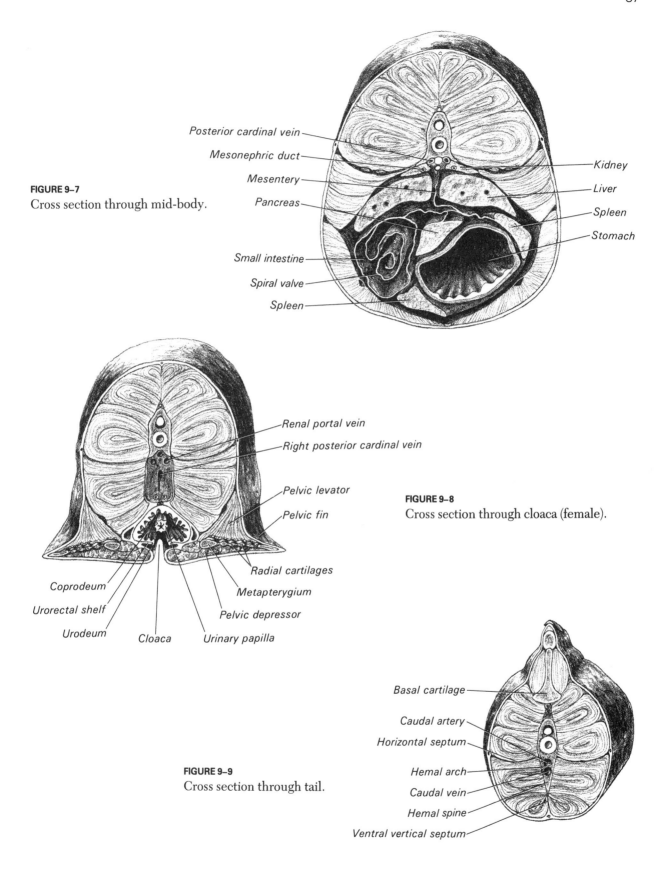

**FIGURE 9–7**
Cross section through mid-body.

Posterior cardinal vein
Mesonephric duct
Mesentery
Pancreas
Kidney
Liver
Spleen
Stomach
Small intestine
Spiral valve
Spleen

**FIGURE 9–8**
Cross section through cloaca (female).

Renal portal vein
Right posterior cardinal vein
Pelvic levator
Pelvic fin
Radial cartilages
Metapterygium
Pelvic depressor
Urinary papilla
Cloaca
Urodeum
Urorectal shelf
Coprodeum

**FIGURE 9–9**
Cross section through tail.

Basal cartilage
Caudal artery
Horizontal septum
Hemal arch
Caudal vein
Hemal spine
Ventral vertical septum

ANATOMY OF THE DOGFISH SHARK

88

**Buccal cavity** The chamber of the mouth, characterized by transverse mucosal folds. It lies beneath the mid-section of the floor of the chondrocranium, known as the **infraorbital shelf** (see also Fig. 2-3).

**Pharynx** This section is a caudal continuation from the mouth and is not sharply demarcated from the buccal cavity. It is the portion of the digestive tract that contains the gill slits and leads into the narrower **esophagus.** The papillae characterize the latter segment.

**Musculature** The hypobranchial muscles may be identified in the region beneath the buccal cavity. These include the **coracohyoid, coracobranchials, coracomandibular,** and **common coracoarcuals** (see also Fig. 3-3). The **thyroid gland** lies above the anterior end of the coracomandibular. The ventralmost muscle is the **first ventral constrictor.**

**Basibranchial cartilages** These median cartilages are the ventral ends of the third to fifth visceral arches. They support the floor of the pharynx.

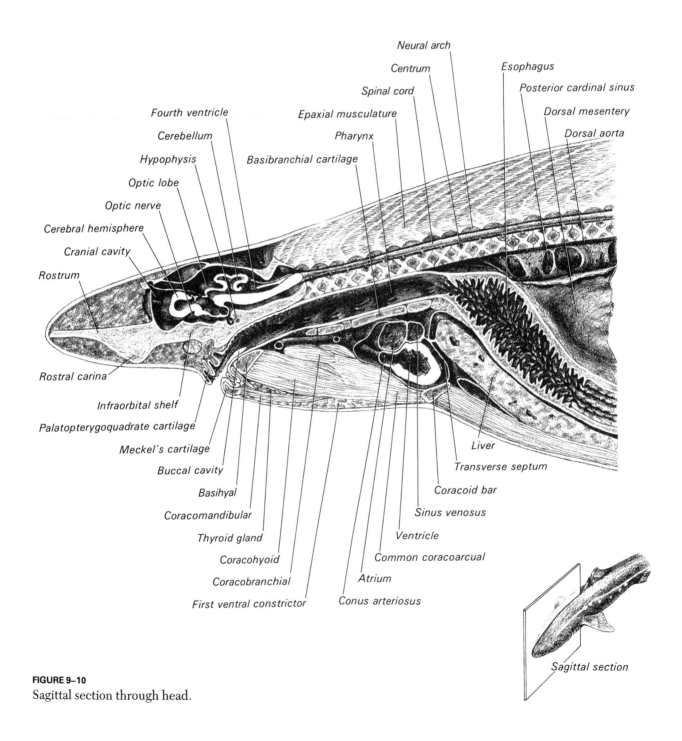

**FIGURE 9–10**
Sagittal section through head.

**Heart**   The three chambers of the heart can be seen in the mid-sagittal section (see also Fig. 5-2). These are the large **atrium** and smaller thin-walled **sinus venosus** beneath the basibranchial cartilages and the thick-walled **ventricle** located ventral to the other two chambers. Extending anteriorly from the ventricle is the **conus arteriosus,** which is characterized by its semilunar valves and which leads to the ventral aorta. The ventricle tip rests on the **coracoid bar.**

**Transverse septum**   The thin partition separating the pericardial from the peritoneal cavities. Opposite the heart is the terminal end of the oviduct. It is separated from the floor of the esophagus by the **liver,** which is suspended by the **dorsal mesentery.**

## FRONTAL SECTION (Fig. 9-11)

*With the use of a hand saw, make a frontal section through the head of a frozen specimen at a level just above the eye. With the aid of forceps, carefully remove the roof of the head as far back as the hyomandibular cartilage to expose the cranial cavity and its contents. One of the eyes can be removed from the orbit.*

**Rostrum**   The scoop-like anterior projection of the chondrocranium filled with a gelatinous material.

**Olfactory capsules**   The rounded swellings located on either side of the base of the rostrum, which house the olfactory sacs. Extending back from the sacs are

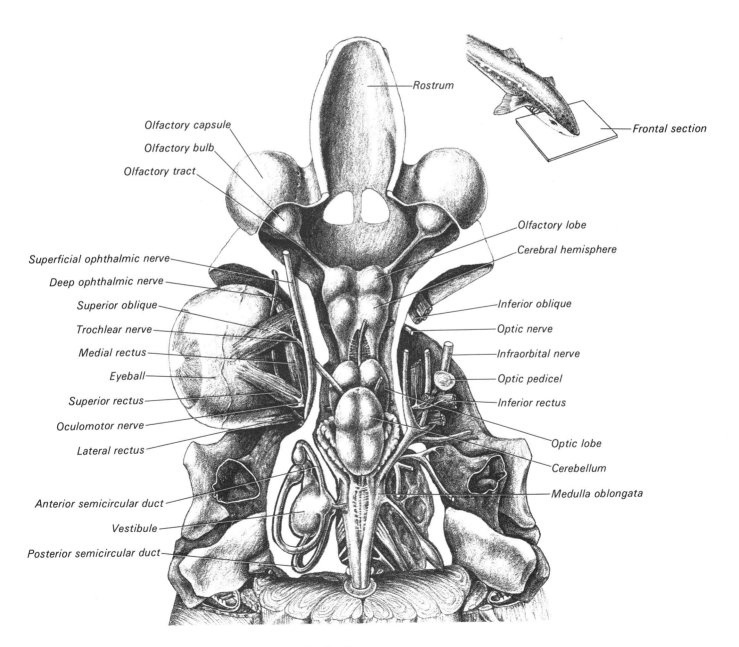

**FIGURE 9–11**
Frontal section through head.

rounded masses, the **olfactory bulbs,** which continue into the **olfactory tracts,** which in turn terminate in the **olfactory lobes** (see also Fig. 8-2).

**Cerebral hemispheres**    The rounded elevations just behind the olfactory lobe.

**Optic lobes**    These eminences are separated from the cerebral hemispheres by the narrow diencephalon.

**Cerebellum**    Because of its large size, which makes it project up at a higher level than the other elements, this oval structure is the most likely part of the brain to have become damaged.

**Medulla oblongata**    This terminal segment of the brain is continuous with the spinal cord.

**Eyeball**    Fills most of the orbit. Several ocular muscles insert on its dorsal surface and several nerves pass behind it.

**Ocular muscles**    Six are present: two **oblique** (superior and inferior) and four **recti** (superior, inferior, medial, and lateral) (see also Fig. 7-2).

**Other orbital structures**    These include the **optic pedicel,** which supports the eyeball; the **optic, oculomotor, trochlear,** and **abducens nerves;** the **superficial** and **deep ophthalmic nerves;** and the **infraorbital nerve.**

**Semicircular ducts**    The vertical **anterior** and **posterior semicircular ducts** may be identifiable lateral to the medulla oblongata. The larger cavity in between the ducts represents the site of the **vestibule.**

# Anatomy of the Dogfish Shark

The principal emphasis of the preceding exercises has been essentially morphological. However the structure and organization of the dogfish, which have been influenced by its evolutionary process, also have functional manifestations. Thus this exercise covers some major evolutionary and functional aspects that are associated with the structural elements of the dogfish shark's various body systems. In this manner we can place the morphological information that has been presented in a more realistic context.

## EXTERNAL MORPHOLOGY

One of the characteristic features of vertebrates, including the dogfish, is a thick epidermis consisting of multiple cell layers. Embedded in the dogfish's skin are placoid scales (see Fig. 1-1) (the most primitive of the four categories of scales). It is the projecting spines of these scales that give the skin its rough texture.

Placoid scales, which differ somewhat in shape in various regions of the dogfish's body, have a common organizational pattern. The tapering spine is capped by enamel which overlies an inner layer of hard, bone-like dentine that extends up through the epidermis from a broad, basal dentine plate (see Fig. 1-2). These scales are particularly interesting because they are very similar to those of early vertebrates, the ostracoderms. The belief is that placoid scales are derived from the large bony scales, or continuous head shield, of some agnathan ancestors but that they have become reduced in size and have increased in number.

The dogfish changes its colors when the pigment in the melanophores, which is found in the dermis of the skin, becomes clumped (see Fig. 1-2). These melanin-bearing chromatophores are similar to those that are responsible for the adaptive color changes that occur in other cold-blooded vertebrates, and they function as protection against predators. In the dogfish *Squalus*, pigment aggregation is apparently controlled by autonomic (adrenergic) nerve fibers.

Modern groups of sharks have apparently survived several hundreds of millions of years, retaining, for the most part, a generalized structural organization. Several hundred species of sharks currently exist. They have a wide range in size, from about one-half foot to about 45 feet. The spiny dogfish lies near the lower end of this spectrum, averaging two to three feet in length.

Sharks exhibit the process of adaptive radiation, which means that they have spread into various habitats with concomitant alterations of behavior and morphological features. Thus we find sluggish forms as well as the rapid-swimming, predaceous ones. The spiny dogfish and some others belong to the latter category. The streamlined body shape of the dogfish and that of similar sharks, which is reflected in its flattened head, fusiform body, and large, strong tail, facilitates its continuous, free-swimming character. Its potential for mobility and energy conservation is reinforced by three additional features: small scales (instead of large bony plates), light weight, all cartilaginous skeleton, and deposition of an oil (squaline) in the liver, which has a specific gravity that is lower than that of water to facilitate the shark's buoyancy.

## Locomotion

Aquatic vertebrates are known as primary swimmers because their ancestors also swam. Although many fishes are inactive or slow moving, others have impressive swimming potential in terms of both endurance and speed.

Four basic requirements that must be satisfied for proficient swimmers are (1) low water resistance, or drag, (2) vertical position control, (3) maintenance of orientation and steering, and (4) propulsion.

The fusiform body of the dogfish shark offers the best solution to both pressure and friction drag as a result of its structural adaptation. Although the small spiny scales represent an adaptation in body surface, they still produce some degree of turbulence and thus some drag.

Forces that act on a swimming fish are traditionally represented in three directions that are at right angles to each other (Fig. 10-1). It *yaws*, or turns its head, from side to side; it may *roll*, or rotate its body on its axis; and it may *pitch*, or dip, up or down. Once begun, the forces that generate yaw or pitch continue to accelerate along with the fish's movements. The unpaired anterior and posterior dorsal fins (see Fig. 1-3) serve to reduce instability of both yawing and rolling, which is generated when the dogfish flaps its powerful tail (see below). By undulating its pelvic fins, the dogfish may enhance the action of its dorsal fins in regaining stability.

Lateral movement of the heterocercal dogfish tail results in forcing water backward and downward. Consequently, the tail is lifted up (i.e., moved in an opposite direction, upward and forward) and, as a result, the head of the swimming dogfish is angled downward. Downward pitch accelerates the natural tendency for the dogfish to sink, because it is heavier than water and lacks a swim bladder. This factor is offset by (1) the lift of the water, (2) the pectoral fins, whose angle can be adjusted to tilt upward at their front edge and thus lift the head, and (3) the ventral flattened snout, which forms a plane that pushes against the water during swimming, thereby directing the head upward.

For most fishes propulsion is brought about by undulation of the axial musculature acting on the flexible axial skeleton, producing waves of curvature that travel down the length of the body (Fig. 10-2). Body undulation is facilitated by the segmental organization of the trunk musculature. Alternating waves that pass down each side of the fish's body produce a backward thrust of the body on the water, inducing forward propulsion.

In the dogfish the back-and-forth movement of the caudal fin is the major factor in propulsion, because it increases the thrust at the posterior end of the body. This fin constantly flaps at an oblique angle against the water (Fig. 10-3), generating a force $(F_t)$, which, in turn, produces an equal force in the opposite direction $(F_w)$. The latter force can be resolved into forward $(F_f)$ and lateral $(F_l)$ components, which are at right angles to each other (Fig. 10-3). Because of the shark's streamlined shape, the water offers little resistance to $F_f$, so propulsion is enhanced. $F_l$ produces yawing, causing the body to pivot around its central axis, which (as noted previously) is offset by dorsal fin stabilization.

## SKELETAL SYSTEM

The head skeleton is the oldest part of the body's entire skeletal system. Although the head skeleton exhibits great structural diversity, it essentially consists of three basic units. (1) The neurocranium is composed of the brain case and its associated sensory capsules; (2) the splanchnocranium is represented by the jaws and gill arches or their derivatives; and (3) the dermatocranium consists of the dermal bony elements that may have evolved from the external armor of the primitive fishes.

Jaw formation was one of the major critical evolutionary developments in vertebrate history and thus merits special attention. The jaw provided a new, highly efficient food-securing mechanism, which facilitated the transition of small, jawless, sedentary fishes to large, active fishes. However, the appearance of jaws in gnathostome fishes did not mark the evolution of an entirely new structure; it merely represented a change in the function of an old structure. Most important evolutionary changes occur in this manner.

The process of jaw formation probably occurred in three phases. It is hypothesized that, in the first phase, the ancestral jawless vertebrates possessed undivided cartilaginous bars, or arches, that supported the gills along the medial walls of the pharynx (Fig. 10-4A). An arch separated each internal pharyngeal slit from its adjacent slit. Since the arches did not enclose the pharynx dorsally, protection was provided there by the axial skeleton. Ventrally, however, the arches met at the

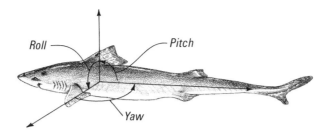

**FIGURE 10–1**
Oscillation of the dogfish during swimming.

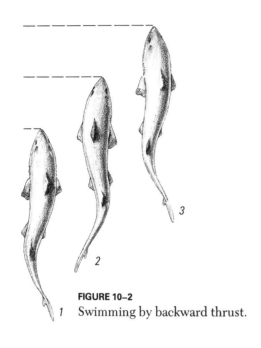

**FIGURE 10–2**
Swimming by backward thrust.

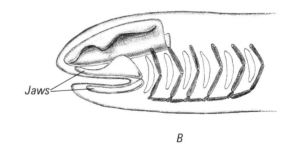

**FIGURE 10–3**
Forces acting at the caudal fin.

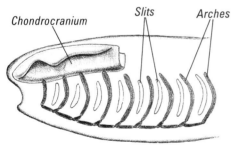

Visceral skeleton
*A*

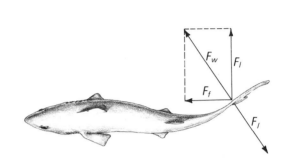

*B*

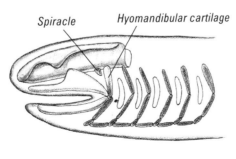

*C*

**FIGURE 10–4**
Evolution of jaws: (A) agnathous vertebrates, (B) ancient gnathostomes, and (C) modern gnathostomes.

midline and were interconnected by a series of longitudinal bars.

The next phase might have occurred in primitive gnathostomes, among an unknown group of placoderms, when the arches became divided into several segments. The major dorsal and ventral segments became the epibranchial and ceratobranchial cartilages, respectively. In addition, a pair of anterior arches that were located opposite the mouth became modified into jaws by enlarging and developing teeth (Fig. 10-4B). The epibranchial components of this arch formed the upper jaw, or fused palatoquadrate cartilages. The lower jaw, or mandibular arch, might have derived from the ceratobranchials of the same arch. At this stage the jaws were not, as yet, supported by the second arch.

In the third phase the upper part of the second arch, or hyomandibular cartilage, became modified to support the posterior end of the jaws, and the rest of the arch became part of the tongue or gill support apparatus (Fig. 10-4C). This type of arrangement, which is present in the dogfish, permits the mouth to be larger and open more widely than it could otherwise. This change, however, brought about the reduction in size of the first gill slit to a small, round spiracle.

## MUSCULAR SYSTEM

The undulatory movements of the segmentally arranged axial musculature that produces body propulsion result from the contraction of red and white muscle fibers. However, at normal cruising speeds, only the red muscles are active. The white fibers are primarily used as a reserve power source to facilitate bursts of speedy swimming, such as in catching a prey or escaping from a predator.

The red muscle fibers are simply organized muscle bands that run longitudinally, parallel to the body axis. The white fibers, however, are folded in a complex manner (Fig. 10-5). They are arranged as $<$-shaped myomeres with the upper edge turned forward. The zigzag angles become increasingly acute in the more caudal myomeres. Since each myomere folds beneath its neighbors, a vertical section at any level will cut through several of them.

The folding of myomeres is considered to serve two possible purposes. First, every myomere extends cranially and caudally over several body segments, which permits even short parallel fibers to act over a relatively long distance. Moreover, the oblique orientation of the myosepta extends the pull over several vertebral joints (Fig. 10-6). Second, as a result of the overlapping of myomeres, any given vertebral joint can be flexed by several different myomeres. This phe-

nomenon facilitates both smoothening out and reinforcement of the contraction movements.

Finally, it should be noted that both the red and white muscle fibers are components of striated somatic musculature. This type of muscle fiber, by virtue of its microscopic and submicroscopic organization, is especially suited for the intense contraction that is needed for locomotion as contrasted with the smooth visceral musculature located in the digestive tract. Branchial muscle, on the other hand, which is associated with the gills and jaws, is structurally similar to skeletal muscle. This arrangement is quite natural, since the branchial muscles are actively involved in contraction during food ingestion and respiration.

From the evolutionary perspective, the organization of the musculature becomes increasingly complex. Thus in cyclostomes, especially the lamprey, the presence of myomeres is evident, but a lateral septum is lacking. Thus the axial musculature is not divided into epaxial and hypaxial divisions. Jawed fishes are more advanced, with myomeres that are positioned at an angle and a massive axial musculature that is divided by a septum into two portions. In tetrapods the limbs carry out propulsion, so their limb muscles enlarge and the axial musculature diminishes.

## DIGESTIVE SYSTEM

In all vertebrates the digestive tract can be divided into three functional parts. The first part is involved primarily in securing and holding the food; the second part is the site of digestion (chemical breakdown) and absorption; and the third part is concerned with water resorption.

If we examine the evolutionary pattern of the digestive tract from cyclostomes to mammals, it becomes obvious that the digestive tract becomes increasingly complex, including progressively more components (Fig. 10-7). The impact of evolutionary change is strikingly evident around the anterior end of the gut, where special structures facilitate feeding. This development is only one reflection of the cephalization process. Thus the dogfish, which has a distinct head, bears a pair of powerful jaws that act as tongs to grasp and handle food. This feature provides jawed fishes with a major advantage over agnathans. The presence of a larger brain, good eyes, and balancing organs, which are associated with this feature, facilitates the coordination that is necessary for a successful predator.

Beyond the pharynx the digestive tract is relatively simple. There is a short esophagus through which food passes to a large stomach; and the stomach is characterized by a high acid environment and the ability to secrete pepsin, which splits protein. It appears that

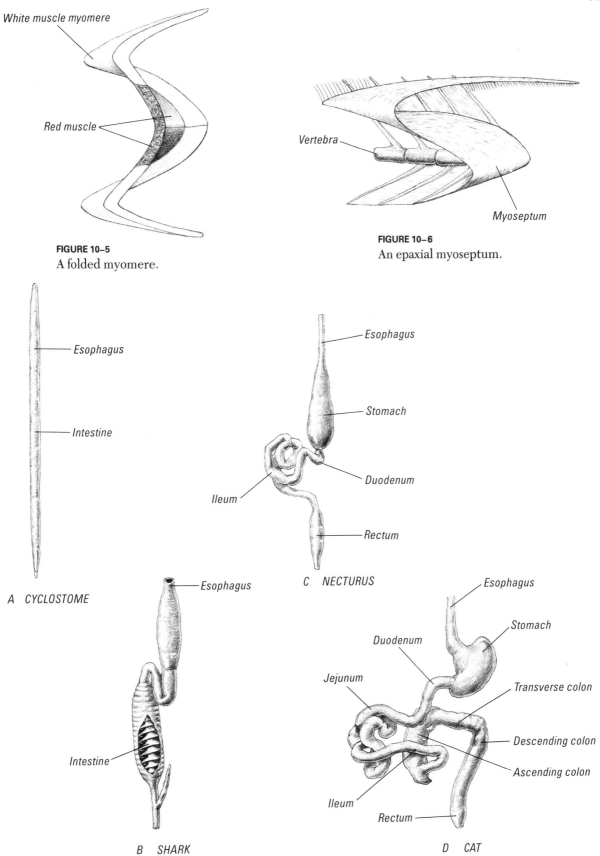

**FIGURE 10–5**
A folded myomere.

**FIGURE 10–6**
An epaxial myoseptum.

*A CYCLOSTOME*

*B SHARK*

*C NECTURUS*

*D CAT*

**FIGURE 10–7**
Digestive tract of representative vertebrates.

the formation of a stomach and the ability to secrete pepsin, which may not have arisen simultaneously, were special gnathostome developments. Those developments facilitated the digestion of the large food masses that could be ingested by the more effective mouth.

The intestine is divided into three segments: duodenum, ileum, and rectum. The duodenum, which is short and undifferentiated, receives secretions from the liver and pancreas. The liver secretes bile, which facilitates the catabolism of fat, while the pancreas secretes juices that contain a variety of enzymes, which aid in the breakdown of all types of nutrients. The pancreas also contains unclustered islet cells, which secrete insulin.

The straight ileum is the longest and widest part of the digestive tract, and its internal epithelial surface area is increased markedly by the presence of a spiral valve. Although an equivalent increase could be achieved by lengthening and coiling the entire intestine, spiralization of the mucosa is more suitable for the long, slender body cavity of the fish.

The rectum presents a single short cecum-like organ, the rectal gland. Many believe that this organ helps to maintain the dogfish in osmotic equilibrium with seawater. Apparently, it does so by extracting excess sodium salts from the blood and secreting them into the rectum, thus supplementing kidney function.

## RESPIRATORY SYSTEM

In the dogfish, the mouth opens into the buccal cavity, which, in turn, leads into the pharynx (see Fig. 4-5). The pharynx has five pairs of gill slits on its sides that open independently to the outside.

During breathing, the muscles that are attached to the basihyal cartilage on the floor of the mouth contract, thus expanding the volume of this chamber. Seawater then flows through the open mouth (as well as the spiracles) but not through the gill slits. The external edges of the gill slits act as passive valves, which come together temporarily, sealing the pharynx into which the water has passed. Next, the muscles that draw the epibranchial and ceratobranchial cartilages closer together contract, and the floor of the mouth is brought up, which results in increased internal pharyngeal pressure. Consequently, water is forced through the gills and out of the pharynx via the gill slits, whose edges have parted, because water pressure is now higher in the mouth.

If we take a closer look at the passage of water through the gills, we can see how gas exchange takes place. The gill, or primary filaments of a gill arch, are stacked on top of each other on opposite sides of a supporting interbranchial septum (see Fig. 4-6). Both sides of each gill filament bear a multitude of secondary lamellae (Fig. 10-8). Thus water that is taken in via the mouth during respiration is channeled through progressively smaller passageways until it reaches the secondary lamellae, at which point the passageway might be as narrow as a fraction of a millimeter. It is there that gas exchange takes place.

Gas exchange is possible because of the ramification of blood vessels within the gills. Thus an afferent branchial artery enters each gill and provides a filamental branch for each gill filament. This fine branch ascends to the apex of the filament and then returns to terminate at the efferent branchial artery, which leads the blood out of the gill (Fig. 10-8). Then capillaries loop across the thin-walled secondary lamellae between pairs of filament vessels (Fig. 10-8, insert). The blood in the lamellae flows from outside the lamellae inward, while the water flows from inside the gills to the outside. Thus, the currents of the two liquids, passing within one micron of each other, travel in opposite directions, resulting in a countercurrent exchange.

Exchange by diffusion occurs when the oxygen-depleted blood that enters the respiratory system comes in close proximity to water that has a somewhat higher oxygen gradient. The continual passage of oxygen from the water to the blood occurs because the imbalance in the diffusion gradient is maintained even when the blood–oxygen content increases, since the water–oxygen content increases simultaneously as "newer" water comes in contact with the lamellae. The diffusion of carbon dioxide from the blood to the water is facilitated by the same mechanism.

## CIRCULATORY SYSTEM

The heart of the dogfish consists of four chambers: sinus venosus, atrium, ventricle, and conus arteriosus. The entire heart is enclosed in a tough, semirigid membrane, the pericardium (Fig. 10-9).

The sinus venosus receives all the venous blood, since it is the site at which the paired common cardinal and hepatic veins terminate (see Fig. 5-7). Its thin walls reflect the sparse amount of musculature present. To be effective, no back flow of blood should occur when the sinus contracts. In typical heart chambers, this phenomenon is prevented by the presence of valves. However, these are absent in the sinus venosus. Nevertheless, back flow is prevented by shortening of the sinus at each contraction, by narrowing of the lumen of the common cardinal veins, and by muscular spincters that are present at the termination of the hepatic veins.

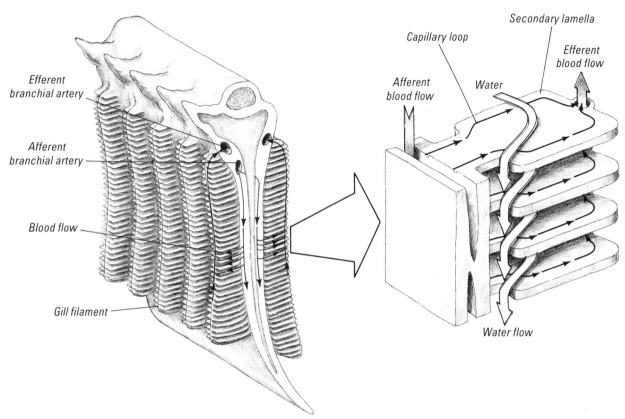

**FIGURE 10–8**
Blood and water flow through the gills.

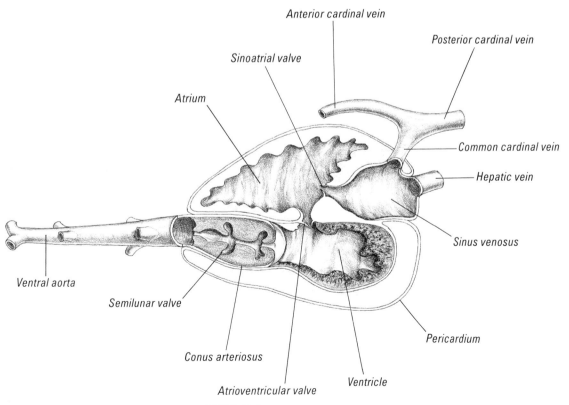

**FIGURE 10–9**
Blood flow through the heart.

The blood propelled by the sinus venosus passes into the atrium, which has valves at both the entrance and exit of its blood flow. The walls of the atrium are relatively thin; its musculature is uniformly distributed; and, internally, it has many small outpocketings in its wall, which serve to expand its total surface area. As a result the atrium can act as an expandable storage chamber, which serves to fill the heart's main pump, the ventricle.

The architecture of the cardiac ventricular musculature in the dogfish is primarily spongy. Thus, the blood, instead of being held in a central cavity, flows into spaces between the loosely organized muscle fibers of the ventricle.

During ventricular contraction the atrioventricular valves close the orifice between the atrium and ventricle. Then the blood from the ventricle passes through the conus arteriosus, which is a continuation of the cardiac musculature. The conus serves to smooth out the pulses in blood pressure, so a relatively even flow reaches the delicate capillary beds in the secondary lamellae of the gills. The conus houses a number of semilunar valves, so it also prevents the back flow of blood into the ventricle (see Fig. 5-2).

Because the pericardium is semirigid, when the ventricle contracts, the pressure between the pericardium and the heart is lowered, which helps to expand the sinus venosus. This action draws the blood from the adjacent sinuses into the heart. The pericardium also suspends the heart in a fluid-containing compartment, thus serving to cushion vibrations generated by the propulsive activities of the animal.

Cardiac muscle is inherently capable of contracting in a rhythmic fashion even though histological continuity is lost as the heart develops. A pacemaker mechanism has evolved to sustain sequential contractions.

In the dogfish the heartbeat is generally initiated in the sinus venosus, and then an electrical wave of excitation spreads over the heart, bringing about the contraction sequence of the atrium, followed by that of the ventricle.

## UROGENITAL SYSTEM

The excretory organs of vertebrates are the paired kidneys, which are drained by ducts. Although several different types of kidneys are recognized, only those relevant to an anamniote, such as the dogfish shark, are discussed here.

The archinephros represents the most primitive type of kidney and is assumed to have been present in ancestral vertebrates (Fig. 10-10A). It apparently extended down the entire length of the body cavity. The many segmentally arranged tubules of an archinephros

presumably opened at their free ends into the coelom by means of ciliated, funnel-shaped apertures, the nephrostomes. Glomeruli, which are suspended in the coelomic cavity, are thought to have been aligned with the nephrostomes. Tissue fluid of ancestral vertebrates is postulated as having filtered out of the glomeruli into the coelom, traveling via the nephrostomes into the tubules. From there the fluid passed down the archinephric ducts to be eliminated from the body. The embryos of hagfishes have an archinephric kidney.

The majority of anamniotes have different types of embryonic and adult kidneys. The anterior portion of the nephrogenic tissue mass is a transitory embryonic organ (pronephros), while the bulk is the functional opisthonephros of adults. Each of its tubules consists of a renal corpuscle, a pair of convoluted tubules, and a collecting tubule. This type of kidney is found in adult cyclostomes, where the duct carries only urine (Fig. 10-10B). Since sperm ducts are not present, the male germ cells are liberated directly into the coelomic cavity and are released to the outside via a pair of genital pores. This system is obviously not a very efficient means for sperm transport.

In primitive gnathostomes the tubules at the anterior end of the opisthonephros become connected with the testis. As a result both sperm and urine are transported by a common duct. However, since this multipurpose duct was not fully satisfactory, alternative modes for separating urinary and sperm transport evolved.

The dogfish shark represents an intermediate step in this direction (Fig. 10-10C). Here too the cranialmost tubules serve to carry sperm to the archinephric duct. The opisthonephric kidney is drained primarily by means of accessory (mesonephric) urinary ducts (see Fig. 6-2). The archinephric duct carries urine only from the anterior part of the kidney and thus is principally concerned with sperm transport. (In the female the ducts are exclusively urinary.) In teleosts a new sperm duct is present, and sperm and urine are eliminated by completely separate pathways. Amniotes have also developed a new urinary duct, but it is not homologous to that of sharks.

## NERVOUS SYSTEM

The brain can be considered an anterior expansion of the spinal cord. After an initial enlargement the developing brain bends ventrally and is subdivided by two constrictions into the prosencephalon, mesencephalon, and rhombencephalon. Somewhat later the prosencephalon develops two pairs of outpocketings, the olfactory lobes and the cerebral hemispheres, which together form the telencephalon, while the re-

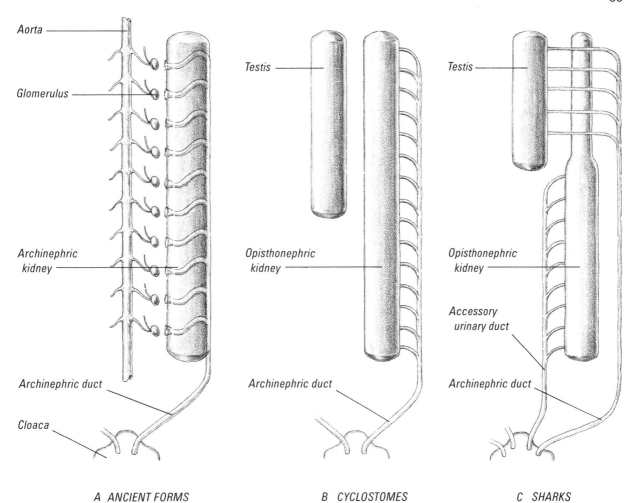

**FIGURE 10–10**
Evolution of male vertebrate kidney and urogenital ducts (*A*, *B*, and *C*).

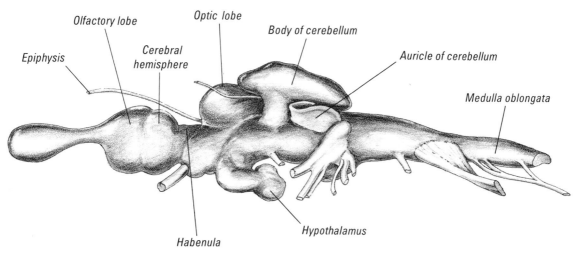

**FIGURE 10–11**
Brain (lateral view).

mainder of the forebrain becomes the diencephalon. With the development of the cerebellum, it and a few other parts become the metencephalon, while the balance of the hindbrain comprises the myelencephalon. In the course of these developments the original flexures in the brain are straightened, so that it assumes a horizontal plane (Fig. 10-11).

The parts of the brain that become especially well developed are those that are used most frequently. In lower vertebrates function is quite localized in the brain compared with that of higher forms. The olfactory lobes, as expected, are well developed in carnivorous predators, such as the dogfish shark. The cerebral hemispheres, on the other hand, are small and are not distinctly demarcated from the olfactory lobes, with which they probably have associated functions, serving to receive and integrate olfactory (and behavioral) information.

The epiphysis may have both endocrine and photo-receptive functions. The habenula, which is located at the rear of the diencephalon, is thought to serve as an integration center for olfactory impulses passed on from the cerebral hemispheres. The function of the hy-pothalmus is uncertain. It has been related to swimming, pressure perception, or as an association center that controls many visceral activities.

The prominent optic lobes, which receive the optic tracts, serve as association centers for vision. The cerebellum is relatively large, projecting anteriorly over the optic lobes and backward over the posterior choroid plexus. It consists of a central body and a pair of auricles. The body of the cerebellum receives proprioceptive impulses from the muscles of the body via the spinal cord and medulla. The auricles receive impulses from the inner ear, lateral line system, and other sensory areas. Both parts of the cerebellum serve to coordinate muscular movements and aid in maintaining equilibrium during swimming.

The medulla serves as a relay center for impulses that pass between the spinal cord and the more anterior parts of the brain. In addition, it is an association center that controls certain visceral activities. The expanded anterior part of the medulla is continuous with the auricles and thus receives cranial nerve fibers from the inner ear, lateral line system, and other receptor areas.

# ANATOMY OF THE MUD PUPPY *NECTURUS*

# External Morphology

# Anatomy of the Mud Puppy *Necturus*

*Necturus* is studied as an example of a primitive amphibian closely related to the forms from which more advanced land vertebrates develop. Its intermediate phylogenetic position as a primitive terrestrial tetrapod is reflected in the fact that it retains some features found in fish and exhibits structural elements corresponding to those in higher vertebrates.

Because the mud puppy has limbs and lungs, it is placed within the superclass Tetrapoda. Because it has a tail, it belongs to the amphibian order Urodela (Caudata). The genus *Necturus* is represented by two species, one of which has, in turn, subspeciated. The most common form, *N. maculosus maculosus,* ranges from the Great Lakes to Louisiana and east to the Atlantic. The other species, *N. punctatus,* is found chiefly in the lower segments of the rivers of North and South Carolina. It is a small, uniformly colored form and is probably a dwarf derivative of *N. maculosus.* Another form found in the Carolina states closely resembles *N. maculosus* but fails to attain its size. It is therefore considered a distinct subspecies, *N. maculosus lewisi* (Fig. 1-1).

*Necturus* is frequently caught by fishermen but is of little commercial value. It searches for food at night and lives on crayfish, aquatic insects, and worms. Like other amphibians, *Necturus* cannot withstand elevated temperatures for prolonged periods. The optimal temperature for its survival is about 18°C. The fact that the mud puppy lives in a wide variety of habitats

suggests that temperature does not significantly control distribution. Rather, the distribution of amphibians, like that of fishes and birds, often depends upon their breeding site preferences. Thus, *Necturus* is most commonly found in streams having a stony bed, since such sites are preferred to any other habitat for nesting.

Although *Necturus* spends its entire life in water, it does not undergo metamorphosis as do most other amphibians, but rather retains its external gills. The persistence of a larval characteristic in an adult animal that has not undergone metamorphosis is known as **neoteny.**

*Necturus maculosus maculosus*

*Necturus maculosus lewisi*

**FIGURE 1–1**
Subspeciation by dwarfing.

## INTEGUMENT (Fig. 1-2)

The skin of *Necturus* is specialized as a highly protective coat. This protective quality is due to a slimy nature and the fact that the body color changes according to variations in environmental light. These properties are the result of the activities of distinct histologic structures in the skin.

*With a compound microscope, study a prepared section of skin from* Necturus *or from another amphibian if one from the mud puppy is not available.*

Like other amphibians, *Necturus* lacks scales. Its skin is smooth and moist because of the presence of numerous glands capable of secreting copious amounts of mucus.

As in other vertebrates, the integument of *Necturus* consists of an epidermis and a dermis. The **epidermis** is made up of a **stratum basalis,** a layer of low columnar cells, upon which rest several layers of increasing numbers of polyhedral cells that grade into flattened cells near the free surface. The outermost squamous layer is only slightly cornified, as with all water-living amphibians. A stratum corneum, or true cornified layer, is present in the land-living amphibians, such as frogs.

The epidermis of *Necturus* is characterized by the presence of numerous club, or flask-shaped, cells, which appear to be of a glandular nature and, together with the mucous glands, are responsible for the slime coat covering the body surface.

The epidermis is separated from the dermis by a basement membrane and a fibrous layer containing pigment cells, which are known as **chromatophores.** They are classified on the basis of the kind of pigment they contain. The general color tone of amphibians is green, owing to the arrangement of the chromatophores in three layers. The melanophores, which contain dark pigment, are the deepest and are covered by a middle layer of iridocytes. The iridocytes are pseudochromatophores, since they lack pigment but contain crystals of guanine that diffract the light. This light has a blue-green color. The most superficial of the three pigment layers is made up of lipophores, which contain a yellowish pigment that filters out the blue so that the overall color effect is green. Changes in body coloration are determined by pigment cell activity. Pigment cells may respond directly or indirectly to incident light. In an indirect response, impulses resulting from stimulation of the eyes are relayed first to the brain and then to the adrenal and pituitary glands, causing hormonal secretions that bring about pigment cell alterations.

The **dermis** is relatively thin and is composed of an outer, looser **stratum spongiosum** and an inner, more compact **stratum compactum.** The dermis contains an abundant number of blood vessels, lymph spaces, nerves, and glands. The glands consist of saccular mucous glands and poison glands, both of epidermal origin. The **mucous glands,** lying just subjacent to the epidermis, have a low cuboidal or almost squamous cell lining. The **poison glands,** containing granular, irregularly spaced secretory cells, are larger and more deeply located and are sheathed in a heavy connective-tissue layer. The poison glands occur on the dorsal surface of the body and function as a deterrent to predators. Mixed mucous and poison glands can occasionally be observed.

## EXTERNAL FEATURES (Fig. 1-3)

**Head**    Flattened dorsoventrally, it contains a terminal mouth.

**Lips**    These are fairly well developed and bound the mouth, which sucks in water through the front.

**External nares**    Widely seperated at the anterior end of the snout, they communicate with the mouth at the internal nares.

**Eyes**    Small, located above the corners of the mouth. They have no lids.

**External gills**    Three in number, they are functional and are located just in front of the forelimbs. Being derivatives of the integument, these gills are not homologous with the internal gills of fishes.

**Gill slits**    Two pairs are present bilaterally between the external gills. They communicate with the pharynx.

**Gular fold**    Transverse fold of skin located at the ventral surface, anterior to the level of the gills. It is the dividing line between head and trunk.

**Trunk**    Compressed dorsoventrally and covered by a smooth skin.

**Pectoral limbs**    The forelimbs are small and weak, and each is divided into an arm, a forearm, and a hand that bears four digits.

**Pelvic limbs**    The hind limbs are similar to the forelimbs, and each is divided into a thigh, a leg, and a foot that bears four digits.

**Cloacal aperture**    This opening is located ventrally in the midline, posterior to the hind limbs. Thus situated, it marks the posterior limit of the trunk. Digestive products, excretory products, and gametes pass through the cloacal aperture.

**Tail**    Compressed laterally, it is well developed.

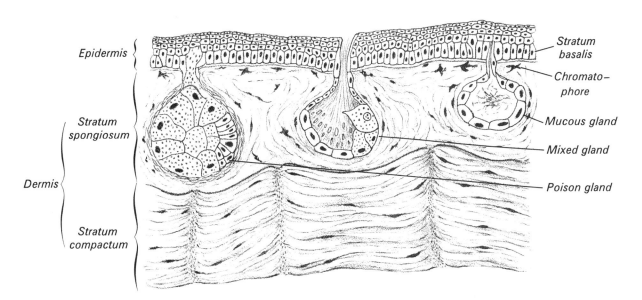

*Epidermis* {

*Dermis* {

*Stratum spongiosum*

*Stratum compactum*

*Stratum basalis*

*Chromato-phore*

*Mucous gland*

*Mixed gland*

*Poison gland*

**FIGURE 1–2**
Skin (cross section).

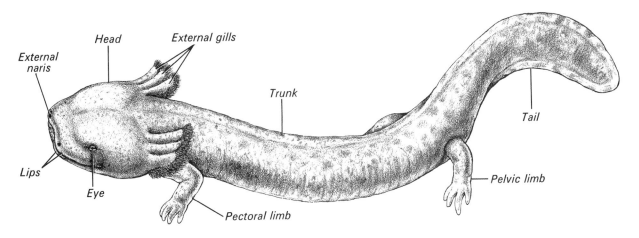

*Head*

*External gills*

*External naris*

*Trunk*

*Tail*

*Lips*

*Eye*

*Pectoral limb*

*Pelvic limb*

**FIGURE 1–3**
External features.

# Anatomy of the Mud Puppy *Necturus*

The skeleton of *Necturus* is a mixture of bone and cartilage. Although it is more complex than that of the dogfish shark, it is still primitive. The skeleton may be divided into axial and appendicular parts.

## AXIAL SKELETON

The axial skeleton consists of the skull, vertebral column, sternum, and ribs.

*Study a special preparation of the chondrocranium from which the dermal bones have been removed. Specially preserved specimens are also necessary for the study of the cartilaginous splanchnocranium.*

### Skull: Chondrocranium, Early Stage (Fig. 2-1)

In the early stages of development the skull is entirely cartilaginous. It consists of a cranium that protects the brain and sense capsules that protect the sense organs. These two cartilaginous parts make up the chondrocranium, which is open on the dorsal side and which consists of the following elements.

**Trabeculae**  A pair of parallel, laterally placed, longitudinal cartilaginous bars that are fused anteriorly.
**Ethmoid plate**  Plate of cartilage, bridging the midline, that results from the fusion of the trabeculae near their anterior ends.

**Trabecular horns**  Two small projections extending anteriorly from the ethmoid plate.
**Antorbital cartilages**  These are short, horn-like processes that project laterally from each of the trabeculae at points where the trabeculae converge toward the ethmoid plate.
**Parachordal plates**  Shelves of cartilages that link the trabeculae with the otic capsules.
**Quadrate cartilages**  On each side they connect the parachordal plate with the quadrate bone (see below).
**Otic capsules**  Prominent bilateral bulbous structures located behind and somewhat above the parachordal plates.
**Supraoccipital arch**  Connects the two otic capsules dorsally.
**Basioccipital arch**  Connects the two halves of the chondrocranium ventrally.

### Skull: Chondrocranium, Intermediate Stage (Fig. 2-1)

As the skull develops, a few of the bilaterally paired cartilages of the chondrocranium are supplanted by cartilage (replacement) bones.

**Quadrate bones**  These articulate dorsolaterally with the quadrate cartilages. They are the sites of articulation for the lower jaw.
**Exoccipital bones**  These provide for articulation of the skull with the vertebral column.

**Prootic bones**    These are the ossified anterior portions of the otic capsule. The middle part of the otic capsule remains cartilaginous.

**Opisthotic bones**    The ossified posterior parts of the otic capsule, located lateral to the exoccipitals.

## Skull: Chondrocranium, Mature Stage (Figs. 2-2, 2-3)

The roof over the greater part of the chondrocranium is formed by a series of dermal (membrane) bones. These bones are derived from mesenchyme that has become ossified without passing through the cartilage stage of development. They are more superficially located than are cartilage bones, and they invest other parts of the chondrocranium as well as its roof. This mode of bone formation is phylogenetically related to that in bony fishes, in which enlarged scales form bony plates in the dermis on the head.

*Dorsally,* the skull presents these bones.

**Parietal bones**    These cover the posterior half of the cranium. Posteriorly, each meets the opisthotic bone, and in the mid-lateral region each parietal bone is in contact with a quadrate cartilage.

**Frontal bones**    These cover a large portion of the anterior half of the cranium and articulate with the parietal bones along two large V-shaped sutures.

**Premaxillary bones**    These two angular tooth-bearing structures are the most anterior bones of the skull. Each consists of a medial and lateral process. The two medial processes join with each other and with the underlying frontal bones. The lateral processes are free and form the anterior border of a space, the external naris.

*Ventrally,* the skull presents these bones.

**Exoccipital bones**    These are located at the posterior end of the cranium, and each bears an occipital condyle, which articulates with the atlas.

**Parasphenoid**    The large flat bone that makes up most of the roof of the mouth.

**Opisthotic bones**    Located on each side at the posterolateral corner of parasphenoids, they extend anteriorly to articulate with the squamosal and parietal bones.

**Prootic bones**    These lie lateral to the parasphenoids.

**Squamosal bones**    Elongated in an anteroposterior direction, they extend along the lateral edge of the posterior third of the skull.

**Quadrate bones**    On each side they articulate posteriorly with the squamosal, dorsolaterally with the quadrate cartilage, and ventrally with the large **palatopterygoid bone.** The palatopterygoid bone bears teeth along its anterolateral border and articulates posterolaterally with the mandible.

**Vomer bones**    These cover the anterior third of the roof of the mouth. They articulate with the parasphenoid and ethmoid bones medially and bear teeth on their lateral borders.

**Ethmoid bone**    This is a small diamond-shaped bone that has its posterior angle wedged into an apical notch in the parasphenoid bone.

**Otic capsules**    These fragile structures are replaced largely by prootic and opisthotic bones. Ventrolaterally the otic capsule presents an oval aperture, the **fenestra ovalis,** which is closed off by the stapedial plate of the **columella** (stapes). The columella is a tiny disc-shaped bone that has a spine-like **stylus** that transmits vibrations to the middle ear.

## Skull: Splanchnocranium (Figs. 2-4, 2-5)

The splanchnocranium consists of the mandibular, hyoid, and three additional visceral arches.

**Mandibular arch**    This, the first arch, articulates with the quadrate bone of the skull at the expanded posterior ends of the mandible. The mandible is formed by **Meckel's cartilage,** which remains unossified. The mandible consists of three curved bones on each side, arranged longitudinally around Meckel's cartilage. The **dentary bone** forms the greater part of the external surface of the lower jaw and bears teeth. The **splenial bone** is located near the middle of the internal surface of the jaw, is small, and also bears teeth. The **angular bone** makes up most of the internal surface of the mandible.

**Hyoid arch**    Lies just behind the mandibular arch and supports the tongue. It consists of the short, paired median **hypohyal** and the long, paired lateral **ceratohyal cartilages.**

**Visceral arches three to five**    Three pairs of gill, or branchial, arches lie directly behind the second, or hyoid, arch, but only the first is well developed. This first pair consists of paired **ceratobranchial** and **epibranchial cartilages.** A **basibranchial cartilage** connects the third visceral arch with the hyoid arch. The fourth and fifth visceral arches are greatly reduced because the ceratobranchial elements are very small or absent and thus these two arches consist primarily of epibranchials. A second basibranchial cartilage projects posteriorly from the place of junction of the first ceratobranchials.

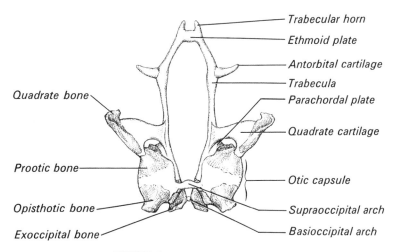

**FIGURE 2–1**
Chondrocranium (dorsal view).

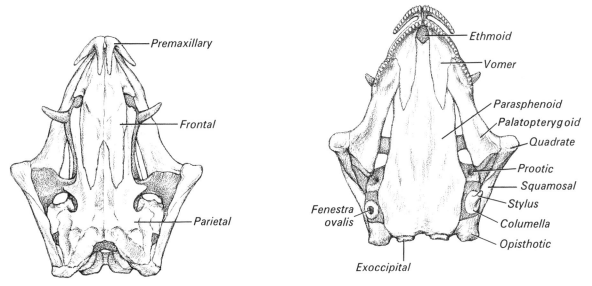

**FIGURE 2–2**
Skull (dorsal view).

**FIGURE 2–3**
Skull (ventral view; columella removed on one side).

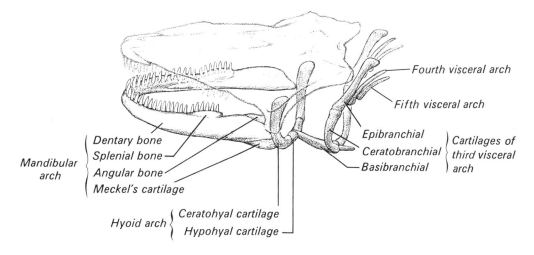

**FIGURE 2–4**
Splanchnocranium (lateral view).

## Vertebral Column (Figs. 2-6, 2-7)

The vertebral column consists of four types of vertebrae. The major portion of each vertebra is its centrum, which is concave on both of its surfaces. Thus, these vertebrae are classified as amphicoelous.

**Cervical vertebra** This single cervical vertebra is characterized by the absence of transverse processes. Its prezygapophyses articulate with the occipital condyles.

**Thoracolumbar vertebrae** There are 17 trunk vertebrae, all of which are associated with ribs. Each vertebra consists of a biconcave **centrum** above which is a **neural arch** enclosing a **vertebral foramen.** The vertebral foramina collectively make up the vertebral canal, which houses the spinal cord. A **neural spine** projects dorsally from the top of the neural arch. The paired **prezygapophyses** project dorsolaterally from the anterolateral surfaces of each neural arch and articulate with a corresponding pair of ventrolaterally directed **postzygapophyses** present on the posterolateral surfaces of the neural arch of the preceding vertebra. Extending laterally from each side of the vertebra is a **transverse process.** It consists dorsally of a **diapophysis,** which originates from the neural arch, and ventrally of a **parapophysis,** which originates from the centrum. The transverse processes articulate with laterally projecting **ribs.**

**Sacral vertebra** This single vertebra is characterized by its stout pair of sacral ribs.

**Caudal vertebrae** Numerous vertebrae that have the same features as trunk vertebrae but do not have ribs. All except the first have **hemal arches** that project from their ventral surface and correspond in position to neural arches. Each arch encloses a **hemal foramen,** which collectively make up the hemal canal that houses the caudal artery and vein. **Hemal spines** project ventrally from the caudal vertebrae.

## Sternum

The sternum consists of a few cartilages embedded in the myosepta of the ventral thoracic region.

## Ribs (Fig. 2-6)

The ribs are small and join the transverse processes of the thoracolumbar vertebrae. Each rib bifurcates proximally and presents two heads: a dorsal **tuberculum** and ventral **capitulum.** The tuberculum articulates with the diapophysis; the capitulum with the parapophysis.

The ribs of the sacral vertebrae are well developed and support the pelvic girdle. The caudal vertebrae have no ribs and their transverse processes become progressively smaller.

## APPENDICULAR SKELETON

The appendicular skeleton consists of the pectoral and pelvic girdles and their associated limbs.

## Pectoral Girdle and Forelimb (Fig. 2-8)

The pectoral girdle is not attached to the vertebral column and is largely cartilaginous. Each half consists of the following.

**Scapula** An irregular ossified mass. On its lateral surface it bears a depression, the **glenoid fossa,** at the site where this bone articulates with the forelimb.

**Coracoid** The broad flat plate that extends anteromedially from the side of the scapula to meet the corresponding cartilage in the mid-ventral line. A foramen, for the passage of blood vessels and nerves, may be seen in the coracoid plate.

**Procoracoid** A slender flat cartilage that extends directly forward from the scapula.

**Suprascapular cartilage** A broad, flat, dorsally projecting plate extending from the scapula.

The pectoral limb consists of the following.

**Humerus** The bone of the upper arm, or brachium.

**Radius** Anteromedially located bone of the forearm, or antebrachium.

**Ulna** Posterolaterally located bone adjacent to the radius.

**Carpals** Six wrist bones that are arranged in three rows.

**Metacarpals** Four long bones of the palm that are just distal to the carpals.

**Phalanges** There are two on each of the digits except the third, which has three.

## Pelvic Girdle and Hind Limb (Fig. 2-9)

The pelvic girdle is mostly cartilaginous. It also consists of the following paired elements.

**Puboischiadic plate** A ventrally placed flat plate. Its anterior part, which is triangular, is formed by fusion of the two **pubes,** whereas its posterior part is formed by the two partially ossified **ischia.**

111

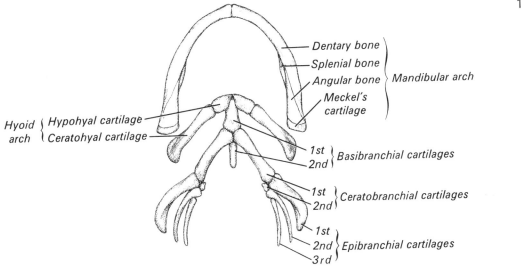

**FIGURE 2–5**
Splanchnocranium (ventral view).

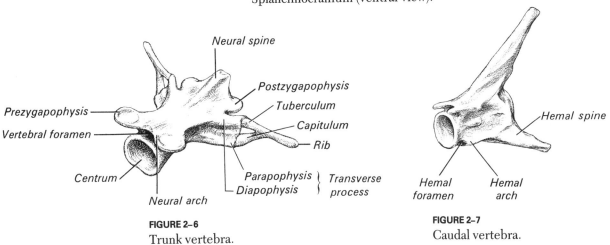

**FIGURE 2–6**
Trunk vertebra.

**FIGURE 2–7**
Caudal vertebra.

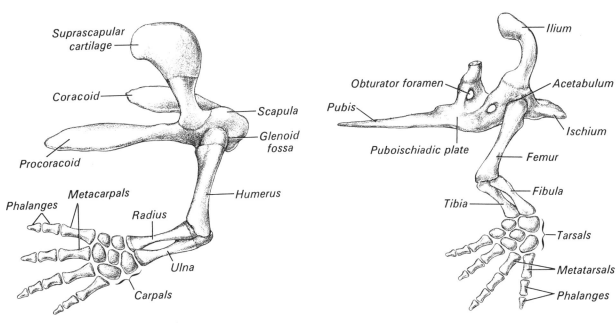

**FIGURE 2–8**
Pectoral girdle and forelimb (lateral view of one-half).

**FIGURE 2–9**
Pelvic girdle and hindlimb (lateral view of one-half).

**Ilium** Elongated bony segment extending dorsolaterally from the puboischium above the site of attachment of the hind limb.

**Acetabulum** Socket providing for the articulation of the pelvic limb.

**Obturator foramen** The open cavity in the side of the ischiopubic plate that is located anterior to the acetabulum.

The pelvic limb consists of the following bones.

**Femur** The heavy bone of the thigh.

**Tibia** The anterior bone of the shank.

**Fibula** The posterior bone of the shank.

**Tarsals** The five very small ankle bones.

**Metatarsals** The four foot bones.

**Phalanges** The bones of the digits. Each digit consists of two phalanges, except the third digit, which consists of three.

# Anatomy of the Mud Puppy *Necturus*

In *Necturus*, the musculature of the trunk and tail is primitive, like that of fish, whereas the musculature of the jaw and appendicular parts is more specialized. The muscular system will be considered here in terms of axial and appendicular parts. The axial musculature will be subdivided into trunk, eye, head, and gill musculature; the appendicular musculature, into the muscles of the pectoral girdle, the forelimb, the pelvic girdle, and the hind limb. The following descriptions of the locations of the muscles will aid in their identification. The origins, insertions, and actions of the muscles are summarized in Table 3-1.

*Skin the animal on one side from the mid-dorsal to the mid-ventral lines, including the gill region and the forelimb and hind limb. Start on the dorsal side and make a longitudinal incision through the skin, extending it from the end of the snout to just posterior to the hind limbs. Take care not to damage the underlying muscles. On the ventral surface make a midline longitudinal skin incision from the anterior end of the lower jaw to the cloaca. Pull off the skin with the fingers or forceps. While removing the skin, note the white connective tissue, or superficial fascia, between the skin and muscles. Take special care in the vicinity of the gular fold, since muscle fibers project into it. Note the large superficial venous sinuses in certain areas of the head.*

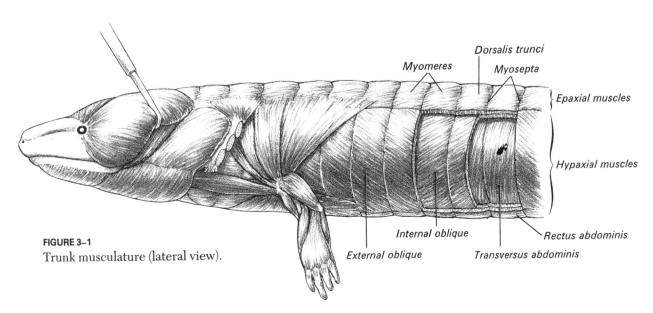

**FIGURE 3–1**
Trunk musculature (lateral view).

## AXIAL MUSCULATURE

### Trunk Muscles (Fig. 3-1)

*Most of the trunk muscles are clearly evident after skinning, but in order to expose the three sheets of hypaxial muscles, cut through successive layers on one side.*

As in the dogfish shark, lateral and median septa divide the trunk musculature into quadrants, with the **epaxial muscles** lying above the horizontal plane and the **hypaxial muscles** below. However, because the lateral septa are more dorsal in *Necturus* than in the dogfish shark, the epaxial musculature is reduced. The trunk and tail muscles consist of a series of **myomeres** separated by connective tissue **myosepta.**

The epaxial musculature is organized into two muscle layers. The superficial layer, which exhibits a metameric arrangement, constitutes the greatest part of the epaxial muscle mass and has become modified into a single longitudinal bundle, the **dorsalis trunci,** on each side. The deep layer of the epaxial muscle forms a series of intersegmental bundles extending between adjacent vertebrae.

The hypaxial muscles are organized into three distinct muscle layers. In addition, a pair of band-like muscles is located in the mid-ventral plane. The three muscle layers are inserted on a median, mid-ventral connective tissue band, the **linea alba.**

**External oblique**   The outermost muscle layer. Its fibers run obliquely in a posteroventral direction.

**Internal oblique**   The middle layer of muscle. Its fibers run obliquely in an anteroventral direction.

**Transversus abdominis**   The deepest layer. Its fibers run in a transverse direction. This muscle is, on its inner surface, in contact with the parietal peritoneum.

**Rectus abdominis**   Located on either side of the linea alba, this muscle extends from the pubis to the pectoral region. The fibers run longitudinally and are interrupted at regular intervals by myosepta.

### Eye Musculature

The extrinsic muscles of the eye are basically similar to those of the dogfish shark. They are innervated by three cranial nerves. The oculomotor nerve (N. III) innervates the **inferior oblique** as well as the **superior, inferior,** and **medial recti** muscles. The trochlear nerve (N. IV) innervates the **superior oblique.** The abducens (N. VI) innervates the **lateral rectus,** which divides

to form an additional muscle, the **retractor bulbi,** which is also innervated by the abducens.

### Muscles of Head and Gill Region (Figs. 3-2, 3-3)

The *dorsal* surface of the head reveals the following muscles.

**Anterior mandibular levator** (or temporal)   This rather broad muscle is located just lateral to the mid-dorsal line fo the head, above the level of the eye.

**External mandibular levator** (or masseter)   This massive muscle is located immediately lateral to each anterior mandibular levator and extends around the lateral surface of the head.

**Mandibular depressor** (or digastric)   Located at the level of the corner of the mouth beneath the lateral part of the masseter.

**Dorsalis trunci**   The cervical part of this muscle mass lies in the dorsal midline posterior to the anterior mandibular levator muscle.

**Branchial levators**   Fan-shaped muscles with slips that extend up into the gills. They are located lateral to the cervical portion of the dorsalis trunci.

The *ventral* surface of the head reveals the following muscles.

**Intermandibular**   This muscle covers the superficial surface of the floor of the mouth from the median raphe to the mandible on each side.

**Interhyoid** (or mylohoid)   This muscle runs just behind and parallel with the intermandibular muscle. The posterior part of the interhyoid is sometimes called the sphincter colli.

**Geniohyoid**   One runs on either side of the mid-ventral line just beneath the intermandibular. (Cut the intermandibular and turn it sideways to disclose the geniohyoid.)

**Branchiohyoid**   This large muscle lies beneath the interhyoid and lateral to the geniohyoid. It runs along the ventral surface of the lower jaw and curves around its angle.

**Rectus cervicis** (or sternohyoid)   This muscle, which is segmented by tendinous inscriptions, is a forward continuation into the cervical region of the rectus abdominis. It covers a large area of the neck between the hyoid apparatus and pectoral girdle. Separate strands of the rectus cervicis may be seen attached to the procoracoid cartilage. These muscle slips may be regarded as a distinct muscle, the **omoarcual.**

115

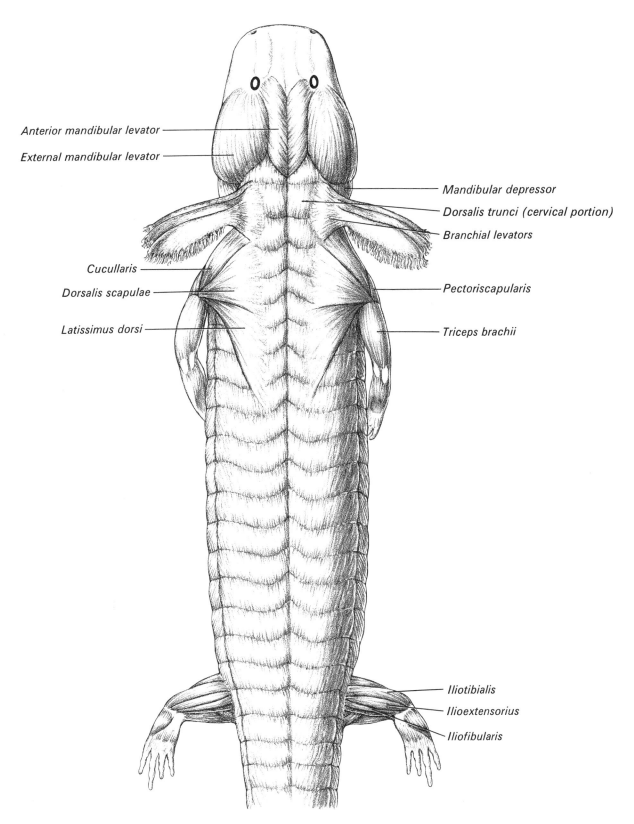

Anterior mandibular levator

External mandibular levator

Mandibular depressor

Dorsalis trunci (cervical portion)

Branchial levators

Cucullaris

Dorsalis scapulae

Pectoriscapularis

Latissimus dorsi

Triceps brachii

Iliotibialis

Ilioextensorius

Iliofibularis

**FIGURE 3–2**
Musculature (dorsal view).

## APPENDICULAR MUSCULATURE

### Muscles of the Pectoral Girdle (Figs. 3-2, 3-3)

The following three muscles can be identified on the *ventral* surface.

**Pectoralis**   A broad fan-shaped muscle extending from the linea alba to the proximal end of the humerus.
**Supracoracoideus**   This smaller fan-shaped muscle is located just anterior to the pectoralis.
**Procoracohumeralis**   A slender muscle that lies on the lateral surface of the procoracoid process antero-lateral to the supracoracoideus.

Four superficial pectoral muscles can be seen on the *dorsal* surface, in addition to the rectus cervicis described previously with the axial musculature.

**Cucullaris** (or trapezius)   Extends from the base of the gills to the scapula near the glenoid cavity.
**Dorsalis scapulae**   This fan-shaped muscle is located just posterior to the cucullaris.
**Latissimus dorsi**   This large fan-shaped muscle lies posterior to the dorsalis scapulae.
**Pectoriscapularis**   A small muscle that lies anteroventral to the cucullaris.

### Muscles of the Forelimb (Figs. 3-2, 3-3)

The forelimb group consists of three muscles on the arm and the numerous small muscles of the forearm. The three muscles of the arm are as follows.

**Triceps brachii**   This large extensor muscle covers the dorsal surface of the arm. Its coracoid and humeral heads of origin are more readily distinguishable than its scapular head.
**Humeroantebrachialis** (or biceps brachii)   This muscle extends along the entire ventrolateral surface of the humerus.
**Coracobrachialis**   This muscle covers the ventromedial surface of the humerus.

The muscles of the forearm consist of several small elements organized into a group of extensors and a group of flexors of the hand. The extensor muscles lie on the dorsal surface of the forearm.

### Muscles of the Pelvic Girdle (Figs. 3-2, 3-3)

On the *ventral* surface five pairs of muscles can be identified. They act chiefly to flex the shank but they also adduct the thigh. In addition, a large structure, the **cloacal gland,** should be noted lying between the skin and the muscles.

**Puboischiofemoralis externus**   This represents the anterior larger part of a triangular muscle mass that extends laterally from the mid-ventral surface of the pelvic girdle to the limb.
**Puboischiotibialis**   This is the posterior third of the aforementioned triangular muscle mass.
**Pubotibialis**   This narrow muscle extends along the lateral two-thirds of the anterior border of the triangular muscle mass.
**Puboischiofemoralis internus**   This muscle lies dorsal to the pubotibialis along the anterior edge of the thigh.
**Ischioflexorius**   This muscle is located along the ventromedial surface of the thigh posterior to the puboischiotibialis.

Three ventral pairs of longitudinal muscles are located on either side of the cloaca. They extend between the pelvic girdle and the tail and serve to flex the latter.

**Ischiocaudalis**   This is the narrow, most medial longitudinal muscle, lying just adjacent to the cloaca.
**Caudopuboischiotibialis (or caudocruralis)**   Strap-like muscle lying lateral to the ischiocaudalis.
**Caudofemoralis**   This is the most lateral of the three muscles.

On the *dorsal* surface of the pelvic girdle, three muscles can be identified in addition to the puboischiofemoralis internus.

**Iliotibialis**   The most anterior of a pair of muscles that pass from the base of the ilium over the knee to the tibia.
**Ilioextensorius**   This muscle is difficult to separate from the iliotibialis. It represents the posterior member of the aforementioned pair of muscles.
**Iliofibularis**   A slender muscle located posterior to the above pair and along the postaxial border of the thigh.

As in the forearm, the muscles of the shank are organized into a dorsal group of extensors and a ventral group of flexors of the foot.

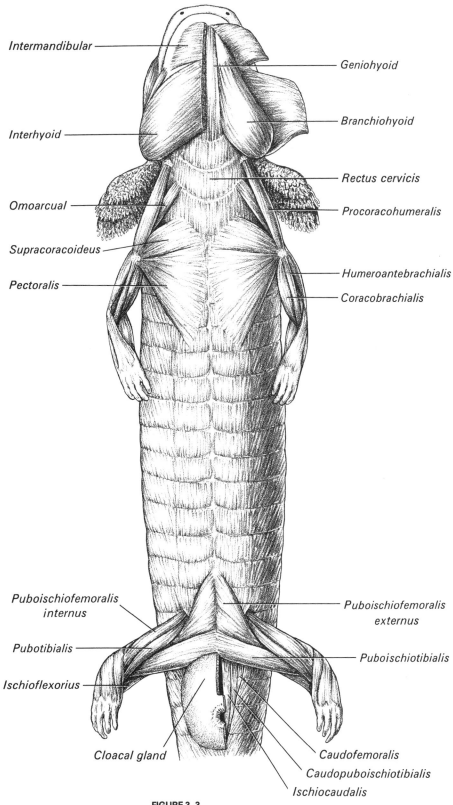

Intermandibular

Geniohyoid

Interhyoid

Branchiohyoid

Rectus cervicis

Omoarcual

Procoracohumeralis

Supracoracoideus

Humeroantebrachialis

Pectoralis

Coracobrachialis

Puboischiofemoralis
internus

Puboischiofemoralis
externus

Pubotibialis

Puboischiotibialis

Ischioflexorius

Cloacal gland

Caudofemoralis

Caudopuboischiotibialis

Ischiocaudalis

**FIGURE 3–3**
Musculature (ventral view).

**TABLE 3-1**
Summary of musculature of the mud puppy.

| Muscle | Origin | Insertion | Action |
|---|---|---|---|
| *Trunk Muscles* | | | |
| Epaxial | | | |
| Dorsalis trunci | Myosepta and transverse processes | Myosepta and transverse processes | Flexes trunk laterally |
| Hypaxial | | | |
| External oblique | Myosepta and horizontal septum | Myosepta and linea alba | Supports and compresses viscera |
| Internal oblique | Myosepta and horizontal septum | Myosepta and linea alba | Supports and compresses viscera |
| Transversus abdominis | Horizontal septum | Linea alba | Supports and compresses viscera |
| Rectus abdominis | Puboischiadic plate | Sternum | Supports and compresses viscera |
| *Head and Gill Region Muscles* | | | |
| Dorsal Group | | | |
| Ant. mandibular levator | Parietal and frontal bones | Dentary bone | Elevates lower jaw |
| Ext. mandibular levator | Parietal bone | Dentary bone | Elevates lower jaw |
| Mandibular depressor | Parietal and squamosal | Angular bone | Depresses lower jaw |
| Dorsalis trunci (cervical part) | Myosepta and transverse processes | Parietal and opisthotic bones | Turns head sideways |
| Branchial levators | Fascia on longissimus dorsi | Epibranchial cartilages | Elevates gills |
| Ventral Group | | | |
| Intermandibular | Dentary and angular bones | Median raphe | Tenses mouth floor |
| Interhyoid | Fascia on branchiohyoid | Median raphe | Tenses throat |
| Geniohyoid | Dentary bone | Second basibranchial | Draws hyoid cranially |
| Branchiohyoid | First gill arch | Ceratohyoid cartilage | Draws hyoid caudally |
| Rectus cervicis | Sternum | Ceratohyal cartilage | Draws hyoid caudally |
| *Pectoral Girdle Muscles* | | | |
| Ventral Group | | | |
| Pectoralis | Linea alba | Medial surface of humerus | Draws humerus medially |
| Supracoracoideus | Linea alba | Medial surface of humerus | Draws humerus medially |
| Procoracohumeralis | Procoracoid cartilage | Lateral surface of humerus | Draws humerus cranially |
| Dorsal Group | | | |
| Cucullaris | Dorsal cervical fascia | Scapula near glenoid cavity | Draws scapula anterodorsally |
| Dorsalis scapulae | Suprascapular cartilage | Anterior border of humerus | Draws humerus cranially |
| Latissimus dorsi | Fascia posterior to scapula | Posterior border of humerus | Draws humerus caudally |
| Pectoriscapularis | Fascia on cervidorsalis trunci | Anterior border of scapula | Draws scapula cranially |
| *Forelimb Muscles* | | | |
| Triceps brachii | Coracoid, scapula, and humerus | Near distal end of ulna | Extends forearm |
| Humeroantebrachialis | Proximal part of humerus | Near proximal end of radius | Flexes forearm |
| Coracobrachialis | Coracoid | Anterior surface of humerus | Flexes forearm |

*(continued)*

**TABLE** 3-1 (*continued*)
Summary of musculature of the mud puppy.

| Muscle | Origin | Insertion | Action |
|---|---|---|---|
| *Pelvic Girdle Muscles* | | | |
| Ventral Group | | | |
| Puboischiofemoralis externus | Pelvic girdle | Crest of femur | Adducts thigh |
| Puboischiotibialis | Ventral surface of ischium | Proximal part of tibia | Adducts thigh, flexes shank |
| Pubotibialis | Pubis anterior to acetabulum | Proximal end of tibia | Adducts thigh, extends shank |
| Puboischiofemoralis internus | Puboischiadic plate | Anterior surface of femur | Draws thigh anteriorly |
| Ischioflexorius | Posterior surface of ischium | Distal end of shank | Flexes shank and foot |
| Ischiocaudalis | Third and fourth hemal spines | Posterior border of ischium | May compress cloacal gland |
| Caudopuboischiotibialis | Third to fifth hemal spines | Puboischiotibialis | Draws thigh caudally |
| Caudofemoralis | Near caudopuboischiotibialis | Crest of femur | Draws thigh caudally |
| Dorsal Group | | | |
| Iliotibialis | Craniolateral surface of ilium | Proximal end of tibia | Extends shank |
| Ilioextensorius | Craniolateral surface of ilium | Proximal end of tibia | Extends shank |
| Iliofibularis | Lateral surface of ilium | Posterior surface of fibula | Flexes shank |

# Digestive and Respiratory Systems

# Anatomy of the Mud Puppy *Necturus*

## DIGESTIVE SYSTEM

The digestive system of the mud puppy differs very little from that of the dogfish shark. There are, however, several evolutionary advances. A spiral valve is no longer present and, to compensate for the resulting reduction in internal surface area, there has been an increase in the length of the small intestine. The terminal portion of the digestive tract, the rectum or large intestine, has become differentiated into a more structurally distinct organ. Thus the digestive tract of the mud puppy illustrates the early tetrapod stage, but it retains many larval features, such as a vertical transverse septum, gill slits, and a poorly formed tongue.

### Mouth and Pharynx (Fig. 4-1)

*Insert one blade of a pair of scissors into the right corner of the mouth and then cut posteriorly along the side of the head to a point just ventral to the gills. Then turn the floor of the mouth toward the right to identify the parts of the oral and pharyngeal cavities.*

**Mouth** The oral cavity is bordered externally by the lips and internally by the pouch between the hyoid and gill arches.

**Teeth** These are fish-like teeth, being small, conical, and homodont. Two V-shaped sets are located on the upper jaw. An outer set of **premaxillary teeth** is present on the premaxillary bones, and an inner set of **vomerine teeth** on the vomer. A very short additional set of **pterygoid teeth** project from the palatopterygoid bones and lies in line with the vomerine teeth, from which it is separated by a short, toothless gap. The lower jaw has a single arch-shaped row of **dentary teeth** and a short set of **splenial teeth** at each end of the arch.

**Internal nares** These are the openings from the nasal cavity to the oral cavity. Each can be seen immediately lateral to the groove separating the vomerine from the pterygoidal teeth. Each naris, which is covered by a fleshy flap, may be explored using a very fine probe.

**Tongue** It is well developed but immovable and is supported by the hyoid arch.

**Pharynx** This chamber lies posterior to the oral cavity.

**Gill slits** A pair on each side open directly to the outside from the pharynx. They are guarded by short projections called gill rakers that prevent large particles from passing through the gill slits.

**Glottis** The very small slit-like opening in the middle of the floor of the pharynx. The glottis opens into the larynx.

**Esophageal opening** The posterior opening from the pharynx.

## Digestive Organs (Fig. 4-2)

*Make two incisions, one on each side of the mid-ventral line, in order to leave a strip of tissue about one-fourth inch wide between the two cuts. Each incision should extend posteriorly from just in front of the pectoral girdle, through the pelvic girdle, to the level of the cloaca. Leave the median tissue strip intact to protect certain blood vessels and points of mesenteric attachment in the mid-ventral line. Now make a lateral cut through the body wall from the middle of each longitudinal incision. This will permit greater access to the viscera.*

**Esophagus**   This part of the digestive tract is short. It extends from the esophageal opening in the pharynx to the stomach. The epithelium that lines the inside of the esophagus is ciliated.

**Stomach**   The beginning of this elongated organ is not sharply demarcated from the esophagus. The cranial two-thirds of the stomach has a greater diameter than the caudal third. The entire organ runs in almost a straight course to terminate at the constricted **pylorus.**

**Duodenum**   The anterior part of the small intestine is about one inch in length. It receives the openings of the bile and pancreatic ducts. The duodenum continues posteriorly for a short distance and then turns forward and to the right to join the next part of the intestinal tract.

**Ileum**   The continuation of the small intestine from the duodenum resembles a loosely coiled tube. It

extends as far as the rectum.

**Rectum**   This short straight terminal part of the gut ends at the cloaca. It is also known as the large intestine. (It will be necessary to cut through the pelvic girdle to see its full extent.)

## Digestive Glands (Fig. 4-2)

**Liver**   The long structure that occupies a large portion of the mid-ventral region of the abdominal cavity. The sides of the liver are somewhat lobulated.

**Gall bladder**   Located on the dorsal aspect of the right side of the liver near its posterior end. The bile duct passes from the gall bladder through the tissue of the pancreas to the duodenum.

**Pancreas**   Irregularly shaped, whitish organ lying along the length of the duodenum.

**Spleen**   Elongated dark-colored organ that has no digestive functions.

## Peritoneum (Fig. 4-2)

The body cavity, or **coelom,** is divided by a partition, the **transverse septum,** into two separate compartments. The anterior **pericardial cavity** contains the heart, and the larger posterior **pleuroperitoneal cavity** contains the remainder of the viscera. The following mesenteries should be identified.

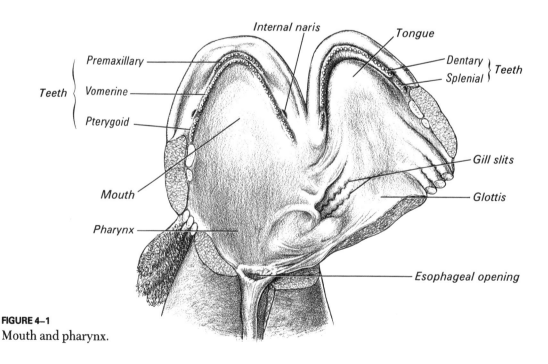

**FIGURE 4-1**
Mouth and pharynx.

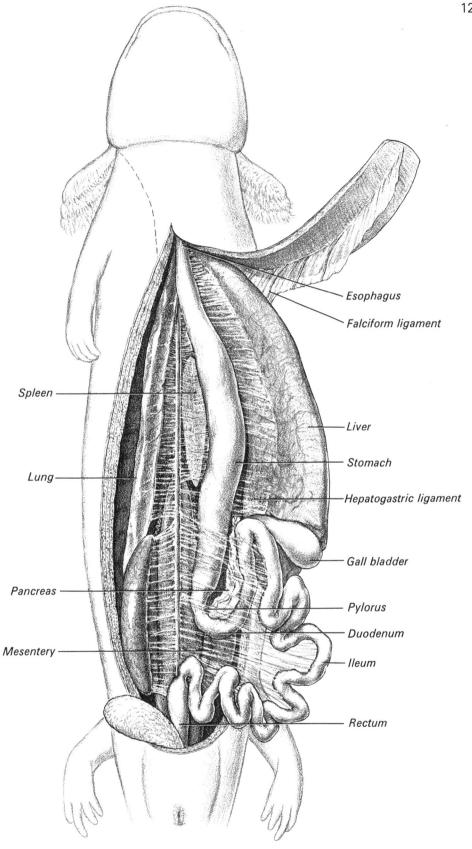

Spleen

Lung

Pancreas

Mesentery

Esophagus

Falciform ligament

Liver

Stomach

Hepatogastric ligament

Gall bladder

Pylorus

Duodenum

Ileum

Rectum

**FIGURE 4–2**
Digestive and respiratory systems.

**Mesogaster**   That part of the dorsal mesentery that passes to the stomach.

**Mesentery**   The mesentery proper is that part of the dorsal mesentery that suspends the small intestine from the dorsal body wall.

**Mesorectum**   That part of the dorsal mesentery extending to the rectum. It may also be called the mesocolon.

**Gastrosplenic ligament**   The peritoneal duplication extending between the stomach and spleen.

**Falciform ligament**   The part of the ventral mesentery extending between the liver and ventral body wall.

**Hepatogastric ligament**   The peritoneal duplication extending between the anterior part of the stomach and liver.

**Hepatoduodenal ligament**   The mesentery between the liver and duodenum.

## RESPIRATORY SYSTEM

*Necturus* has three modes of respiration that have varying degrees of importance: cutaneous, branchial, and pulmonary. Cutaneous respiration, which is aided by the moist nature of the skin, appears to play a relatively minor role. The most important site of gaseous exchange appears to be the external gills. The importance of the lungs under ordinary conditions appears to be negligible since they are poorly vascularized, contain no alveoli, and receive blood that has already been oxygenated in the gills. Thus, it is possible that the lungs are effective respiratory organs only if the other two modes of respiration cannot provide adequate amounts of oxygen.

**Larynx**   The very small chamber extending posteriorly from the glottis. Its walls are supported by a pair of lateral cartilages.

**Trachea**   A very short tube that leads directly into the lungs without the intervention of bronchi.

**Lungs**   Two long thin-walled, sac-like structures extending posteriorly and laterally to the stomach. They are not subdivided. One lung may be cut open and its smooth inner surface noted. Since the lungs are not enclosed in a separate part of the coelom, as are those in mammals, the body chamber in which they lie is appropriately called the pleuroperitoneal cavity.

# Anatomy of the Mud Puppy *Necturus*

*Make a shallow longitudinal incision extending be-tween the pleuroperitoneal incision and the posterior ends of the geniohyoid muscles to reveal the pericardial sac enclosing the heart. Slit the sac, expose the heart, and identify its parts. Make a longitudinal cut extend-ing along the ventral surface of the ventricle, the conus arteriosus, and the bulbus arteriosus to reveal their internal structure.*

**Ventricle**   The large thick-walled posterior chamber of the heart. It is a common receiving chamber for blood entering from both atria. Blood empties from it into the conus arteriosus. **Atrioventricular valves** are present between both atria and the ventricle.

**Atria**   A pair of thin-walled sacs that appear on either side of the bulbus arteriosus. They are separated structurally by a perforated **interatrial septum.** The right atrium receives blood from the sinus venosus and empties into the ventricle. The left atrium re-ceives blood from the pulmonary venous trunk, formed by both pulmonary veins, and it too empties into the ventricle.

**Conus arteriosus**   The single vessel that extends an-teriorly from the ventricle. A row of **semilunar valves** are located at the junction of the conus and ventricle.

**Bulbus arteriosus**   This represents the portion of the ventral aorta lying within the pericardial cavity and is a continuation of the conus arteriosus. The bulbus

is divided internally by a longitudinal septum into two channels. Outside of the pericardial cavity it branches to distribute blood to the gills.

**Sinus venosus**   This thin-walled chamber lies dorsal to the heart. It receives blood from all parts of the body except the lungs. Blood passes from this cham-ber through the **sinoatrial valves** into the right at-rium.

The circulatory system of the mud puppy differs significantly from that found in lower forms. These differences include (1) the addition of a pulmonary circulation and an associated three-chambered heart;

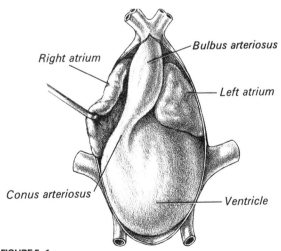

Right atrium · Bulbus arteriosus · Left atrium · Conus arteriosus · Ventricle

**FIGURE 5–1**

Heart (ventral view).

(2) the absence of first and second aortic arches and their associated afferent and efferent branchial arteries; (3) the development of a postcaval vein, which is the principal vessel draining the body; and (4) the presence of cutaneous vessels that supplement the gill and lung respiratory systems.

## ARTERIAL SYSTEM

The arterial system consists of the two aortae and their branches.

### Ventral Aorta and Its Branches (Fig. 5-2)

*Carefully dissect away the muscle tissue in front of the pericardial cavity. Then, after the afferent branchial and external carotid arteries have been identified, cut through the angle of the mouth on the intact side to just in front of the gills. On the other side, which was dissected in Exercise 4, extend the cut further posteriorly, ventral to the base of the gills and medial to the base of the forelimbs. Continue this posterior incision until the lower jaw and the posterior ventral region can be pulled to one side, exposing the roof of the mouth and pharyngeal cavities. Next, with forceps pull away the membrane that covers the roof of the mouth and the pharynx.*

**Afferent branchial arteries**   From the bulbus arteriosus the ventral aorta passes anteriorly and divides almost immediately into right and left branches. Each of these branches then subdivides into two rami. The smaller, anterior ramus is the first afferent branchial artery; the larger, posterior vessel is a common trunk that divides to form the second and third afferent branchial arteries. These three pairs of branchial arteries extend laterally to the gills where they ramify to form the capillaries of the gill filaments where respiration occurs.

**External carotid artery**   This is a branch of the first efferent branchial artery and takes origin just before the latter vessel passes into the first gill. It supplies the floor of the mouth, including the tongue.

**Efferent branchial arteries**   These drain the gills, one arising from each gill. The second and third efferent branchial arteries unite (as a posterior branch) immediately upon leaving the gills. The first, or anterior, efferent joins this common trunk a short distance from the origin of the latter. Cross-connections between the afferent and efferent branchial arteries are present in the base of each gill.

**Internal carotid artery**   Arises from the first efferent branchial artery near its point of junction with the common trunk from the second and third efferent branchials. It runs anteromedially along the roof of the mouth.

**Radix aorta**   Each is formed by union of the first efferent branchial artery with the common trunk from the second and third efferent branchial arteries. The radices aortae unite in the midline to form the dorsal aorta.

**Vertebral artery**   Each arises from a radix aorta shortly before the aortae unite in the midline. The vertebral arteries run cranially to disappear into the base of the skull.

**Pulmonary artery**   Arises on each side from the vessel formed by union of the second and third efferent branchial arteries (posterior branch). It runs to the dorsal side of the lung.

### Dorsal Aorta and Its Branches (Fig. 5-2)

*The opening made previously in the pleuroperitoneal cavity makes the abdominal arteries accessible. Now, identify the branches of the dorsal aorta.*

**Dorsal aorta**   Formed as a result of the union of the two radices aortae, it extends into the tail where it is known as the **caudal artery.**

**Subclavian artery**   Arises as a bilateral pair of vessels posterior to the junction of the radices. It runs laterally and gives off a large **cutaneous artery** that extends caudally along the length of the body wall on its inner surface. Along its course the cutaneous artery gives off **segmental arteries** that anastomose with parietal arteries from the dorsal aorta. Posteriorly the cutaneous artery anastomoses with the epigastric branch of the iliac artery. The subclavian artery continues across the shoulder as the **axillary artery.** It gives off no branches but passes into the arm where it becomes the **brachial artery.** This artery gives branches to the muscles of the forelimb and terminates in fine branches that extend to the terminal part of the limb and its digits.

The unpaired abdominal arteries are as follows.

**Gastric artery**   This is the most anterior unpaired artery. It supplies the digestive tract. Near the stomach it divides into a dorsal and a ventral branch. The ventral branch sends rami to the spleen.

**Celiacomesenteric artery**   Arises about an inch and a half posterior to the gastric artery. It gives rise to several **intestinal arteries.** It then divides into a **splenic artery** that runs to the posterior part of the spleen; a **pancreaticoduodenal artery** that runs to the pancreas, the duodenum, and the pyloric part

127

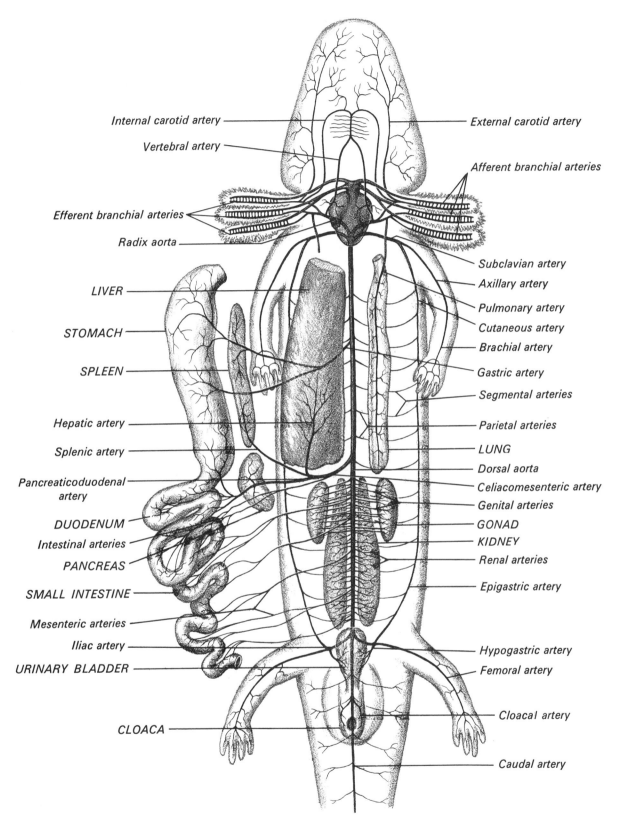

**FIGURE 5–2**
Arterial system (ventral view).

of the stomach; and a **hepatic artery** that runs alongside the hepatic portal vein to the liver.

**Mesenteric arteries**   A series of vessels arising directly from the dorsal aorta that traverse the dorsal mesentery to supply the small intestine.

The paired abdominal arteries are as follows.

**Parietal arteries**   These are segmentally paired dorsal branches of the dorsal aorta that run to the body muscles. They anastomose with comparable segmental branches from the cutaneous artery.

**Genital arteries**   A series of paired vessels extending laterally from the dorsal aorta as spermatic arteries to the testes in the male and as ovarian and oviducal arteries to the ovaries and oviducts in the female.

**Renal arteries**   These numerous pairs of vessels extend from the dorsal aorta to the kidneys.

**Iliac arteries**   A single pair of large vessels that arise just posterior to the last mesenteric artery. Each gives rise to an **epigastric artery,** which ascends along the ventral body wall to anastomose with the cutaneous artery, and to a **hypogastric artery,** which sends branches to the cloaca and bladder. The iliac artery then continues as the **femoral artery** into the hind limb, where it branches supply the muscles of the thigh and shank. The femoral artery terminates in branches to the digits.

**Cloacal arteries**   The first pair of vessels stemming from the caudal artery. They pass into the cloacal region. (Additional paired arteries arise from the caudal artery in the tail.)

## VENOUS SYSTEM (Fig. 5-3)

The venous system is composed of three sets of vessels: two portal systems and the systemic circuit. All of the vessels drain into the sinus venosus via the common cardinal veins. The venous system will be found in the opened pleuroperitoneal cavity.

### Hepatic Portal System

The hepatic portal system drains the digestive tract and spleen. It breaks up into capillaries within the liver.

**Hepatic portal vein**   This vessel ascends in a cleft along almost the entire length of the dorsal surface of the liver, which it enters near the site of emergence of the bile duct. This vein is formed by union of the mesenteric and the gastrosplenic veins within the pancreas.

**Mesenteric vein**   Runs longitudinally in the mesentery where it is formed by the union of numerous **intestinal veins** from the small intestine. As it passes through the pancreas it receives a number of small **pancreatic veins.**

**Gastrosplenic vein**   This vein is formed by the union of the **gastric vein** from the stomach and the **splenic vein** from the spleen. The gastrosplenic vein joins the mesenteric vein within the pancreas.

**Ventral abdominal vein**   A tributary of the hepatic portal vein that will be discussed in connection with the renal portal system. It unites the two portal systems.

### Renal Portal System

The vessels that make up this system drain blood from the posterior appendage, tail, and pelvic region and carry it to the kidney.

*Extend the longitudinal abdominal incision caudally until the entire kidney can be exposed on either side, if this has already not been done. Try to locate as many of the veins belonging to this system as possible. Some of them are quite difficult to find.*

**Caudal vein**   A single median vein that drains blood from the tail. It is located posterior to the kidney and divides, just caudal to that organ, into two renal portal veins.

**Renal portal veins**   These veins ascend along the lateral borders of the kidneys. In their course they give off numerous medial branches, **afferent renal veins,** to the kidneys and receive small lateral branches, **parietal veins,** from the body wall. In addition, each renal portal vein gives off a single lateral branch, the **iliac vein,** opposite the posterior limb. At the anterior end of the kidneys the renal portals continue cranially as the posterior cardinal veins, which belong to the systemic circuit.

**Pelvic veins**   Each arises as a result of the union between an iliac vein, which originates from the renal portal vein near the posterior end of the kidney, with a **femoral vein** from the hind limb. The two pelvic veins meet in the mid-ventral line to form the ventral abdominal vein.

**Ventral abdominal vein**   This prominent vessel ascends along the ventral abdominal wall and receives as tributaries small **vesicular veins** from the urinary bladder and **parietal veins** from the ventral abdominal wall. The ventral abdominal vein enters the posterior tip of the liver and, after giving off branches to the liver, joins the hepatic portal vein (see definition).

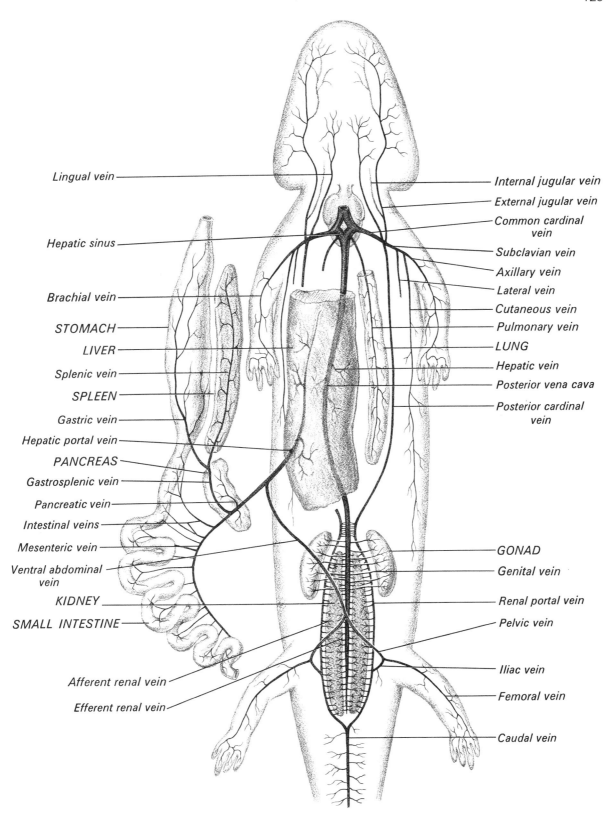

Lingual vein

Hepatic sinus

Brachial vein

STOMACH

LIVER

Splenic vein

SPLEEN

Gastric vein

Hepatic portal vein

PANCREAS

Gastrosplenic vein

Pancreatic vein

Intestinal veins

Mesenteric vein

Ventral abdominal vein

KIDNEY

SMALL INTESTINE

Afferent renal vein

Efferent renal vein

Internal jugular vein

External jugular vein

Common cardinal vein

Subclavian vein

Axillary vein

Lateral vein

Cutaneous vein

Pulmonary vein

LUNG

Hepatic vein

Posterior vena cava

Posterior cardinal vein

GONAD

Genital vein

Renal portal vein

Pelvic vein

Iliac vein

Femoral vein

Caudal vein

**FIGURE 5–3**
Venous system (ventral view).

## Systemic Veins

Most of the systemic veins are drained by the common cardinal veins. The **common cardinal veins** receive blood from both the posterior systemic veins from the abdominal viscera and hind limbs as well as from the anterior systemic veins from the head and forelimbs.

The posterior systemic veins are as follows.

**Posterior vena cava**  This large vessel arises from between the kidneys where it receives **efferent renal veins** from them and **genital veins** from the gonads. Slightly anterior, the posterior vena cava turns ventrally and runs forward through the dorsal mesentery to enter the liver. Within the liver it ascends in a straight line, receiving several **hepatic veins** along the way. It then passes forward to the septum transversum to divide into the **hepatic sinuses,** which empty into the common cardinal veins. A pair of **pulmonary** veins from the lungs passes dorsal to the hepatic sinuses and unites to form a single vessel that enters the left atrium.

**Posterior cardinal veins**  These are continuations of the paired renal portal veins (see definition). Each begins at the point at which the posterior vena cava turns ventrally toward the liver. They run cranially alongside the dorsal aorta to penetrate the transverse septum and join the common cardinal veins. They drain the dorsal body wall as they ascend by means of parietal veins.

The anterior systemic veins, which empty eventually into the common cardinal veins, are as follows.

**External jugular vein**  This is formed by numerous small veins draining the head. As the external jugular vein descends to empty into the subclavian vein, it is joined by the **internal jugular vein** from the brain.

**Subclavian vein**  Formed by union of the **axillary vein,** which is a continuation of the **brachial vein** from the forelimb, and the **cutaneous vein** from the skin of the lateral body wall. The subclavian vein empties into the common cardinal vein.

**Lingual vein**  A small vessel coming from the tongue that empties into the subclavian vein.

**Lateral vein**  Formed from tributaries coming from the muscles of the lateral body wall.

# Urogenital System

# Anatomy of the Mud Puppy *Necturus*

The breeding habits of *Necturus* are not completely understood, but presumably the male deposits a **spermatophore** that is picked up by the female. The spermatophores are clusters of sperm enveloped by a gelatinous substance, which is secreted by a large superficial cloacal gland and by several smaller dispersed pelvic glands, all of which open up into the cloaca. A third set of glands, the abdominal, are rudimentary and nonfunctional and do not contribute to the formation of spermatophores (as they do in salamanders).

The spermatophores are probably shaped by the posterior ventral portion of the cloaca. The means by which the spermatophore is transferred to the female is unclear. Two methods of transference have been observed. The first method involves direct physical contact; in the second, the male deposits spermatophores in a stream bed and the female retrieves them with the aid of muscular movement of the lips of her cloaca. Thus, neither amplexus nor copulation occurs in *Necturus* (or in other salamanders).

A dorsal diverticulum of the cloaca, the **spermatheca,** serves as a receptacle for the spermatozoa. Thus, when the ova pass into the cloaca from the oviducts, the sperm are already present.

Although the female *Necturus* receives spermatophores in the fall, eggs are not laid until the following spring. The female hollows out a space in the sand underneath a submerged log or rock, then turns her body upside down and deposits her eggs on the underside of the solid object.

The eggs, which appear as pale yellow spheres, are deposited individually. As many as 150 may be laid. The female remains in this nest guarding the eggs (Fig. 6-1). This brooding instinct, in which one or both of the parents remains with the eggs, appears to be an important component of the reproductive mechanisms in higher vertebrates. In *Necturus*, however, brooding appears to be a part of a prolonged behavioral syndrome, since it has been observed that females occupy the "nests" after the young have departed. Because some adults use these nests as shelters throughout the year, the brooding of *Necturus* may be merely the result of the adult's disinclination to leave its favorite shelter.

The eggs take from six to nine weeks to hatch, after which time the larvae are less than one inch in length. It takes as long as about eight years for the animals to reach maturity. They can live for as long as 25 years.

**FIGURE 6–1**

Female *Necturus* brooding her eggs.

## MALE UROGENITAL SYSTEM (Fig. 6-2)

*The components of the urogenital system are evident in the dorsal part of the exposed pleuroperitoneal cavity.*

**Testes**   Elongated organs, located on each side of the mid-dorsal line above the kidneys.

**Mesorchium**   The mesentery that attaches each testis to the dorsal surface. By holding up the mesorchium to the light, a number of small white tubules, the **efferent ductules,** can be seen passing from the testes into the kidney. These tubules convey sperm.

**Kidneys**   A pair of brown, narrow, elongated organs that lie in the posterior part of the body cavity close to the dorsal body wall. They are intraperitoneal rather than retroperitoneal structures, being entirely enclosed by visceral peritoneum. The more slender anterior parts of the kidneys are genital in function, since the tubules in this region of the kidney are modified for sperm passage. The broader posterior parts are urinary in function.

**Mesonephric ducts**   These lie on the lateral border of each of the kidneys. Each is tightly coiled in the genital region, is straight in the urinary region, and terminates in the cloaca. These ducts convey both sperm and urine.

**Cloaca**   The common terminal sinus of the digestive, reproductive, and excretory systems. The mesonephric ducts open dorsally into the cloaca on either side of the midline, just posterior to the opening of the large intestine. Papillae are present on the posterior ventral wall of the male cloaca.

**Urinary bladder**   This structure arises as a ventral outpocketing from the wall of the cloaca. Since the mesonephric ducts do not terminate in the bladder, the urine passes into this organ by gravitational flow.

The bladder is probably of greatest importance during the breeding season, when the cloaca is plugged by the swollen cloacal gland and its gelatinous secretions. The gland opens into the cloaca by means of several very small ducts. (If these small ducts cannot be found, it is probable that the gland was removed during dissection of the pelvic muscles.)

## FEMALE UROGENITAL SYSTEM (Fig. 6-3)

**Ovaries**   Paired sac-like organs covering the anterior end of the kidneys. They consist of masses of ova.

**Mesovarium**   The mesenteries that serve to suspend the ovaries from the mid-dorsal body wall.

**Oviducts**   Pair of large coiled ducts just lateral to each kidney and ovary. The funnel-shaped opening of the oviduct, the **ostium,** is located at the anterior end of the pleuroperitoneal cavity dorsal to the lungs. The ostium is ciliated. The posterior end of each oviduct is somewhat expanded to form a **uterus.** The uteri open into the sides of the cloaca close to the opening of the urinary bladder.

**Mesotubarium**   The mesentery that attaches each oviduct to the mid-dorsal body wall.

**Kidneys**   Similar to those in the male.

**Mesonephric ducts**   These are relatively narrow uncoiled tubes that run along the lateral margin of the kidneys. Each duct empties separately into the cloaca and serves a purely excretory function.

**Cloaca**   Similar to that in the male. Its posterior ventral wall lacks the numerous papillae that characterize the male cloaca. The papillae are replaced by smooth folds.

**Urinary bladder**   Like that in the male, this is a cloacal outpocketing. It opens into the ventral surface of the cloaca by means of a wide mouth.

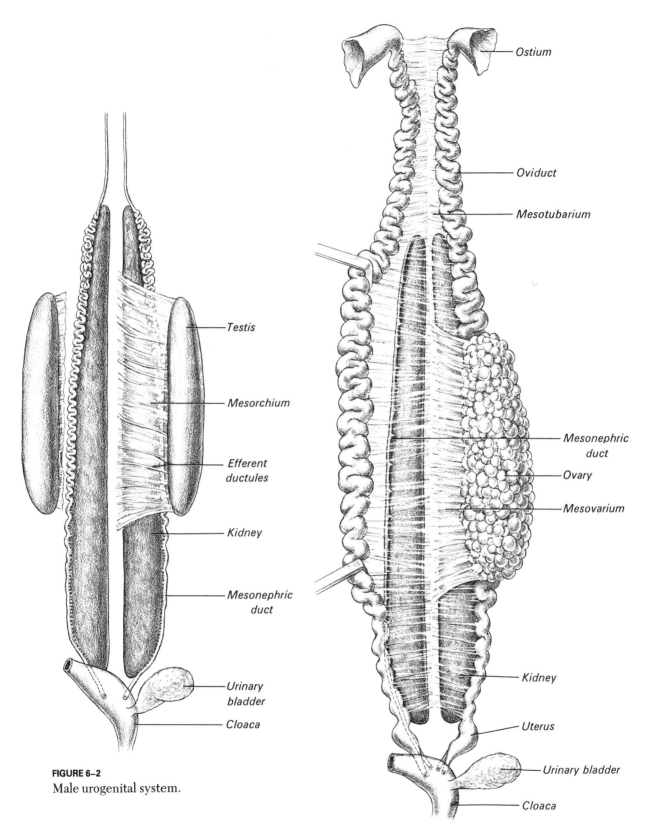

**FIGURE 6–2**
Male urogenital system.

Testis

Mesorchium

Efferent
ductules

Kidney

Mesonephric
duct

Urinary
bladder

Cloaca

Ostium

Oviduct

Mesotubarium

Mesonephric
duct

Ovary

Mesovarium

Kidney

Uterus

Urinary bladder

Cloaca

**FIGURE 6–3**
Female urogenital system.

# Sense Organs and Nervous System

# Anatomy of the Mud Puppy *Necturus*

## SENSE ORGANS

The small size of the mud puppy precludes a detailed study of its sense organs in the laboratory. The sense organs will, however, be described briefly.

### Lateral Line System

This system consists of minute pits in the skin that are distributed on the head and along the body. Ampullae of Lorenzini are absent. The lateral line organs are innervated primarily by cranial nerves from the facial (N. VII), glossopharyngeal (N. IX), and vagus (N. X) nerves.

### Olfactory Organs

A specialized nasal epithelium lines the passageways between the external and internal nares. This layer is innervated by the olfactory nerve (N. I). The nasal passages function for both olfaction and respiration.

### Eyes

The eyes are poorly developed and are generally similar in structure to those of other vertebrates. The functional capabilities of the eyes in *Necturus*, however, are not well known.

### Ears

The mud puppy has neither a tympanic membrane nor a middle ear cavity. It does have an inner ear that is apparently sensitive to waterborne vibrations.

The inner ear, or **membranous labyrinth,** is well developed in the mud puppy. The labyrinths are embedded in the otic capsules of the skull and function as organs of equilibration. Each membranous labyrinth is composed of three **semicircular ducts:** anterior, posterior, and horizontal. Each of these thin ducts exhibits an enlargement, the **ampulla,** at its lower end. The semicircular ducts on each side communicate with a sac-like **vestibule** that is composed of an upper chamber, the **utriculus,** and a lower chamber, the **sacculus.** These chambers are not sharply separated from each other. The sacculus contains a prominent **otolith,** made up of crystalline granules, and exhibits a small pocket, the **lagena,** at its posterior end. The lagena is the evolutionary precursor of the cochlea, the end organ for hearing in higher vertebrates.

### Taste Buds

Taste buds are restricted to the tongue, mouth, and pharynx and can be seen only by means of the microscope.

## NERVOUS SYSTEM

The nervous system consists of a cerebrospinal division (the brain and spinal cord), a peripheral division (the cranial and spinal nerves), and an autonomic division.

## Brain

*Expose the dorsal surface of the cranium by removing the overlying skin and muscles, and then carefully chip away the bone that forms its roof.*

The brain is an elongated, somewhat lobulated organ that is enclosed by two membranes. The thicker external membrane, or **dura mater,** lies flattened against the inner surface of the skull. The thinner internal membrane, or **pia mater,** is adherent to the brain surface.

On the *dorsal* surface (Fig. 7-1) the brain presents the following structures.

**Telencephalon** The forebrain consists of two paired parts, the olfactory lobes and the cerebral hemispheres. The **olfactory lobes,** which make up the anterior third of the large forebrain, are not distinctly separated from the cerebral hemispheres. Olfactory nerves extend forward from these lobes. The **cerebral hemispheres** make up the posterior two-thirds of the telencephalon.

**Diencephalon** The 'tween-brain lies just in back of the posterior end of the cerebral hemispheres. The thin vascular tela choroidea forms the roof of the diencephalon. A choroidal sac, the **paraphysis,** projects forward from the tela choroidea. A projection from the posterior part of the roof, the **epiphysis,** or pineal body, is also present.

**Mesencephalon** The midbrain is located posterior to the telencephalon and consists chiefly of paired **optic lobes.**

**Metencephalon** This is represented by the **cerebellum** and appears as a narrow transverse ridge that lies behind the optic lobes.

**Myelencephalon** Consists of the **medulla oblongata** extending posteriorly from the cerebellum. Within the medulla is a cavity, the **fourth ventricle,** that is also covered by a thin vascular tela choroidea.

*Carefully free the brain from its surroundings. To facilitate this, cut the cranial nerves coming out of the brain and then transect the brain at the end of the medulla.*

On the *ventral* surface (Fig. 7-2) only the diencephalon exhibits structures that have not been described above. This region of the brain consists of the **optic chiasma,** the site at which the optic nerves cross from one side of the body to the other, and the **infundibulum,** a broad stalk located just posterior to the chiasma. The infundibulum supports the **hypophysis,** or pituitary gland, at its end.

## Spinal Cord

The posterior end of the medulla oblongata merges imperceptibly with the spinal cord. The spinal cord lies within the neural canal and is basically uniform in diameter throughout all but its posterior end, which tapers to a fine thread, the **filum terminale.** Prominent longitudinal grooves are present in the mid-dorsal and mid-ventral lines.

## Cranial Nerves

There are eleven cranial nerves in *Necturus.* Their small size makes dissection impractical. In number, distribution, and general function the cranial nerves are similar to those of the dogfish shark. Their characteristics are summarized in Table 7-1.

## Spinal Nerves

Metamerically arranged, paired spinal nerves emerge from the spinal cord. Each nerve is formed by union of its ganglion-bearing **dorsal root** with the **ventral root** in the intervertebral foramen. A short distance after being formed, each nerve gives off a **ramus communicans,** or communicating branch, to an adjacent sympathetic ganglion and then divides into **dorsal** and **ventral rami.**

In the regions of the limbs some of the ventral rami of the spinal nerves intercommunicate to form plexuses. Anteriorly, the ventral rami of the second, third, fourth, and fifth spinal nerves interconnect on each side to form the **brachial plexus,** which is distributed to the forelimb. Posteriorly, the ventral rami of the nineteenth, twentieth, and twenty-first nerves interconnect to form the **lumbosacral plexus,** which passes into the hind limb.

## Autonomic Division

The autonomic division is difficult to study because of the widespread distribution of its thin nerve fibers. It innervates smooth muscle fibers of viscera, cardiac muscle, and glands. It has two components.

**Parasympathetic system** Made up of a group of nerve fibers carried by cranial nerves (N. III, VII, IX, X) and by sacral nerves.

**Sympathetic system** Made up of nerve fibers in two slender sympathetic trunks that extend the length of the body, parallel to the dorsal aorta. They bear metamerically arranged sympathetic ganglia that receive the communicating branches from the spinal nerves. The sympathetic fibers emerging from the ganglia are distributed in a pattern similar to that observed in higher forms.

**TABLE 7-1**
Cranial nerves in the mud puppy.

| No. | Name | Site of Emergence from Brain | Site of Innervation | Function |
|-----|------|------------------------------|---------------------|----------|
| 0 | Terminal | Ventral border of olfactory lobe | Sensory endings of snout | Sensory |
| I | Olfactory | Anterior end of olfactory lobe | Nasal mucosa | Sensory |
| II | Optic | Optic chiasma | Retina of eye | Sensory |
| III | Oculomotor | Posterior part of mesencephalon | Eye muscles (except two) | Motor |
| IV | Trochlear | Between mesen- and metencephalon | Superior oblique | Motor |
| V | Trigeminal | Anterolateral border of medulla | Face and jaws | Mixed |
| VI | Abducens | Anteroventral end of medulla | Lateral rectus | Motor |
| VII | Facial | Lateral border of medulla | Face and jaws | Mixed |
| VIII | Statoacoustic | Lateral border of medulla (with VII) | Inner ear | Sensory |
| IX | Glossopharyngeal | Posterolateral border of medulla | Pharynx and tongue | Mixed |
| X | Vagus | Behind glossopharyngeal | Viscera and lateral line | Mixed |

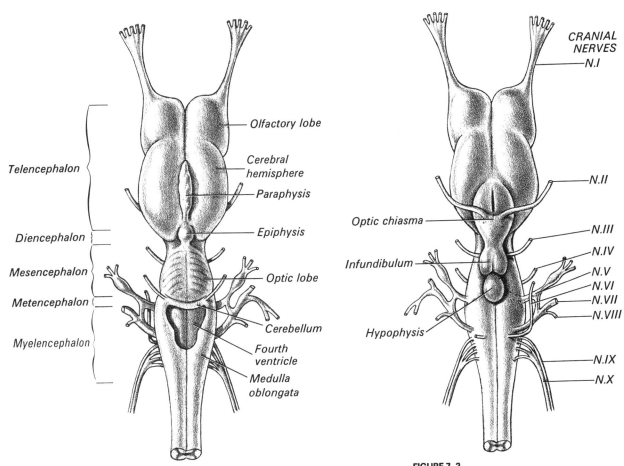

**FIGURE 7–1**
Brain and cranial nerves (dorsal view).

**FIGURE 7–2**
Brain and cranial nerves (ventral view).

# Evolution and Function

# Anatomy of the Mud Puppy *Necturus*

This exercise seeks to integrate the anatomical information on the mud puppy *Necturus* in an evolutionary and physiological context. Adaptation to life on land is of major significance, and amphibians underwent considerable functional morphological changes during the transition from an aquatic to a terrestrial habitat.

Amphibians, although common in such habitats as ponds, nevertheless are quite inconspicuous animals. Modern amphibians are descendants of the first land vertebrates, the labyrinthodonts. The labyrinthodonts evolved from crossopterygean fishes and enjoyed a period of great expansion followed by a marked decline. Relatively few kinds of amphibians have survived to the present. They are distributed in both temperate and tropical parts of the world. While

three subclasses are recognized, two are extinct; thus the subclass *Lissamphibia* contains all surviving amphibians. This subclass has three orders of animals: URODELA, or tailed forms, ANURA, or tailless amphibians, and APODA, or legless burrowing animals (Fig. 8-1). The mud puppy *Necturus* belongs to the order URODELA.

Most amphibians are characterized by three major features. First, their eggs do not develop fetal membranes and therefore must be deposited in water. Second, amphibians have a two-stage life cycle, typically a larval aquatic stage that metamorphoses into an adult terrestrial form. Third, the skin of amphibians is usually moist. This characteristic is discussed below in detail.

*Order ANURA: Frog*

*Order APODA: Caecilian*

**FIGURE 8–1**

Modern amphibians.

*Order URODELA: Salamander*

It should be noted at this point that *Necturus* is used as a generalized amphibian because it retains a tail and has four functional limbs. It was therefore also considered a terrestrial tetrapod, when in reality it is an aquatic one. *Necturus* has not developed adaptations to terrestrial life because it fails to undergo metamorphosis and thus is an atypical amphibian. Nevertheless, the mud puppy is more common and larger than most metamorphosing salamanders and thus it is used for comparative anatomy studies. This exercise therefore focuses on *Necturus*, but special features of other amphibians are noted where appropriate.

## EXTERNAL MORPHOLOGY

The skin of most amphibians is not able to withstand prolonged exposure to dry air. Consequently, most terrestrial forms have means to keep the skin moist or to prevent desiccation, for example, the thick, wavy, impermeable skin of many toads.

In vertical section seen under the microscope (Fig. 8-2), the thin skin of amphibians, like cyclostomes and fishes, consists of epidermal and dermal layers. The outermost layer of the epidermis, the very thin stratum corneum, consists of dead keratinized cells. In response to hormonal stimulation, the dead cells are shed every few days in small numbers, but sometimes in large patches after longer periods. The bulk of the epidermis consists of several layers of stratified squamous epithelium. The glands that lie in the dermis are more complex and are under nervous or hormonal control. In the mud puppy these glands consist of the smaller, more numerous mucous glands, whose secretion insures that the skin remains moist, and the larger poison glands. The poison glands provide a protective function since they secrete irritating or toxic substances. Thus an uninitiated dog that attempts to chew a toad (genus *Bufo*) gets not dinner but a severely irritated mouth and may become ill. The dermis of the mud puppy is characterized by an abundance of blood vessels, which provide the potential for respiratory exchange (especially $CO_2$) that takes place through the skin.

## LOCOMOTION

Land vertebrates are faced with two major locomotion problems: how to support their weight in air and how to transport it over the ground. The problem of support is solved with the firm connection of the vertebral column and limbs. In all true land vertebrates, the pectoral and pelvic girdles are attached directly or indirectly to the vertebral column.

The problem of animal transport is solved by the axial arrangement of the tetrapod limb bones. The basic pattern for such a limb involves one bone (humerus or femur) articulating with the girdle at its proximal end and with two bones at its distal end. These two bones in turn articulate with several rows of small bones, the most distal row of which articulates with the digits. Such an arrangement, which permits independent joint movement and which has a complex musculature associated with it, allows for the limb to function as a system of levers for support and propulsion.

The multijointed tetrapod limb organization is radically different from that of fish, where usually neither of the two girdles is attached to the vertebral column. Moreover, the axial limb arrangement of tetrapods differs significantly from the structure of the fins of bony fishes. Most aquatic forms do not need their limbs for support and use their massive axial musculature to provide undulatory body movement that ensures propulsion. To adapt to terrestrial life habitats, amphibians acquired muscular and skeletal modifications in the form of four limbs and complex limb musculature. This enables the body to be lifted off the ground and permits a walking gait.

When a urodele such as a salamander is at rest, the entire weight of the body is placed on the abdomen. When the animal initiates movement (Fig. 8-3), its body weight is transferred to the limbs, which carry through a standard multiphasic cycle of movement. The motion of a salamander has three major features: (1) the plantar surface of a foot is placed on the ground some distance away from the side of the body; (2) the order in which the limbs leave or make contact with the ground exhibits a diagonal pattern whereby the right forelimb arches upward and forward followed very shortly by the left hind limb, and at the same time the left forelimb and right hind limb propel the animal forward by applying force to the ground; (3) during most of the cycle, three feet are on the ground simultaneously, but in some phases only two or all four are in contact with the ground.

During each movement cycle, the body curvature is altered by contraction of the vertebral (axial) muscles, which results in lengthening the stride. The synchronized movement of diagonal limbs is characteristic of the relatively rapid progression of an undisturbed animal moving at a deliberate speed. The rapid movement of a frightened salamander occurs by violent undulation of the unelevated body as the animal slides along on its ventral surface.

The movement pattern of the salamander ensures that the forces acting on the feet are kept in equilibrium with the body weight throughout the cycle. This is because usually only one foot at time is in motion,

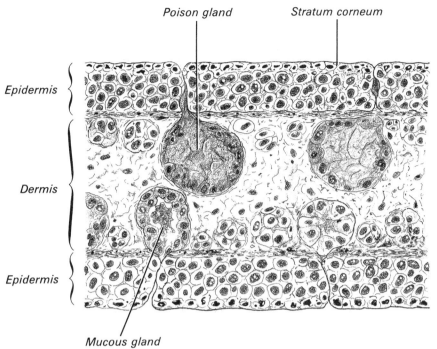

**FIGURE 8–2**
Skin of *Necturus* (section through tail).

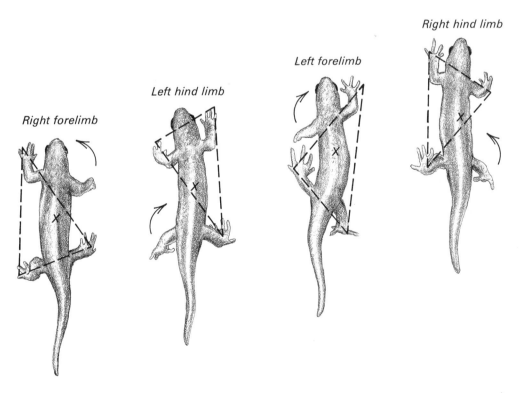

**FIGURE 8–3**
Stages of locomotion of the salamander (half a cycle).

with the other three grounded. The movement cycle occurs at a rapid enough pace to ensure against unwanted body displacement occurring between one footfall and the next. In addition, the movement of the feet in the proper sequence ensures stability, since the animal's center of gravity (see the *X* on Figure 8-3) is kept within the confines of the triangle formed by the three stationary feet.

## SKELETAL SYSTEM

In the course of adaptation to a terrestrial habitat, the fish-like characteristics of the progenitors of amphibians, such as gill bars and operculum, regressed or disappeared. Amphibians are the first vertebrates to evolve a neck. The craniovertebral joint is quite flexible and enables the animals to raise and move the head without moving the rest of the body. While other important changes also took place, a trend toward stability of skull form was initiated by crossopterygeans and continued through their descendants. Living amphibians have highly specialized skulls.

The most significant transitional skull changes in amphibians occurred in the visceral skeleton, especially the hyoid arch. In fish, the elements making up the hyoid arch, particularly the hyomandibular, are responsible for moving the operculum and supporting the skull (Fig. 8-4A). In some fossil amphibians the operculum has already disappeared, and the quadrate bone of the upper jaw now articulates with the squamosal bone of the skull, with the hyomandibular no longer participating in jaw suspension. Rather, this bone now serves as a brace between the braincase and cheek. Subsequent hyomandibular changes are related to hearing.

In order for terrestrial forms to receive sound waves that move through air, two adaptations are required. First, they need bone that bridges a middle-ear cavity. Second, they must possess a tympanic membrane (eardrum). Both conditions were eventually met in the course of amphibian evolution. Thus modern Anura possess a middle-ear bone, the stapes (columella), which is formed from the hyomandibular. It extends between the inner ear and the tympanic membrane (Fig. 8-4B). The transformation of the hyomandibular from jaw suspension and gill ventilation to braincase support and then to sound reception is a good example of functional change during evolution.

Urodeles, like *Necturus*, have a rudimentary stapes and lack a tympanic membrane. Sound reception apparently occurs by means of contact between the shoulder girdle and the ground.

The structures of the vertebrae were gradually radically altered in order to accommodate the change from aquatic to terrestrial life. This was an adjustment to the stress of gravity. Also, the effect of support of the body axis transmitted by the lengthening limbs had to be accommodated. Finally, land animals required a regionally differentiated vertebral column to provide increased mobility to the head and support for the pelvic girdle, rather than being uniformly flexible throughout its length. The requirements were met with the development of vertebrae with (1) firm, heavily ossified centra, (2) processes that increase muscle leverage, (3) intervertebral joints with specialized articular processes that can enhance or restrict motion, depending on location, and (4) a more intricate relation to the girdle.

## MUSCULAR SYSTEM

Several trends can be identified in the evolution of the axial musculature of tetrapods. In fishes the axial musculature is massive to provide them with the propulsive power for locomotion. With the limbs assuming this function in land-living forms, the limb muscles enlarge while the axial musculature declines. Also, in tetrapods the axial skeleton assumes a new supportive role by becoming firmer, while the diminished axial musculature assumes a more intimate relationship to the skeleton. In addition, myosepta regress or disappear and muscle fibers lengthen to span several vertebrae, contributing to the potential for flexion of the vertebral column. Some muscles (external oblique) assume a sheet-like character, while others (dorsal scapular) become associated with the pectoral girdle.

The axial musculature of amphibians is transitional between that of fishes and reptiles. The dorsal epaxial muscles, although segmented by nearly vertical (instead of angled) myosepta, can be considered a single muscle, the dorsalis trunci (Fig. 8-5). The hypaxial muscles of the trunk are placed in three groups: (1) the subvertebral group, a small mass located beneath the transverse processes of the vertebrae that serves to flex it, (2) the rectus abdominis, which runs lengthwise along the ventral body wall between both girdles, and (3) the lateral group, located between the two other groups. This flank musculature is the most differentiated segment of the hypaxial muscles. It consists of three superimposed sheet-like layers: external oblique, internal oblique, and transversus abdominis, each with fibers running in opposing directions, thus reinforcing the strength of the abdominal wall.

The pectoral girdle does not articulate with the vertebral column. Rather, several muscles hold this girdle to the trunk. In contrast, the pelvic girdle gains an

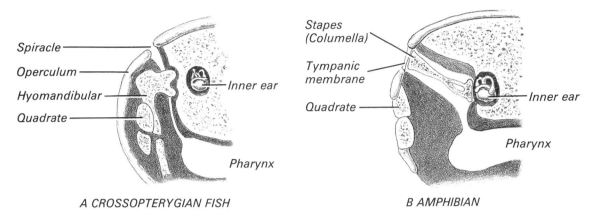

**FIGURE 8–4**
Evolutionary changes in the skull.

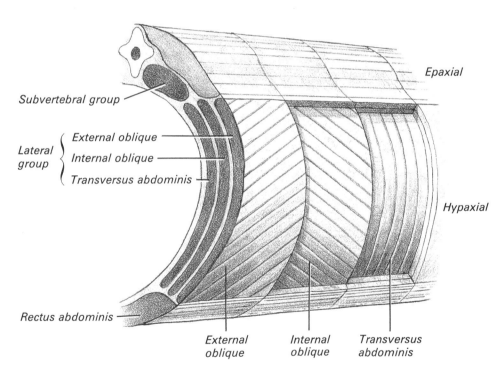

**FIGURE 8–5**
Trunk musculature of a urodele.

articulation with the vertebral column and hence does not require as extensive muscular support.

Finally it should be noted that the muscles of the mud puppy *Necturus* are specialized, and thus some people prefer to study the muscles of another urodele, such as the tiger salamander *Ambystoma tigrinum*.

## DIGESTIVE SYSTEM

Adult amphibians have relatively short and simple digestive tracts (Fig. 8-6), in which a coiled small intestine is demarcated from a shorter large intestine (rectum). In some forms a small cecum is present at the origin of the large intestine.

In most amphibians much of the buccal cavity is lined with ciliated epithelium. The cilia, which continuously beat towards the pharynx, carry food particles and secretions toward the stomach. By means of complex muscular contractions involving the floor of the mouth and the tongue, ingested food is thrust into the pharynx. This expanded chamber is also lined with ciliated, mucus-secreting epithelium and leads into the esophagus, a relatively short, thin-walled tube that is also lined with ciliated epithelium. The musculature of the esophageal wall contracts rhythmically, moving the food toward the stomach. A muscular sphincter is located at the pharyngoesophageal juncture. A similar sphincter is present at the junction of esophagus and stomach.

The stomach is simply an expanded chamber with a contractible pyloric sphincter at its end. The small intestine of *Necturus* and other urodeles is a short coiled tube with a small internal diameter (see Dogfish Shark, Fig. 10-7). The small intestine tends to be longer and more coiled in larval anurans, becoming relatively short in adult anurans. Intestinal shortening is associated with the change in food habits. Tadpoles consume plant food (algae), whereas adult frogs and toads are exclusively carnivorous.

From the small intestine, digested food is passed on by contraction (peristalsis) of the muscular walls into a short, wide-bored large intestine. The remnants of the digestive process are then passed to the cloaca, from whence they exit the body.

## RESPIRATORY SYSTEM

Dry air provides terrestrial vertebrates with potentially more oxygen than that available to fishes. What is essential in utilizing this resource is a moist respiratory membrane, so that the oxygen can be in solution and thus can diffuse into the blood. The exposure and ventilation of the respiratory membrane without excess loss of body water is achieved by lungs. The rate of ventilation is low, and not all the air in the lungs is exchanged in each breathing cycle. Lungs also solve the problem of supporting the body and organs in air, which is less dense than water. Lungs usually contain internal partitions, which increase the surface area and provide support.

Amphibian lungs are greatly diverse in structure. Diversity presumably is correlated with multiple modes of breathing. In addition to lungs, such supplementary structures as gills, skin, and even buccal mucosa can serve as respiratory elements in amphibians. In some amphibians, one method of breathing is so dominant that another has been eliminated. Thus aquatic lungless salamanders, which inhabit swift mountain streams, rely on skin or gills for respiration. Terrestrial lungless salamanders depend solely on the skin for respiration.

Lung structure in amphibians in general is not significantly advanced beyond that of lungfish. The internal lung lining may be smooth throughout, there may be simple sacculations in only the proximal portion, or the entire lining may be pocketed by the presence of ridges or septa. Internal septal arrangements increase the internal surface area of the lung (Fig. 8-7). A trachea is present, but a tracheobronchial tree that can provide direct, narrow passageways from the outside to the sites of gas exchange is absent.

The total surface area of amphibian lungs is proportional to the extent that the lung participates in the total body gas exchange. In terrestrial amphibians, the surface area is increased by enlargement of the primary alveolar septa and their subdivision.

Basically the lungs of amphibians are two simple sacs that conform to the shape of the pleuroperitoneal cavity. Lungs are bulbous in anurans and elongated in urodeles (see Fig. 4-2). The left lung of the legless caecilians is rudimentary.

## CIRCULATORY SYSTEM

Animals using lungs as respiratory organs have hearts whose anatomy is different from that of the simple two-chambered organ found in fishes. The atrial chamber receives not only systemic venous unoxygenated blood but also oxygenated blood from the lungs. In a fish heart the two types of blood would mix, consequently significantly diminishing their physiological effectiveness. In amphibians, however, the right and left sides of the heart evolved towards separation, with the initial changes represented by the formation of a median partition dividing the atrium into two distinct chambers. This separation diminishes the

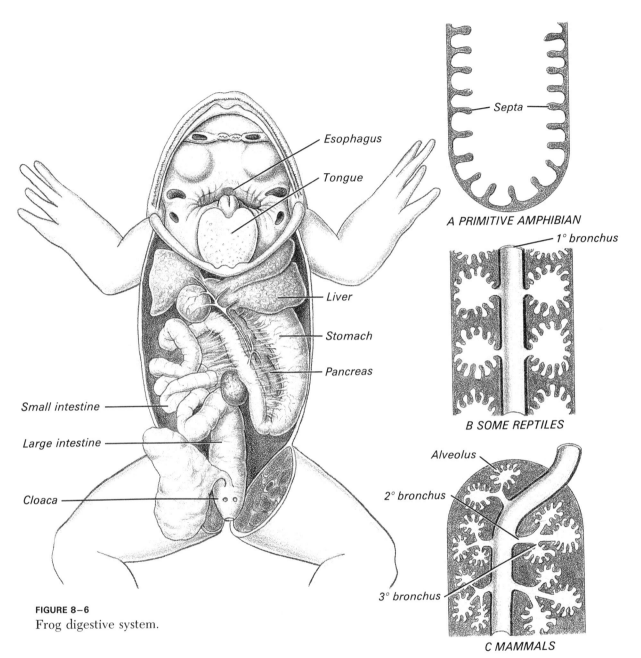

**FIGURE 8–6**
Frog digestive system.

Esophagus

Tongue

Liver

Stomach

Pancreas

Small intestine

Large intestine

Cloaca

Septa

*A PRIMITIVE AMPHIBIAN*

1° bronchus

*B SOME REPTILES*

Alveolus

2° bronchus

3° bronchus

*C MAMMALS*

**FIGURE 8–7**
Evolutionary changes in internal lung surface.

negative effect of a mix of the two different types of blood.

The circulatory system of amphibians shows little advance in complete separation of systemic and respiratory heart circuits over that of the lungfish. The heart of amphibians (as well as dipnoans and reptiles) thus is transitional between single-circuit and double-circuit pumps. Because the lower tetrapods are less active than birds or mammals, they can function with incomplete double-circuit systems. In amphibians breathing is supplemented by gills, skin, and buccal mucosa. On the other hand, active vertebrates, like birds and mammals, that respire exclusively with lungs have hearts that pump one stream of oxygenated blood and a completely independent stream of unoxygenated blood.

In amphibians (and dipnoans) the atrium is partially (dipnoans and urodeles) or completely (anurans) divided. Systemic venous blood enters the sinus venosus, which is an antechamber of the right atrium (Fig. 8-8). Oxygenated pulmonary blood returns from the lungs to the left atrium. The ventricle of amphibians is undivided (partially divided in dipnoans). Although the ventricle is single, its spongy walls temporarily suspend blood inside many small crypts, thereby minimizing the mixing of the two types of blood as they enter the ventricle. An additional feature that impedes mixing is the laminar outflow of the bloodstream upon ventricular contraction.

The conus is large, and in anurans and terrestrial urodeles it is subdivided by a partition, the spiral valve. This creates a ventral passage that leads to the carotid and systemic arteries and a dorsal passage leading to the pulmonary artery. Aquatic urodeles and caecilians lack a spiral valve.

Oxygenated blood returning to the heart from the skin is not, as would be expected, directed into the left atrium like that carried from the lungs by the pulmonary veins. Rather, cutaneous blood is mixed with systemic venous blood and then returned to the right atrium. Thus the skin merely serves to elevate the oxygen content of systemic venous blood rather than to provide oxygen-rich blood directly to the tissues.

## UROGENITAL SYSTEM

The opisthonephric type of kidney is formed in many adult fish and amphibia (see Dogfish Shark, Fig. 10-10). It developed from the main mass of nephrogenic tissues (except the portion that gave rise to the pronephros). Many tubules generally develop from each nephrogenic segment. Since the early evolution of the urinary system is discussed in Exercise 10 of the Dogfish Shark section, the balance of this section will be devoted to the reproductive system.

The basic organizational plan of the vertebrate reproductive tract is outlined in Fig. 8-9. It consists of the primary sex organs, or gonads, in which the gametes (and sex hormones) are formed, and the sex ducts and accessory sex organs, which convey the gametes (eggs and sperm) or embryo to the outside. Species utilizing internal fertilization also have external genitalia to facilitate copulation; those in which fertilization is external usually do not have external genitalia.

The testes generate sperm and steroid hormones. The hormones regulate the function of the male reproductive tract and other sexual accessory organs and also sexual behavior. The testis consists of two major components: spermatogenic and steroidogenic. In anurans (and amniotes) the spermatogenic component is in the form of elongated convoluted seminiferous tubules. These contain the stem spermatogonia or primordial germ cells, as well as cells undergoing spermatogenesis, and supporting Sertoli cells. The endocrine component consists of interstitial tissue, the connective tissue filling the spaces between the tubules and the glandular Leydig cells.

A different pattern of organization is present in urodeles (and other lower amniotes). The spermatogenic component consists of lobules whose stem spermatogonia are discharged along with all other sperm cells at each spawning. The lobules collapse to become reconstituted during the next reproductive cycle. The permanent stem cells may lie in one of several places outside the lobules. When the spermatogenic cycle is resumed, spermatogonia migrate to the lobules, where they reproduce to form clusters of spermatocytes that differentiate into mature sperm cells.

The ovaries of vertebrates are located in the peritoneal cavity, attached to the body wall on either side of the dorsal mesentery near the kidney. In tetrapods, the ovaries do not elongate proportionately to body growth and develop into compact bodies. The basic histological organization consists of two elements: a supportive connective tissue bed, the stroma, and the germinal cells, or follicles, which are embedded in the stroma. The follicles develop as a consequence of the migration of primordial germ cells from the yolk sac into the gonad during embryonic development, followed by their proliferation and growth over one or more seasons, during which their cytoplasm fills with yolk. The yolk will serve to nourish the developing embryo. The ovaries of amphibians (and most teleosts and some reptiles) continue to produce new generations of oocytes after maturity.

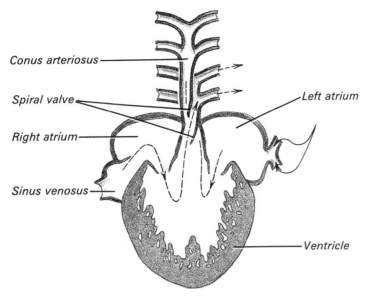

**FIGURE 8–8**
Amphibian heart (ventral view of front section).

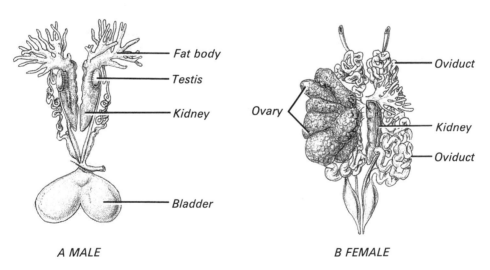

*A MALE*

*B FEMALE*

**FIGURE 8–9**
Frog urogenital system.

## NERVOUS SYSTEM

The brain of amphibians is not specialized, especially in urodeles (see Figs. 7-1, 7-2). Moreover, it is not even more advanced than that of cartilagineous fishes and dipnoans. In spite of the greater variety of the amphibians' mode of movement (swimming, leaping, or walking), they do not require the same degree of coordination as does a fish for its active mobility in water.

The cerebral hemispheres of amphibians are not well developed and are more separate from one another than those of fishes (Fig. 8-10). The olfactory lobes are not easily distinguished from the hemispheres. The corpus stratum is small, and primitive hippocampal and pyriform areas have formed. The epithalamus is smaller than that in fish, but anurans have well-developed pineal bodies. The hypothalamus gives evidence of incipient mammillary bodies while the epithalamus is beginning to enlarge. The cerebellum is poorly developed and auricles are absent. The absence of auricles is probably a consequence of the reduction in size of the lateral line system. However, the small size of the cerebellum is puzzling, for even though amphibians are not highly mobile, the jumping capability of anurans does require some degree of functional integration. Thus a larger cerebellum, the brain's coordination center, at least in anurans would be expected to be present. It is possible that a spinal reflex largely accounts for anurans' locomotory potential, since a decapitated frog can be induced to jump if appropriately stimulated.

The optic lobes are small in urodeles and moderate to large in anurans, reflecting the increasing impor-

tance of sight in land animals. As in fishes, olfactory stimuli are the major sensory input into the cerebral hemispheres. Cerebral efferent pathways extend down to the hypothalamus, the ventral thalamic nuclei, and the reticular formation of the brain stem, but none go directly down to the spinal cord. The tectum is the primary somatic integration center. Fibers from the tectum project down to the reticular formation, but tracts have evolved that extend directly down to the motor columns in the spinal cord (tectospinal tracts).

In the autonomic nervous system (ANS), a sequence of two sets of nerve fibers (preganglionic and postganglionic) forms a pathway from the central nervous system to smooth muscle or gland cell effectors. It appears that the ANS of nonmammalian vertebrates is basically similar to that of mammals, but much information about its anatomy and physiology is incomplete.

All vertebrates, starting with bony fish, have a bilateral pair of well-defined paravertebral sympathetic chains. The autonomic nerve fibers that emerge from the spinal cord pass in higher vertebrates through the ventral roots, but in amphibians some pass through the dorsal roots. In all vertebrates the cranial parasympathetic outflow passes through cranial nerves III, VII, IX, and X (and XI in mammals). The sacral parasympathetic outflow is rudimentary in anurans and absent in urodeles (and teleost fish). Finally it may be noted that in amphibians (and in fish) many sympathetic fibers are present in N.X, which is considered a parasympathetic nerve in mammals. The separation of the two types of fibers is more complete in higher vertebrates.

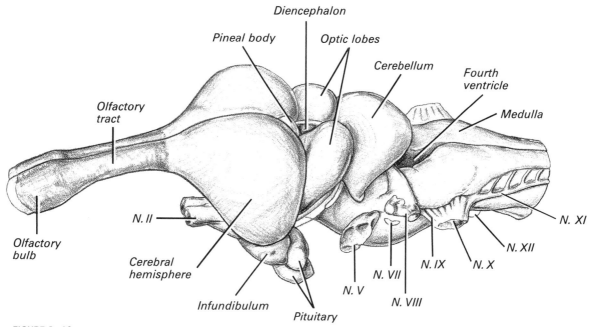

**FIGURE 8–10**
Frog brain (lateral view).

# ANATOMY OF THE CAT
## (INCLUDING THE SHEEP HEART, EYE, AND BRAIN)

# External Morphology

# Anatomy of the Cat

The domestic cat, *Felis domestica*, is descended from multiple crosses between the European wildcat and the Egyptian race of the African wildcat. The somewhat uncertain phylogeny of the cat is outlined in Table 1-1. Like other members of the class Mammalia, the cat has hair, and the female has mammary glands. Because they are flesh-eaters, cats belong to the order Carnivora; because their toes are separated, they belong to the suborder Fissipedia. The domestic cat will be studied as the typical mammalian form.

Carnivores can be traced as far back as the Eocene epoch of the Cenozoic period. The cat is the most specialized of the carnivores, especially in the development of its incisor and premolar teeth and its retractile claws. The habits of the cat are characteristic of carnivores. Cats rarely associate in packs and rarely hunt in open ground. Although they have assimilated themselves well into our civilization, they retain many feral instincts. They easily revert to a wild state and are more self-sufficient when left to shift for themselves than any other domestic animals.

When reverting to its natural state the cat makes use of its exceptional capabilities for seeing in dim light, stalks its prey stealthily on cushioned paws, crouches before the final spring, and kills with its sharply pointed teeth. The cat loves to play with or torment its victim before killing it and is an enemy to all bird life. It is, however, of economic value as a destroyer of insect pests and rodents.

The entire body of the cat—except the nose, lips, and foot pads—is covered with fur, which is kept well-groomed by frequent licking with its rough tongue. The length, quality, and color of the fur varies with the variety of the cat. The mombus cat of Africa has short, stiff hair, but Angora and Persian cats are remarkable for the length and delicacy of their soft fur. The coloration may be solid or a mottling of gray or yellow, with some breeds showing a distinct transverse striping.

The cat has a very delicately adjusted mechanism for excluding light from its eyes. In bright light the pupil narrows to a vertical slit, which may be drawn so tight that only pinpoint openings at its extremities admit light. The choroid, or middle vascular coat, of the eye is also modified to form a brilliant layer that makes the eyes shine in darkness because it reflects any available light.

**TABLE 1-1**
Phylogeny of the cat.

| Geologic Epoch | Genus |
|---|---|
| Recent | *Felis* (true cat) |
| Pleistocene | *Felis* |
| Pliocene | *Felis* |
| Miocene | *Pseudaelurus* (false cat) |
| | *Nimravus* (hunting cat) |
| Oligocene | |
| Upper | *Dinictus* (terrible weasel) |
| Lower | *Aelurietis* (cat weasel) |
| Eocene | Undiscovered creodont |

## INTEGUMENT (Fig. 1-1)

*With a compound microscope examine prepared slides of the skin of the cat or some other typical mammal.*

**Epidermis** The outer region is made up of a number of layers of stratified epithelial cells. The deepest layer, the **stratum basalis,** consists of columnar-shaped cells that are mitotically active. The daughter cells produced by this layer move toward the surface to form several adjacent layers of polygonal cells. These cells, which constitute the **stratum spinosum,** have a prickly appearance and are held together by minute intercellular bridges. The stratum basalis and the stratum spinosum are collectively referred to as the **stratum Malpighii.** Several layers of cells above the stratum spinosum contain highly refractile granules and make up the **stratum granulosum.** Above this are a few layers of cells that appear structureless and constitute the **stratum lucidum.** The most superficial layer, the **stratum corneum,** is made up of cornified cells that are continually being worn away at the surface and replaced from below. The outermost, chronologically oldest cornified cells are responsible for the protective and impermeable qualities of mammalian skin.

**Dermis** This is a fibrous connective tissue layer that supports and attaches the epidermis and conveys blood vessels and nerves to it. Sebaceous (oil) glands, sudoriferous (sweat) glands, and hair follicles are also present in the dermis.

## EXTERNAL FEATURES (Fig. 1-2)

The body is made up of a head, neck, trunk, and tail. The trunk consists of the thorax and abdomen. Each forelimb consists of the arm (brachium), forearm (antebrachium), wrist, palm, and digits; each hind limb consists of the thigh, shank, ankle, foot, and digits. Three other areas that can be recognized are the axilla, the area between the thorax and arm; the inguinal region, which is between the abdomen and the thigh; and the perineum, the area around the urogenital and anal openings.

The head of the cat is large and the neck movable. The body is rather slender and its average length is about two and one-half feet, with the tail making up about one-third of the total length of most breeds. Some forms, such as the Manx, are tailless.

The limbs of mammals have a position markedly different from that of appendages of lower vertebrates. Fish, for example, have laterally projecting fins. The position in higher forms is the result of a 90° rotation of all four limbs, which project ventrally (and are elongated).

The cat walks with its wrist and heel raised off the ground, a posture known as digitigrade. This posture is in contrast to plantigrade (as in man), where the sole of the foot is flat on the ground, or unguligrade (as in the horse), in which the animal walks on the tips of its toes. When stalking its prey or when frightened, the cat crouches and slinks in a posture between that of digitigrade and plantigrade. There are five toes on the forefoot but only four on the hind foot (the great toe, or hallux, being missing). The thumb, or pollex, of the forefoot is situated above the other toes so that it does not touch the ground in walking.

The following specific structures should be identified.

**Eyes** These are guarded by upper and lower eyelids and a nictitating membrane. The nictitating membrane is a transparent layer that extends laterally over the eyeball from the medial angle of the eye.

**Pinna** This is the prominent soft-tissue part of the external ear.

**Vibrissae** These are stiff, bristle-like, tactile hairs around the mouth, on the cheeks, and above the eyes.

**Claws** These epidermal derivatives are present at the terminal end of the digits. They are modifications of the stratum corneum of the skin at these sites.

**Tori** (Fig. 1-3) These are epidermal thickenings of the skin, known also as foot or friction pads, which form cushions on the walking surfaces of the feet. Seven of these pads are present on each forepaw, and five on each hindpaw.

**Anus** Lies in the midline, ventral to the base of the tail.

**Urogenital aperture** This opening in the female is located just anterior to the anus.

**Nipples** These are projections on the ventral side of the body through which the terminal ducts of the mammary gland pass to open onto the skin surface. There are four or five linearly arranged nipples on either side of the midline. They are less conspicuous in males.

**Scrotum** The sac in the male that contains the testes. It is located on each side a short distance anterior to the anus.

**Prepuce** Slight elevation located immediately in front of the scrotum. The penis is normally retracted and thus not evident.

155

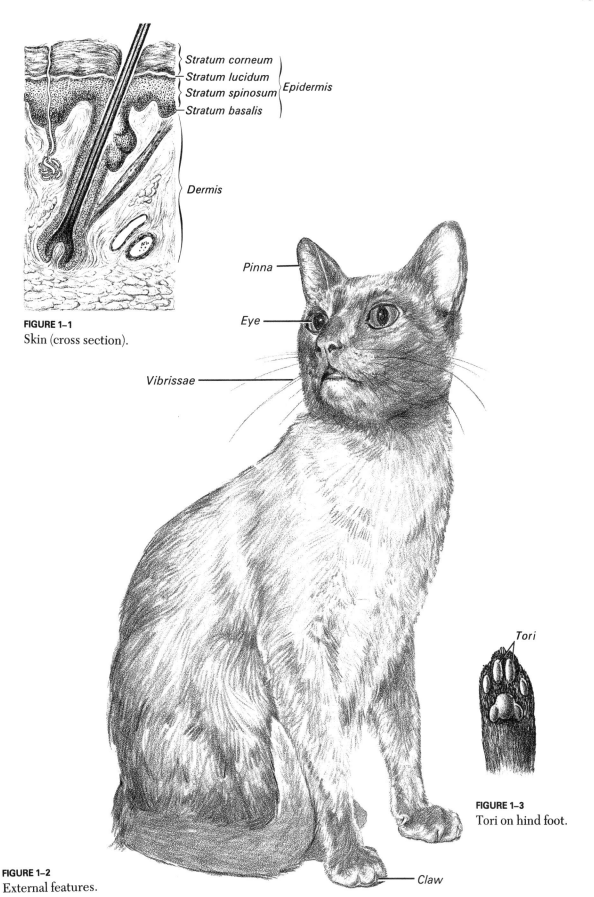

*Stratum corneum*
*Stratum lucidum*
*Stratum spinosum*
*Stratum basalis*
*Epidermis*

*Dermis*

**FIGURE 1–1**
Skin (cross section).

*Pinna*

*Eye*

*Vibrissae*

*Tori*

**FIGURE 1–3**
Tori on hind foot.

**FIGURE 1–2**
External features.

*Claw*

EXERCISE 2

# Skeletal System

# Anatomy of the Cat

The skeleton of the cat (Fig. 2-1) is bony, except for the costal cartilages making up the sternal ends of the ribs. It is divided into axial and appendicular parts and should be studied with the aid of both an articulated and a disarticulated specimen. A disarticulated specimen may be made available for home study in the form of kits that contain, in addition to a skull, a representative bony segment from each body part (such as a femur and a cervical vertebra). This approach will save laboratory time and provide a sound basis for study of the muscle system.

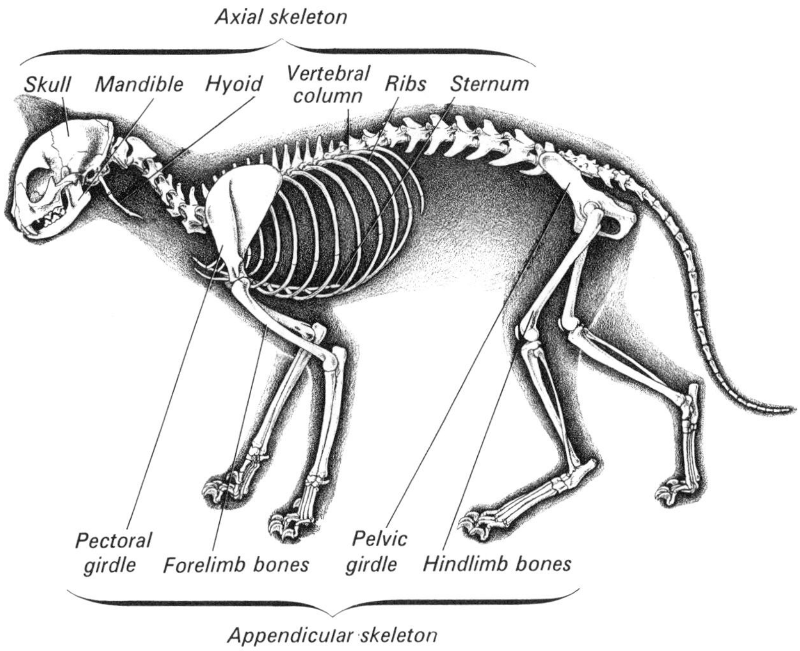


**FIGURE 2–1**
Profile of entire skeleton.

## AXIAL SKELETON

The axial skeleton consists of the skull, mandible, hyoid apparatus, vertebral column, ribs, and sternum. The skull exhibits numerous sutures that are sites of articulation between adjacent bones and aid in demarcating their boundaries.

### Skull (Figs. 2-2 to 2-5)

On the *dorsal* surface (Fig. 2-2) the following bones and structures should be identified.

**Premaxillary bones**   A pair of bones forming the lower boundary of the external nares. Each bone bears three small incisor teeth.
**Nasal bones**   A long pointed bone located posterior to each external naris.
**Maxillary bones**   Each of these two bones lies lateral to a nasal bone and bears five teeth. The maxillary bones make up the greater part of the upper jaw and palate.
**Frontal bones**   A pair of large bones forming the anterior part of the skull and a large portion of the medial wall of each orbit. Each bone has a laterally projecting **postorbital process.**
**Malar (jugal) bones**   Each of these two bones bounds the orbit laterally and forms a part of the zygomatic arch. Each malar bone has a **postorbital process** that

extends upward toward the postorbital process of the frontal bone on that side and a ventrocaudally directed **zygomatic process.**
**Parietal bones**   This pair of bones is located in the posterolateral region of the skull. Each articulates anteriorly with a frontal bone and meets its mate in the mid-dorsal line.
**Interparietal bone**   A single median triangular bone located between the posterior parts of the two parietal bones.
**Temporal Bones**   The most prominent segment of each bone is the large, flat squamosal part. Located on each side, the squamosal is one of four parts that form the temporal bone. The other three parts are the tympanic, mastoid, and petrous parts. Each large, flat squamosal segment gives rise to a **zygomatic process,** which is directed toward the malar bone to complete the **zygomatic arch** (see also Fig. 7-2).
**Supraoccipital bone**   The part of the occipital bone located posterior to the parietals.
**Orbit**   This is the anterior of the two large cavities formed between the frontal bone and zygomatic arch.
**Temporal fossa**   The posterior of the two cavities. It is partially separated from the orbit by the postorbital processes of the malar and frontal bones.
**Lacrimal bones**   A small bone on each side that makes up the anteromedial wall of the orbit.

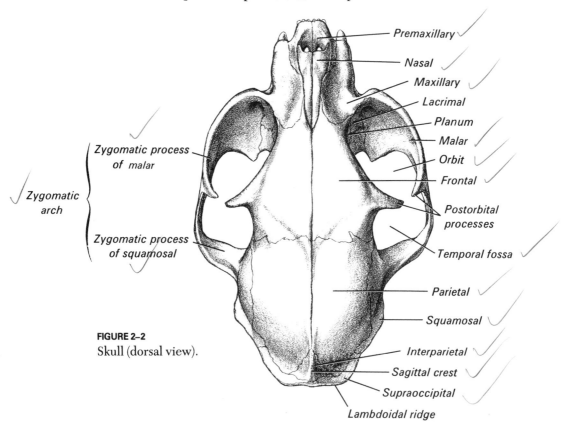

**FIGURE 2–2**
Skull (dorsal view).

**Planum bones**   Each of these two small bones is a part of the ethmoid. When present, each lies between the lacrimal and frontal bones.

**Lambdoidal ridge**   A transverse crest extending across the upper part of the occipital bone.

**Sagittal crest**   A midline ridge that extends vertically along the interparietal and supraoccipital bones.

On the (left) *lateral* surface (Fig. 2-3) identify the following additional paired structures.

**Tympanic bulla**   Thin-walled rounded bony eminence located ventral to the squamosal. It is the tympanic part of the temporal bone (see also Fig. 7-2).

**Mastoid bone**   This is the outer region of the petrous portion of the temporal bone. This bone is somewhat triangular and projects anteroventrally adjacent to the tympanic bulla, terminating just posterior to its opening. The ventral tip of the mastoid bone forms the **mastoid process.**

**Teeth**   The dental formula for each half upper jaw is three **incisors,** one **canine,** three **premolars,** and one **molar** (Fig. 2-4).

On the *ventral* surface (Fig. 2-4) identify the following bones and structures.

**Palatine bones**   Each bone of this pair lies behind the maxillary and, along with part of the maxillary, forms the hard palate, which makes up most of the roof of the mouth.

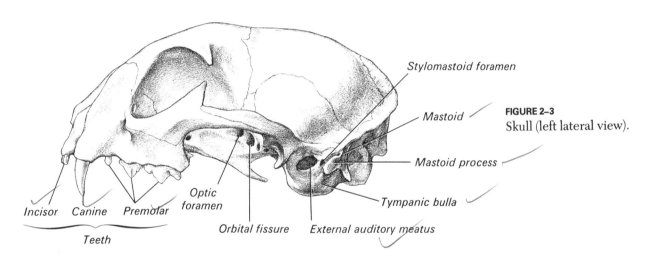

FIGURE 2–3
Skull (left lateral view).

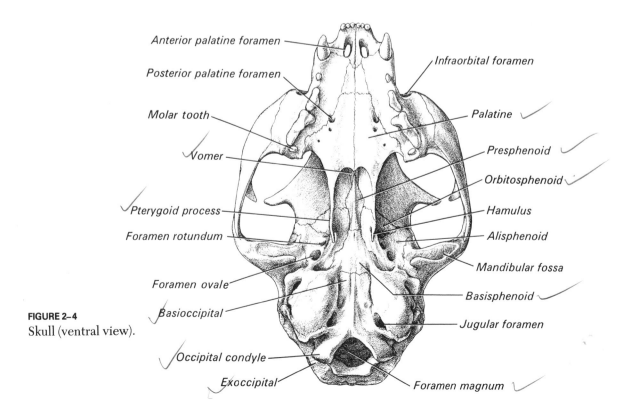

FIGURE 2–4
Skull (ventral view).

**Vomer**   Unpaired bone that forms the posterior part of the nasal septum and the roof of the posterior part of the nasal cavity.

**Presphenoid bone**   A small elongated bone that lies on the midline behind the vomer and forms part of the roof of the pharynx.

**Orbitosphenoid bone**   Each of these is a process of the presphenoid bone, which can be seen in the posteromedial wall of the orbit.

**Basisphenoid bone**   An unpaired midline bone lying behind the presphenoid.

**Alisphenoid bone**   Wing-like part of the sphenoid bone located on the medial wall of the temporal fossa. It projects onto the ventral surface of the skull and there gives rise to the **pterygoid process.** Projecting posteriorly from the pterygoid process is the **hamulus.**

**Mandibular fossa**   Transverse depression lateral to the alisphenoid. It is the concavity on the medial surface of the zygomatic process of the temporal bone.

**Basioccipital bone**   A broad unpaired bone that is present along the midline and extends from the basisphenoid to the foramen magnum.

**Exoccipital bone**   A part of the occipital bone. It extends medially to form the anterolateral boundary of the foramen magnum. Each bears an ovoid prominence, the **occipital condyle,** that articulates with the atlas, or first cervical vertebra.

On a skull that has been sectioned *sagittally* (Fig. 2-5) identify the following bones and structures.

**Ethmoid bones**   The relatively large bone occupying the nasal cavity. It is made up of a series of scroll-shaped bones, the **ethmoturbinals,** which are on each side; a medial unpaired vertical plate, the **mesethmoid,** that forms part of the nasal septum; and a horizontal **cribriform plate** that separates the nasal and cranial cavities.

**Tentorium**   A shelf of bone projecting into the cranial cavity from the parietal bone. It partially separates the cerebral and cerebellar fossae.

**Hypophyseal fossa**   A depression in the basisphenoid bone that houses the pituitary gland, which is bounded in front and in back by bone plates. These plates, together with the bony floor of the fossa, constitute the **sella turcica.**

**Sphenoid sinus**   The cavity within the basisphenoid located anterior to the hypophyseal fossa.

**Frontal sinus**   The cavity within each frontal bone.

The foramina and other openings of the skull are summarized in Table 2-1. Most of these are identified on the illustrations of the cat's skull.

**TABLE 2-1**

Foramina in the skull of the cat.

| Name | Location | Contents |
|---|---|---|
| Anterior palatine | Between pre-maxillary and maxillary ventrally | Nasopalatine nerve and nasal artery |
| Condyloid canal | Exoccipital | Vein from transverse sinus |
| Ethmoid | Orbitosphenoid | Ethmoid nerve and artery |
| External auditory meatus | Lateral surface of tympanic bulla | ——— |
| Facial canal | Petrous region of temporal | N. VII |
| Hypoglossal | Exoccipital bone | N. XII |
| Infraorbital | Maxillary near orbital margin | Infraorbital nerve and artery |
| Internal auditory meatus | Petrous region of temporal | N. VIII |
| Jugular | Junction of tympanic bullae with ex- and basioccipitals | N. IX, X, XI Inferior cerebral vein |
| Magnum | Occipital bone | Spinal cord, vertebral arteries |
| Nasolacrimal | Lacrimal inside orbit | Tear duct |
| Olfactory | Cribriform plate of ethmoid | N. I |
| Optic | Orbitosphenoid | N. II and ophthalmic artery |
| Orbital fissure | Junction of orbito- and alisphenoid bones | N. III, IV, $V^1$, and VI |
| Ovale | Alisphenoid | N. $V^3$ |
| Posterior palatine | Palatine | Palatine nerve and artery |
| Rotundum | Alisphenoid | N. $V^2$ |
| Sphenopalatine | Vertical plate of palatine | Sphenopalatine nerve and artery |
| Stylomastoid | Mastoid region of temporal | N. VII |

## Mandible (Fig. 2-6)

The lower jaw consists of two halves united firmly at their anterior ends. Each half bears three incisors, one canine, two premolars, and one molar.

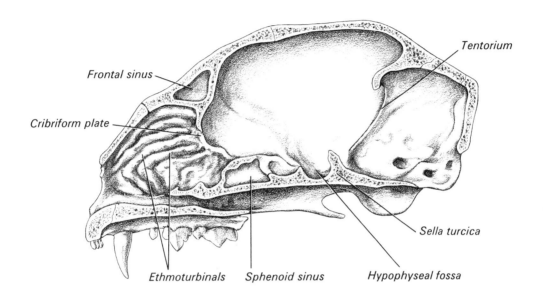

**FIGURE 2–5**
Skull (sagittal section).

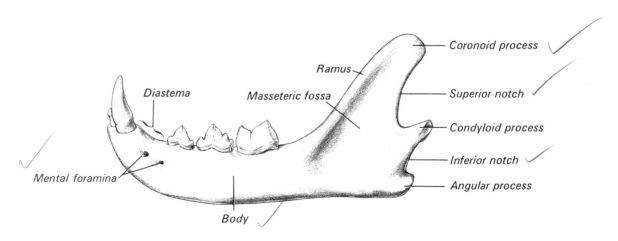

**FIGURE 2–6**
Mandible (lateral view).

**Body**    Anterior horizontal part of the mandible in which the teeth are embedded.

**Ramus**    Posterior and somewhat vertical part of the mandible.

**Mental symphysis**    The site of articulation of both halves of the mandible.

**Coronoid process**    This large projection on the ramus extends up into the temporal fossa of the skull.

**Superior notch**    The concavity between the coronoid and condyloid processes.

**Condyloid process**    This knob-shaped projection at the base of the superior notch articulates with the mandibular fossa of the temporal bone.

**Inferior notch**    Concavity below the condylar process.

**Angular process**    Projection at the posteroventral angle of the mandible.

**Masseteric fossa**    Concavity on the lateral surface of the ramus.

**Diastema**    The gap between the canine and premolar teeth on the mandible.

**Mental foramina**    Two openings on the anterolateral surface of the body beneath the diastema.

**Mandibular foramen**    Opening on the medial surface near the base of the coronoid process.

## Hyoid Apparatus (Figs. 2-7, 2-8)

A hyoid apparatus is present in mammals but is not as large as in more primitive tetrapods, since it is formed from only the ventral parts of the hyoid and first branchial arches. Its size, however, varies greatly among mammals.

The hyoid apparatus supports the tongue and structures of the larynx. It consists of a series of bones extending from the thyroid cartilage to the tympanic bulla. Dissection of the hyoid apparatus may be postponed until after the respiratory system has been studied (Exercise 4).

**Body**    Single median bone, positioned anterior to the upper end of the larynx. It is also known as the **basihyal.**

**Cornua**    Two pairs of bony projections, or horns, that articulate with the body of the hyoid. Each of the larger, more cephalic horns consists of four long slender bones: the **ceratohyal, epihyal, stylohyal,** and **tympanohyal.** The ceratohyal bone extends ventrolaterally, and the other three bones are directed posteriorly. Each of the posterior horns consists of a single **thyrohyal bone.**

## Vertebral Column (Figs. 2-9 to 2-16)

*Examine a set of disarticulated vertebrae and learn to identify a vertebra selected at random as cervical,* *thoracic, lumbar, sacral, or caudal. Referring to a mounted skeleton, arrange the disarticulated vertebrae in their proper order.*

The vertebral column consists of a series of bones called vertebrae, which occupy the mid-dorsal line of the animal. Its cranial end articulates with the occipital condyles of the skull and its caudal end terminates in the tip of the tail.

The vertebral column connects the anterior and posterior segments of the trunk, provides attachments for the ribs, and protects the spinal cord. It consists of 53 vertebrae: 7 cervical, 13 thoracic, 7 lumbar, 3 fused sacral, and 23 caudal vertebrae.

The vertebrae may be divided into typical and atypical forms.

A typical vertebra (Figs. 2-9, 2-10) consists of the following three elements.

**Body**    A short cylindrical part that is also known as the **centrum.**

**Vertebral arch**    The bony arch surrounding the spinal cord that consists of **pedicles,** which project back from the body, and **laminae,** which are flat sloping plates that meet in the median plane.

**Processes**    There are three typical types. **Spinous processes** project backward from the junction of the laminae; **transverse processes** project laterally from the site of junction of the pedicles with laminae; and **articular processes** project cranially, as prezygapophyses, and caudally, as postzygapophyses, from the site of junction of the pedicles and laminae.

The characteristics of typical vertebrae in each of the major regions are listed in Table 2-2. The major segments are as follows.

**Cervical vertebrae** (Fig. 2-11)    These are characterized by the presence of **foramina transversaria,** each of which lies between the two roots of a transverse process.

The atypical cervical vertebrae are the atlas and axis.

**Atlas** (Fig. 2-12)    The first cervical vertebra differs from all others in that it lacks a spinous process and a body. It consists of a **ventral arch** that is thickened at its sides to form the **lateral masses.** Each lateral mass bears at its anteromedial surface a cranial articular facet that articulates with the occipital condyle of the skull. Dorsal to each facet is the **atlantal foramen.** Wing-like transverse processes extend sideways from the lateral masses. A foramen transversarium is located at the base of each transverse process. The **dorsal arch** is much

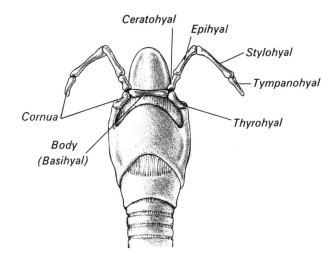

**FIGURE 2–7**
Hyoid (ventral view).

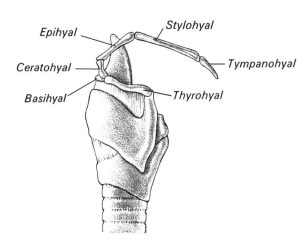

**FIGURE 2–8**
Hyoid (left lateral view).

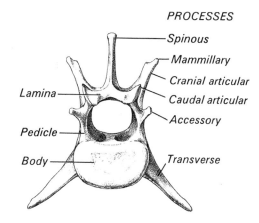

**FIGURE 2–9**
Typical lumbar vertebra (caudal view).

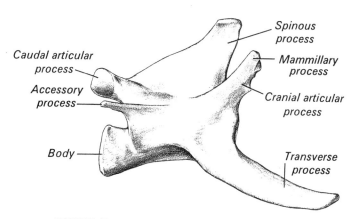

**FIGURE 2–10**
Typical lumbar vertebra (lateral view).

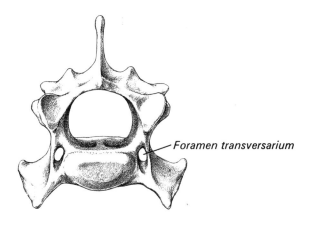

**FIGURE 2–11**
Typical cervical vertebra (anterodorsal view).

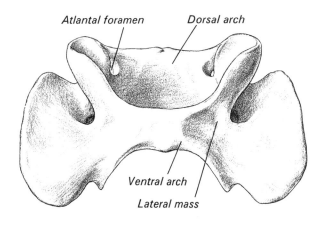

**FIGURE 2–12**
Atlas (ventral view).

**TABLE 2-2**
Characteristics of typical vertebrae in the cat.

| Component | Cervical (3–7) | Thoracic (2–10) | Lumbar (1–5) |
|---|---|---|---|
| Body | Oval | Heart-shaped | Kidney-shaped |
| Vertebral foramen | Triangular | Circular, small | Triangular |
| Spine | Short | Long | Thick |
| Transverse process | With foramina transversaria | Thick | Bears accessory processes |

broader than the ventral arch. The caudal articular facets, which are located on the posterior surface of the atlas, articulate with the second cervical vertebra, the axis.

**Axis** (Fig. 2-13)  The second cervical vertebra is characterized by its **dens,** or odontoid process, a prominent projection extending from the anterior end of the body. It probably represents the body of the atlas that has fused with the axis during embryonic development.

**Thoracic vertebrae** (Fig. 2-14)  These are characterized by **costal demifacets** on the sides of the centra for articulation with the heads of the ribs and **costal facets** on the transverse processes for articulation with the tubercles of the ribs.

The four atypical thoracic vertebrae are the following.

   **First thoracic vertebra**  It resembles the cervical vertebrae but in addition has a costal facet at its anterior end and a costal demifacet at its posterior end.
   **Eleventh, twelfth, thirteenth thoracic vertebrae**  Each bears a costal facet on each side of the body and lacks facets on the transverse processes.

**Lumbar vertebrae** (see Figs. 2-9, 2-10)  These vertebrae are characterized by their large size and by the absence of both costal facets and foramina on their transverse processes. Small **accessory processes** extending from the neural arch are present on most lumbar vertebrae, but they may be lacking on the sixth and seventh vertebrae. **Mammillary processes** project from the dorsolateral surfaces of the cranial articular processes.

**Sacrum** (Fig. 2-15)  Consists of a median portion of three fused vertebrae and a pair of lateral bony masses composed of fused costal elements and transverse processes. It is triangular in shape, having a base, apex, and pelvic and dorsal surfaces. Two pairs of **sacral foramina** are present on both surfaces.

**Caudal vertebrae** (Fig. 2-16)  These vary from 21 to 23 in number, and they become progressively smaller toward the tail. The first eight vertebrae have a distinct vertebral arch, a spine, and transverse processes. These structures disappear, starting with the ninth vertebra, and more caudal vertebrae consist largely of centra. **Hemal processes** project from the ventral surfaces of the vertebrae, beginning with about the third. Together with small **chevron bones** (which are missing in prepared skeletons) these hemal processes form **hemal arches,** which collectively form a **hemal canal.**

## Ribs (Fig. 2-17, 2-18)

*Select a rib from the middle of the thoracic cage (e.g., sixth) to use in identifying the characteristic elements of a typical rib. Then fit two thoracic vertebrae together with a rib and observe how the head of the rib articulates with facets on both vertebral bodies. Also note that the tubercle of the rib fits into an articular facet on the transverse process of the more caudal of the two vertebrae.*

There are 13 ribs, or costae, on each side of the body. The first nine, the **true ribs,** are joined to the sternum by **costal cartilages.** The remaining four are **false ribs.** The costal cartilages of the tenth, eleventh, and twelfth ribs join that of the ninth and thereby become attached to the sternum. The thirteenth is a **floating rib,** since its costal cartilage does not reach the sternum. A rib consists of the following.

**Head**  The head is wedge-shaped. It presents two facets that articulate with the demifacets of the bodies of successive thoracic vertebrae.
**Neck**  The constricted region behind the head.
**Tubercle**  The elevated area on the lateral surface behind the neck. It has a facet that articulates with the costal facet of the transverse process of a thoracic vertebra.
**Shaft**  The thin, long, curved body that exhibits a sharp bend at the **angle.**

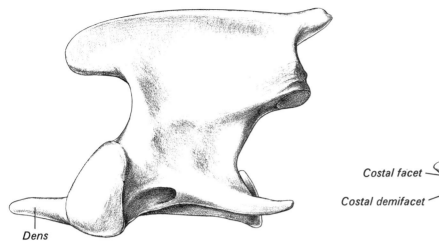

**FIGURE 2–13**
Axis (lateral view).

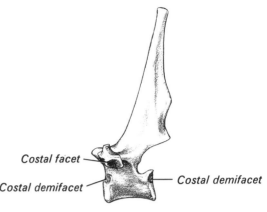

Costal facet
Costal demifacet
Costal demifacet

**FIGURE 2–14**
Typical thoracic vertebra (lateral view).

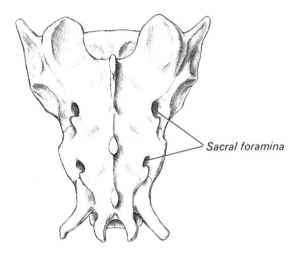

Sacral foramina

**FIGURE 2–15**
Sacrum (dorsal view).

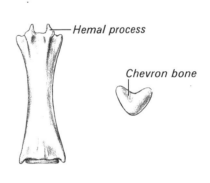

Hemal process

Chevron bone

**FIGURE 2–16**
Caudal vertebra (dorsal view).

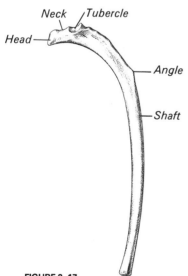

Neck  Tubercle
Head
Angle
Shaft

**FIGURE 2–17**
Rib.

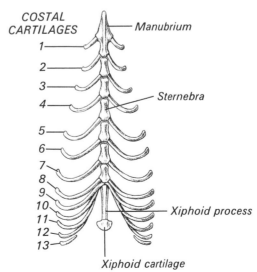

COSTAL
CARTILAGES
1
2
3
4
5
6
7
8
9
10
11
12
13

Manubrium

Sternebra

Xiphoid process

Xiphoid cartilage

**FIGURE 2–18**
Sternum and costal cartilages.

## Sternum (Fig. 2-18)

The sternum, to which the costal cartilages are attached, consists of a row of eight bones, or **sternebrae,** and is divided into three parts.

**Manubrium**    The first sternebra. This bone has a pointed anterior end and extends laterally in the form of a pair of processes.
**Body**    This is composed of the next six sternebrae.
**Xiphoid process**    The last sternebra, which has attached at its posterior end a broad **xiphoid cartilage.**

## Thoracic Skeleton

Several of the elements noted above make up the skeletal framework of the walls of the thorax. It is formed ventrally by the sternum and costal cartilages, dorsally by the bodies of the 13 thoracic vertebrae and the corresponding intervertebral discs, and by the ribs, from their heads to their angles. It is formed laterally by the shafts of the ribs, from their angles to their cartilages. Confirm these facts by referring again to a mounted skeleton.

The osseocartilagenous thoracic cage is conical in shape. It is open at its truncated apex to provide for passage of the trachea, esophagus, large blood vessels of the neck, and thoracic duct as well as the vagus, phrenic, and sympathetic nerves. It is closed at its wider base by the diaphragm. The thoracic cage provides protection for the heart and its great vessels as well as for the lungs.

## APPENDICULAR SKELETON

The appendicular skeleton consists of the pectoral girdle, forelimbs, pelvic girdle, and hind limbs.

## Pectoral Girdle (Fig. 2-19)

The pectoral girdle is represented by the scapulae and clavicles. A third element, the coracoid, found in the pectoral girdle of lower forms as a separate entity, is developed as the coracoid process of the scapula in the cat.

The **scapula** is a large triangular flat bone that exhibits the following features.

**Glenoid fossa**    Concavity at the ventral apex (head) of the scapula for articulation with the humerus.
**Coracoid process**    A short curved projection arising from the anteromedial surface of the head.
**Spine**    A keel-like structure that divides the lateral surface of the scapula into two fossae.
**Metacromion**    A process projecting posteriorly from the spine near its ventral end.
**Acromion**    A short blunt process that extends from the ventral end of the spine.
**Supraspinous fossa**    The concavity on the cranial side of the spine.
**Infraspinous fossa**    The concavity caudal to the spine.
**Subscapular fossa**    The large shallow depression on the medial surface of the scapula.

The **clavicle** is a slender curved rod embedded in the muscles of the shoulder. It is connected by fibrous tissue to the apex of the acromion. The lateral end is slightly enlarged.

## Forelimb (Figs. 2-20 to 2-23)

The skeleton of the forelimb consists of the bones of the arm, foreleg, and forefoot. The longest and largest forelimb bone is the humerus, located in the arm. The foreleg has two bones, the radius and ulna. The forefoot consists of the carpals, metacarpals, and phalanges.

The **humerus** (Fig. 2-20) has the following features.

**Head**    The large convex, smooth, and hemispherical protuberence at the end of the humerus.
**Greater tubercle**    A projection extending forward from the lateral border of the head.
**Lesser tubercle**    A smaller projection extending anteromedially from the head.
**Intertubercular sulcus**    The groove located between the tubercles.
**Pectoral ridge**    The crest extending distally from the ventral end of the greater tubercle along the shaft of the humerus.
**Deltoid ridge**    The crest extending down obliquely, from the posterior end of the greater tubercle, along the lateral side of the shaft of the humerus. Near the middle of the ventral surface of the shaft it meets the pectoral ridge.
**Shaft**    The long body of the humerus. It contains nutrient foramina.
**Capitulum**    This rounded eminence forms the lateral part of the distal articular surface. It articulates with the head of the radius.
**Trochlea**    The spool-shaped medial part of the distal articular surface. It articulates with the ulna.
**Medial epicondyle**    The roughened prominence medial to the trochlea.

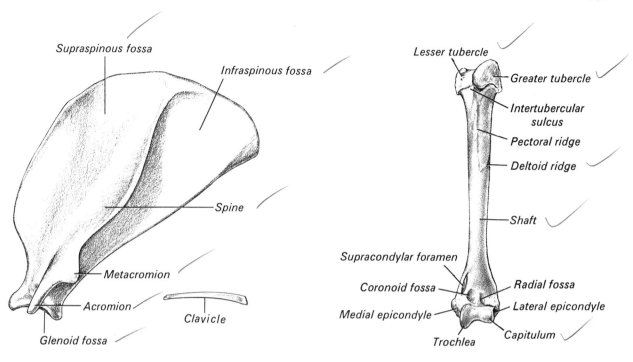

**FIGURE 2–19**
Scapula (lateral view) and clavicle.

**FIGURE 2–20**
Humerus (ventral view).

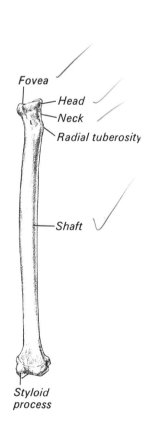

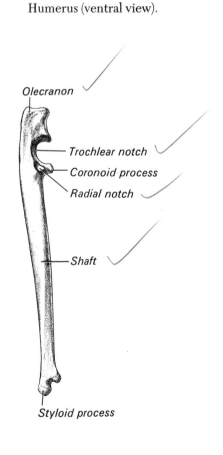

**FIGURE 2–21**
Radius (dorsolateral view).

**FIGURE 2–22**
Ulna (ventromedial view).

**Lateral epicondyle** The elevation lateral to the capitulum.

**Olecranon fossa** The depression above the trochlea on the posterior surface.

**Coronoid fossa** The depression above the trochlea on the anterior surface.

**Radial fossa** The depression over the capitulum adjacent to the coronoid fossa.

**Supracondylar foramen** An oval aperture in the bone, proximal to the medial epicondyle.

The **radius** (Fig. 2-21), located on the lateral side of the forearm, has the following features.

**Head** A disc whose upper surface has a concave **fovea.**

**Neck** The constricted portion below the head.

**Radial tuberosity** Protuberance below the neck of the radius.

**Shaft** The body of the radius. It is convex dorsally and concave ventrally.

**Styloid process** The wedge-shaped projection extending from the medial side of the distal end of the radius.

The **ulna** (Fig. 2-22) is a long slender bone that has the following features.

**Olecranon** The projection from the proximal end of the bone.

**Trochlear notch** The wide concavity indenting the olecranon. At this site the olecranon articulates with the trochlea of the humerus.

**Coronoid process** The projection at the base of the trochlear notch.

**Radial notch** The concave facet on the lateral surface of the coronoid process.

**Shaft** The body of the ulna. It tapers gradually toward the carpus.

**Styloid process** The projection from the distal end of the ulna.

The wrist, or **carpus** (Fig. 2-23), consists of seven bones arranged in two rows. The bones of the proximal row, from medial to lateral, are the scapholunar, cuneiform, and pisiform. Those of the distal row, from medial to lateral, are the trapezium, trapezoid, capitate, and hamate bones. If the carpal bones of a disarticulated skeleton are available, the individual characteristics (and mode of articulation) of each can be determined.

**Scapholunar** A quadrangular bone with its ventral radial angle shaped in the form of a blunt process. Its proximal surface is smooth; its distal surface is marked by oblique ridges.

**Cuneiform bone** Shaped in the form of a flattened pyramid, its proximal radial surface is a smooth facet.

**Pisiform** This bone is about twice as long as broad and has enlarged ends.

**Trapezium** Has the form of a triangular prism curved into a semicircle.

**Trapezoid** Somewhat wedge-shaped, with the apex of the wedge pointing ventrad.

**Capitate** An oblong plate bearing on its proximal surface a semicircular ridge crossing it diagonally.

**Hamate** A wedge-shaped bone with the apex of the wedge directed proximally.

The **metacarpus** (Fig. 2-23) consists of five metacarpals set between the carpal bones and the digits. Each metacarpal is composed of a **base** at its proximal end, a slender **shaft,** and a distal convex **head.** All the digits have three **phalanges,** except the first digit, which consists of only two phalangeal bones. The proximal end of each phalanx is notched, and its distal end is pulley-shaped. The distal phalanges form the skeleton of the **claw.**

## Pelvic Girdle (Fig. 2-24)

The pelvic girdle is made up of two **innominate bones** (Fig. 2-24) that articulate dorsally with the sacrum and ventrally at the midline with each other at the pubic symphysis. Each innominate consists of three bony parts, the ilium, ischium, and pubis, which fuse during development at the **acetabulum,** a depression that receives the head of the femur. The large opening in each innominate bone between the ischium and pubis is the **obturator foramen.**

**Ilium** This part is long and has a flattened expanded anterior surface, a rough medial surface that articulates with the sacrum, and a concave lateral surface.

**Ischium** Has the shape of a triangular prism that is contracted in the middle.

**Pubis** A flat, curved part, contracted in the middle and expanded at the ends.

## Hind Limb (Figs. 2-25 to 2-28)

The skeleton of the hind limb consists of the bones of the thigh, hind leg, and hind foot. The thigh contains a single long bone, the femur; the hind leg skeleton consists of two bones, the tibia and fibula. The hind foot is supported by the tarsus, metatarsals, and phalanges.

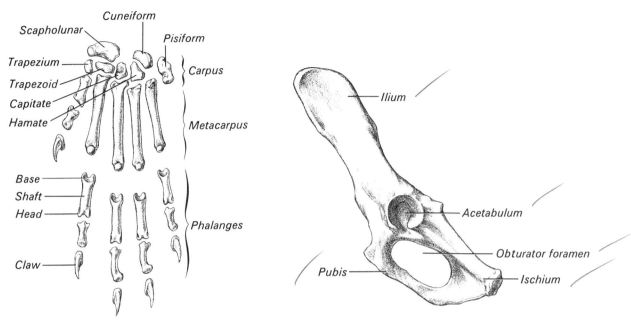

**FIGURE 2–23**
Bones of the forefoot (dorsal view).

**FIGURE 2–24**
Innominate bone (ventrolateral view).

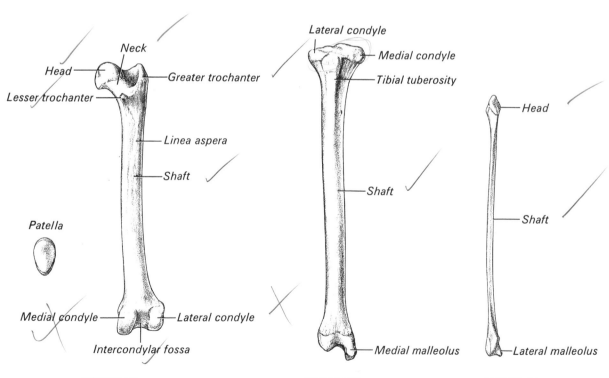

**FIGURE 2–25**
Femur and patella (dorsal view).

**FIGURE 2–26**
Tibia (ventral view).

**FIGURE 2–27**
Fibula (ventral view).

The **femur** (Fig. 2-25) has the following features.

**Head** The hemispheric protuberance that articulates at the acetabulum with the innominate bone.
**Neck** The thick bar connecting the head with the trochanters.
**Greater trochanter** The eminence projecting from the lateral side of the proximal end of the femur, opposite the head.
**Lesser trochanter** A small tubercle on the postero-medial side, a short distance below the head.
**Shaft** The body of the femur. It displays on its ventral surface a rough line, the **linea aspera.**
**Lateral** and **medial condyles** These are eminences present on either side of the distal end of the femur.
**Intercondylar fossa** The concavity separating the projecting condyles.
**Patella** This small triangular sesamoid bone lies against the articular surface at the distal extremity of the femur. Sesamoid means formed in a ligament.

The **tibia** (Fig. 2-26), the larger of the two leg bones, has the following features.

**Lateral** and **medial condyles** Projections on either side of the proximal end of the bone.
**Tibial tuberosity** A small tubercle that is located dorsally where the condylar margins converge downward.
**Shaft** The long body of the tibia. It is triangular in section.
**Medial malleolus** The medial extension at the distal end of the tibia.

The **fibula** (Fig. 2-27) lies on the lateral side of the hind leg. It exhibits the following features.

**Head** The small proximal end that exhibits a facet for articulation with the tibia.
**Shaft** The slender body of the fibula.
**Lateral malleolus** The expanded distal end of the fibula.

The **tarsus** (Fig. 2-28) consists of seven tarsal bones. The longest, the calcaneus, lies on the lateral side. It articulates distally with the cuboid. Between the calcaneus and tibia is the talus, whose distal end articulates with the navicular, or scaphoid. The navicular articulates distally with the three cuneiform bones. The individual tarsal bones have the following characteristics.

**Calcaneus** This bone lies on the lateral side and forms the heel. It is the largest bone of the foot and articulates with the cuboid.
**Cuboid** Has the approximate form of a cube.
**Talus** May be divided into a body, neck, and head: the latter articulates with the navicular.
**Navicular** (scaphoid) Boat-shaped. Its proximal surface has one facet, its distal surface three. It articulates with the three cuneiform bones.
**Lateral cuneiform** Wedge-shaped, with a hook-like process extending from its ventral angle.
**Intermediate cuneiform** Small and wedge-shaped.
**Medial cuneiform** Flat, triangular, and broader at its proximal end.

The **metatarsus** consists of five metatarsal bones (Fig. 2-28), the first of which is rudimentary. They closely resemble the metacarpals.

There are three **phalanges** on each digit (Fig. 2-28). They are similar to those in the hand. Since the first digit is not present on the hind foot, the digits are numbered two to five.

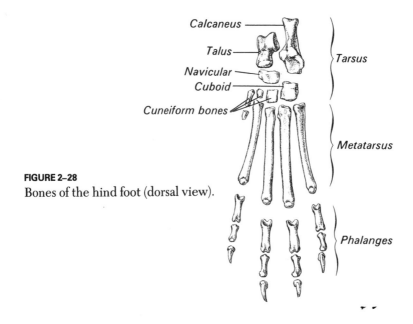

**FIGURE 2–28**
Bones of the hind foot (dorsal view).

# Anatomy of the Cat

The muscles are divided structurally (histologically) and functionally into two groups. The *voluntary* muscles, which are responsible for all coordinated movement, are innervated by branches from the cerebrospinal nerves. The *involuntary* muscles are the contractile tissue in the walls of the alimentary tract, blood vessels, urinary system, reproductive ducts, respiratory passageways, and other viscera. The musculature of the heart is a special type of involuntary muscle tissue. The involuntary musculature is innervated by branches of the autonomic nervous system.

The most important voluntary muscles will be described in terms of their sites of attachment, from which their function can usually be deduced. Each voluntary muscle is attached at both ends, usually to bone. The site of attachment on a fixed bone is the **origin;** the site on a bone that moves is the **insertion.** The most common attachment of a muscle is a glistening white cord of fibrous connective tissue known as a **tendon.** A few tendons are broad and flat; such a tendon is known as an **aponeurosis.** The same muscle from both sides of the body may join in the mid-dorsal or mid-ventral lines, where the line of fusion forms a connective tissue ridge known as a **raphe.**

The function of the muscles is to move skeletal elements that together act as a system of levers. Muscles can perform this activity because most of them are sus-

pended across joints and are capable of contraction, or shortening. A **flexor** is a muscle that by its contraction bends a limb or any portion of it. Its antagonist, the **extensor,** straightens out the limb or brings the bones into line. An **abductor** is a muscle that draws laterally from the axis of the limb or the sagittal plane of the body the bone to which it is inserted. Its antagonist is known as the **adductor.**

Muscles may be named according to their function, shape, number of heads of origin, or a combination of these factors. They are covered and bound together by sheets of connective tissue known as **fascia.** The integument is connected to the muscles by superficial fascia that may contain fat deposits. Deep fascia envelops both individual muscles and groups of muscles.

Terrestrial mammals show a substantial increase in the appendicular musculature, associated with the extensive enlargement of their appendages. Only the intercostal, intervertebral, and rectus abdominis muscles of the trunk have preserved their metamerism. Also, as a result of the migration of the muscles from their initial positions, they have lost their original natural grouping. With the reduction of the tail, the caudal muscles have dwindled. The locations of the muscles will be described here to aid in their identification. Their origins, insertions, and actions are summarized in Table 3-1 at the end of this exercise.

*Pinch up the skin of the abdomen and snip it with scissors. While lifting up the skin, extend the incision upward along the mid-ventral line to the lower lip and downward to the genital region. Make similar incisions along the inside of the forelimbs and hind limbs and then around the paws. Loosen the skin and carefully separate it from the underlying muscles by tearing through the superficial fascia with a scalpel handle or dental probe. In freeing the skin from the body, start ventrally at the edges of the incision and work backward toward the mid-dorsal wall. The skin may be left attached to the mid-dorsal wall in order to facilitate wrapping the exposed body between laboratory periods. If plastic bags are used for storage, the skin may be removed in its entirety.*

*In the dissection of muscles it is essential that the fat and connective tissue be removed from their exposed and deep surfaces. This greatly facilitates their identification. Wherever possible, separate the muscles along their natural boundaries. When transection of a muscle is necessary, separate it from underlying structures and cut it cleanly to leave the origin and insertion intact.*

## CUTANEOUS MUSCLES

Because the cutaneous muscles are usually not easily identifiable they are difficult to study.

**Cutaneous maximum**   This very large thin muscle covers almost the whole side of the body.

**Platysma**   This thin muscle covers the sides of the neck and face.

## SHOULDER MUSCLES (Fig. 3-1)

### Superficial Group

**Acromiotrapezius**   The thin, flat muscle extending from the middle of the back to the shoulder. It is the middle of the three trapezius muscles.

**Spinotrapezius**   This muscle is located posterior to the acromiotrapezius, by which it is partly covered.

**Clavotrapezius**   The large muscle extending from the neck to the shoulder. It lies just anterior to the acromiotrapezius.

**Clavobrachialis**   This triangular muscle is considered a continuation onto the shoulder of the clavotrapezius. (The clavicle can be felt embedded in the muscle at its site of origin.)

*Cut through the middles of the spino-, acromio-, and clavotrapezius muscles and lay back the cut halves so that the underlying muscles are exposed.*

**Levator scapulae ventralis**   This slender muscle extends caudally from the atlas to the scapula, passing beneath the clavotrapezius.

**Acromiodeltoid**   This short thick muscle is visible superficially between the levator scapulae ventralis and the clavobrachialis.

**Spinodeltoid**   This short thick muscle lies ventral to the acromiotrapezius and posterior to the levator scapulae ventralis and acromiodeltoid.

**Latissimus dorsi**   This is a large triangular muscle sheet that extends from the middle of the back to the humerus.

### Deep Group

**Rhomboideus**   This muscle lies beneath the medial portions of the acromio- and spinotrapezius muscles and extends from the border of the scapula to the mid-dorsal line.

**Rhomboideus capitis**   Known also as the occipito-scapularis, this narrow muscle apparently represents the most anterolateral portion of the rhomboideus. It extends from the back of the skull to the scapula.

**Splenius**   This broad muscle lies beneath the clavotrapezius and covers the dorsal side of the neck.

**Supraspinatus**   Lying beneath the acromiotrapezius, this muscle occupies the supraspinous fossa of the scapula.

**Infraspinatus**   This muscle fills the infraspinous fossa.

**Teres major**   This thick muscle lies posterior to the infraspinatus and extends to the humerus.

**Teres minor**   This small muscle can be seen if the infraspinatus and teres major are separated near their insertions. It can be more fully exposed if the acromiodeltoid is cut across its middle. (See Fig. 3-6.)

**Subscapularis**   This muscle occupies the subscapular fossa. (See Fig. 3-2.)

## MUSCLES OF THE BACK (Fig. 3-1)

**Serratus dorsalis superior**   A thin sheet of muscle and tendon located on the dorsal part of the thorax and neck.

**Serratus dorsalis inferior**   This thin muscle lies just caudal to the serratus dorsalis superior and is located over the dorsal part of the lumbar region.

*On one side make a longitudinal incision through the latissimus dorsi somewhat lateral to the mid-dorsal line. Reflect the cut ends of this muscle. Then very carefully remove as much of the lumbodorsal fascia as*

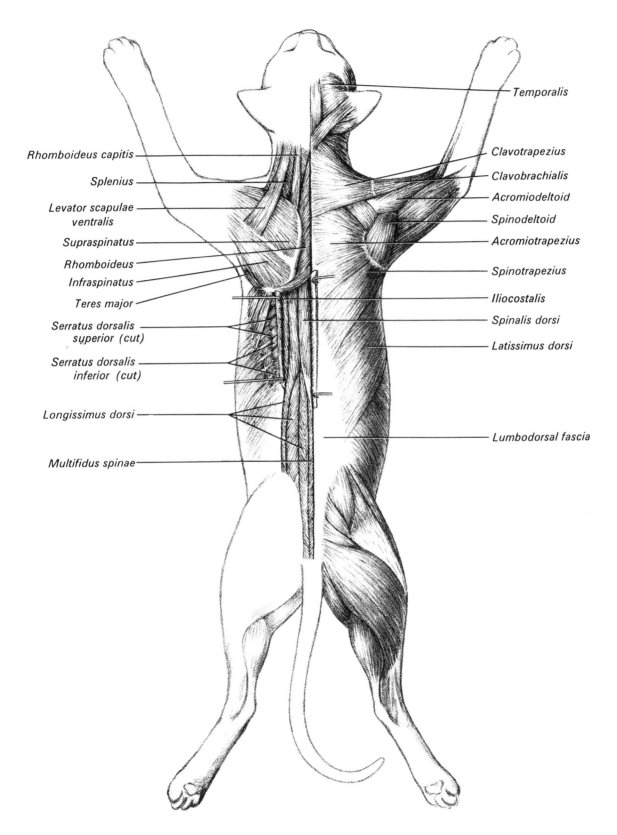

**FIGURE 3–1**
Muscular system (dorsal view).

*possible and delineate the underlying muscles along their natural boundaries.*

**Spinalis dorsi**  The most medial of the deep muscles of the back. It lies in the mid-dorsal thoracic line.

**Longissimus dorsi**  The thick muscle mass lying just lateral to the spinalis dorsi. This muscle fills the space between the spinous and transverse processes of vertebrae in the thoracolumbar region. In the lumbar region the longissimus dorsi can be separated into three distinct muscle bundles.

**Iliocostalis**  The lateralmost of the three longitudinal muscles of the back. It lies on the dorsal portion of the ribs and is covered by the serratus dorsalis muscles.

**Multifidus spinae**  This muscle is best seen in the sacral region adjacent to the mid-dorsal line. In the thoracic region it is covered by the other deep back muscles noted above. It extends onto the neck where it continues as the semispinalis cervicis, a muscle that lies under the splenius.

## THORACIC MUSCLES (Fig. 3-2)

### Pectoral Group

The muscles on the ventral surface of the thorax consist of two large triangular masses, each of which consist of four muscles. Each muscle mass arises from the sternum and passes forward to insert primarily along the humerus. The major action of each group of muscles is to pull the humerus toward the chest (adduction). The muscles are not very sharply separable from one another.

*Turn the cat on its back and expose the chest by spreading the forelimbs. Clean the surface pectoral muscles so that the direction of their fibers can be seen.*

**Pectoantebrachialis**  The most anterior of the four pectoral muscles. It extends from the ventral midline of the chest to the anterior surface of the forelimb.

**Pectoralis major**  This muscle consists of two parts, both of which extend from the sternum to the humerus. The narrower cranial part of the pectoralis major is, for the most part, overlapped by the pectoantebrachialis. The cranial part can be distinctly separated from the much larger caudal part.

**Pectoralis minor**  This small fan-shaped muscle is partly overlapped cranially by the pectoralis major.

**Xiphihumeralis**  The most posterior muscle of this series. It is long, thin, and narrow.

### Deep Thoracic Group

*Transect the pectoral group of muscles in the middle and reflect both halves. Deep to these muscles considerable fat and connective tissue surround the blood vessels and nerves that supply the forelimb. If these structures are not studied at this time, they may be excised on one side. Then pull the scapula away from the body wall and clean the fascia from the deep thoracic muscles.*

**Serratus ventralis**  A fan-shaped muscle extending obliquely upward from the ribs to the scapula.

**Levator scapulae**  This is the anterior continuation of the serratus ventralis. Their contiguous borders cannot readily be separated.

**Transversus costarum**  This short thin muscle on the cranioventral surface of the thorax covers the cranial end of the rectus abdominis (see definition).

**Scalenus**  This is a large and complex muscle that extends longitudinally along the ventrolateral surface of the neck and thorax. It is divisible into three separate muscles: the scalenus anterior, medius, and posterior. The scalenus medius is the largest and most readily identifiable of the three.

**External intercostals**  A series of muscles whose fibers are directed caudoventrally. They close the intercostal spaces.

**Internal intercostals**  A series of muscles, whose fibers are directed caudodorsally, lie beneath the external intercostals. The intercostal muscles close the intercostal gap.

## ABDOMINAL MUSCLES (Fig. 3-2)

*After cleaning and examining the external oblique, carefully cut through its middle for about two inches at a right angle to the direction of the fibers. When the underlying internal oblique is reached, expose it by reflecting part of the external oblique. Note the direction of its fibers. To get at the deepest layer, the transversus abdominis, make a longitudinal cut through the fleshy portion of the internal oblique and reflect a part of this muscle.*

Three of the four abdominal muscles are present as overlapping sheets. The other one has a strap-like appearance. The aponeuroses of the muscles meet in the mid-ventral line to form the **linea alba.**

**External oblique**  This broad flat muscle extends over the entire abdomen and ventral part of the thorax. Its fibers run anterodorsally.

**Internal oblique**  This smaller sheet of muscle is lo-

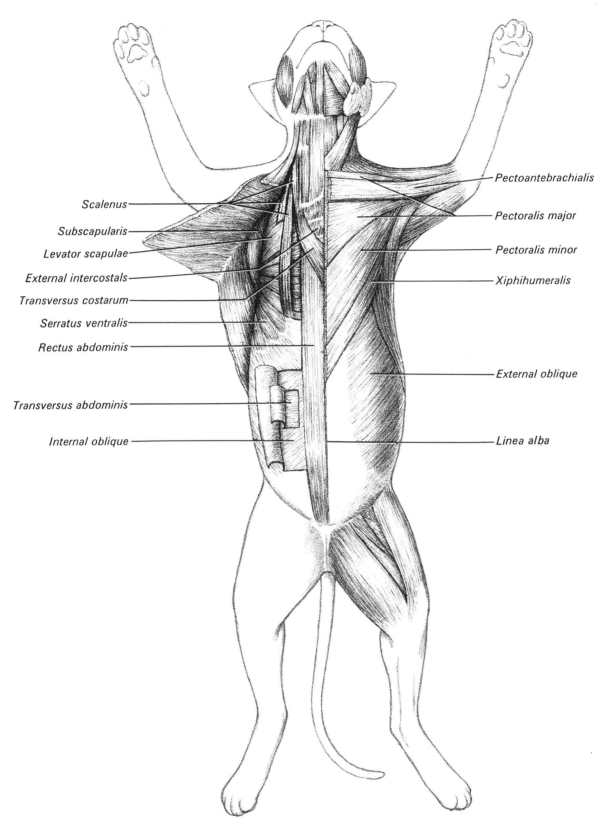

Scalenus

Subscapularis

Levator scapulae

External intercostals

Transversus costarum

Serratus ventralis

Rectus abdominis

Transversus abdominis

Internal oblique

Pectoantebrachialis

Pectoralis major

Pectoralis minor

Xiphihumeralis

External oblique

Linea alba

**FIGURE 3–2**
Muscular system (ventral view).

cated under the lateral part of the external oblique. Its fibers run at right angles to the external oblique.

**Transversus abdominis**    The very thin third layer of muscle of the abdominal wall. Its fibers run transversely and are best seen ventrolaterally.

**Rectus abdominis**    The strap-like muscle on either side of the linea alba that extends longitudinally from the sternum to the pubis.

## MUSCLES OF THE NECK AND HEAD

### Superficial Group (Fig. 3-3)

*Three large salivary glands are located in this region. They will be studied in more detail in Exercise 4, Digestive and Respiratory Systems. They may, however, be identified now and removed on one side, if necessary, in order to expose the underlying muscles. The parotid gland lies anteroventral to the external ear; the submandibular gland can be found at the angle of the jaw; and the sublingual gland is located anterior to the submandibular. After locating these glands, study the paired sternomastoids, the first muscles of the group. Then free one of the sternomastoids at its origin and reflect it laterally to expose the other muscles more clearly.*

**Sternomastoid**    A wide flat muscle that extends obliquely from the cranial end of the sternum to the side of the skull. Together, the two sternomastoid muscles form a V-shaped muscular collar at the base of the throat.

**Sternohyoid**    This narrow muscular band lies along the mid-ventral line of the neck and is partly covered by the sternomastoid. (Retracting both muscles from the midline will expose the bilobed thyroid gland.)

**Digastric**    This thick prismatic muscle can be exposed after the fascia on the undersurface of the mandible is removed. This muscle lies just medial to the body of the mandible and connects it with the base of the skull.

**Mylohyoid**    The thin sheet-like muscle seen after reflection of the digastric. Together, both mylohyoids close off the V-shaped gap formed by the bodies of the mandible.

**Stylohyoid**    This ribbon-like muscle extends across the external surface of the digastric, parallel to the margin of the mylohyoid.

**Masseter**    The large muscle mass located at each angle of the jaw, behind and below the eyes.

**Temporalis**    This muscle is located above and just posterior to the eye and medial to the ear (see Fig. 3-1).

### Deep Group (Fig. 3-4)

*To expose the deep muscles of the neck, it is necessary to transect and reflect the digastric and mylohyoid muscles.*

**Sternothyroid**    This muscle lies on the side of the trachea beneath the sternohyoid.

**Thyrohyoid**    A small muscle lying on the lateral side of the thyroid cartilage of the larynx.

**Cricothyroid**    This muscle, with its fellow, covers the ventral surface of the cricoid cartilage.

**Cleidomastoid**    This narrow flat muscle is located just beneath the clavotrapezius. It extends from the head to the clavicle.

**Geniohyoid**    A strap-like muscle that lies deep to the mylohyoid on either side of the mid-ventral line.

**Hyoglossus**    A short, obliquely directed muscle located lateral to the geniohyoid.

**Styloglossus**    A relatively long muscle that runs parallel to the inside of the body of the mandible.

**Genioglossus**    This muscle lies beneath the geniohyoid, which can be transected to expose the genioglossus more fully.

## MUSCLES OF THE UPPER ARM

### Medial Surface (Fig. 3-5)

*Remove the brachial fascia from the muscles of the upper arm. In doing so, avoid removing the narrow brachioradialis, a forearm muscle (see the next section) that may at times be mistaken for a fascial strand. Cut through the middle of the pectoral muscles and reflect them in order to expose the coracobrachialis, epitrochlearis, and the biceps brachii.*

**Coracobrachialis**    A very short muscle covering the medial surface of the capsule of the shoulder joint.

**Epitrochlearis**    Known also as the extensor antebrachii, this flat muscle lies along the medial surface of the upper arm.

**Biceps brachii**    A spindle-shaped muscle located along the ventral surface of the humerus.

### Lateral Surface (Fig. 3-6)

*Identify the lateral head of the triceps. Then cut it near its insertion and reflect it to expose the other heads of the triceps, anconeus, and brachialis.*

**Triceps brachii**    This is the largest muscle of the upper arm and is made up of three heads. The **long head**

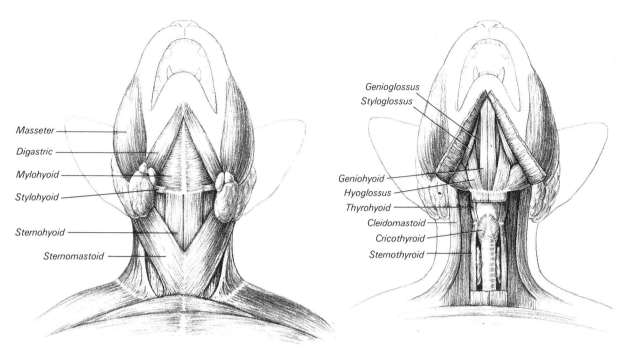

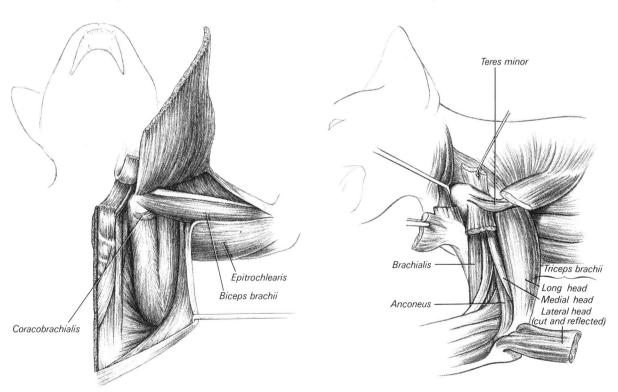

**FIGURE 3–3**
Superficial muscles of the neck.

Masseter

Digastric

Mylohyoid

Stylohyoid

Sternohyoid

Sternomastoid

**FIGURE 3–4**
Deep muscles of the neck.

Genioglossus
Styloglossus

Geniohyoid
Hyoglossus
Thyrohyoid
Cleidomastoid
Cricothyroid
Sternothyroid

**FIGURE 3–5**
Muscles of the upper arm (ventral view).

Epitrochlearis
Biceps brachii
Coracobrachialis

**FIGURE 3–6**
Muscles of the upper arm (dorsal view).

Teres minor

Brachialis

Anconeus

Triceps brachii
Long head
Medial head
Lateral head
(cut and reflected)

is the spindle-shaped muscle mass on the dorsolateral aspect of the arm. The **lateral head** is flat and lies on the outer aspect of the arm just in front of the long head. The **medial head** has three parts and can be seen when the other two heads are separated.

**Anconeus**   A small triangular muscle on the lateral surface of the arm in the region of the elbow joint.

**Brachialis**   This relatively small muscle lies on the outer surface of the arm just anterior to the lateral head of the triceps. The insertion of the pectoralis major on the humerus separates the brachialis from the biceps brachii (the brachialis is on the lateral side, the biceps on the medial side).

## MUSCLES OF THE FOREARM

*To expose the muscles of the forearm it is necessary to remove the tough (antebrachial) fascia that encases them. The long tendons from the muscles thus exposed will be seen to be held in place by fascial ligaments at the level of the wrist. These ligaments must be removed if the tendons are to be traced.*

The muscles of the forearm can be divided into two groups: dorsal extensors and ventral flexors. Each of the two groups can in turn be subdivided into superficial and deep muscles.

### Superficial Dorsal Group (Fig. 3-7)

Most of these muscles arise from or near the lateral epicondyle of the humerus. The tendons of these muscles are held in place at the wrist by a **dorsal carpal ligament**. The muscles can be identified in the following radial to ulnar order.

**Brachioradialis**   A narrow band that extends down from the lateral side of the upper arm. It stands away from the underlying extensors.

**Extensor carpi radialis longus**   A spindle-shaped muscle that underlies the brachioradialis and extends into the medial surface of the arm. It is the longer portion of the two-part extensor carpi radialis.

**Extensor carpi radialis brevis**   This is a deeper and more posterior lying muscle than the longus.

**Extensor digitorum communis**   Thick at its proximal end, this muscle flattens at its distal end, where it breaks up into four tendons that insert into the middle and distal phalanges of digits two to five.

**Extensor digitorum lateralis**   Located adjacent to the extensor digitorum communis, this divides into three or four tendons that join those of the latter muscle.

**Extensor carpi ulnaris**   A prismatically shaped muscle having a single tendon that inserts on the fifth digit.

### Deep Dorsal Group (Fig. 3-8)

*Separate the extensor digitorum lateralis from the extensor carpi ulnaris to expose the underlying muscles.*

**Extensor indicis proprius**   A narrow muscle arising deep to the extensor carpi ulnaris.

**Extensor pollicis brevis**   Also known as the abductor pollicis longus or extensor carpi obliquus, it runs obliquely beneath the tendons of the extensor digitorum communis. It converges to form a tendon that passes over the tendons of the extensor carpi radialis and inserts on the thumb.

**Supinator**   If the superficial muscles are not transected and reflected, this short muscle can be seen only in part. It passes obliquely across the proximal third of the radius, deep to the belly of the extensor digitorum communis.

### Superficial Ventral Group (Fig. 3-9)

The muscles in this group should be identified from radial to ulnar side. The tendons of these muscles are held in place at the wrist by the **transverse carpal ligament**.

**Pronator teres**   A short, wedge-shaped muscle that extends diagonally from the elbow region to the medial border of the radius.

**Flexor carpi radialis**   A spindle-shaped muscle that extends down the radial border of the forearm. It has a long tendon that passes deep to other tendons of the hand.

**Palmaris longus**   The widest of this superficial group of muscles. It forms a broad tendon that passes beneath the transverse carpal ligament and then divides into four tendons.

**Flexor carpi ulnaris**   A rather thick muscle having both humeral and ulnar heads of origin. It extends along the ulnar border of the forearm to the pisiform bone.

### Deep Ventral Group (Fig. 3-10)

*To expose this group, transect and reflect the palmaris longus and flexor carpi ulnaris and pull aside the other superficial muscles.*

**Flexor digitorum superficialis**   This muscle has two muscular heads of origin. One is a small muscle belly arising from the ulnar side of the tendon of the palmaris longus near the wrist. The other, which is larger and deeper, arises beneath the palmaris longus

179

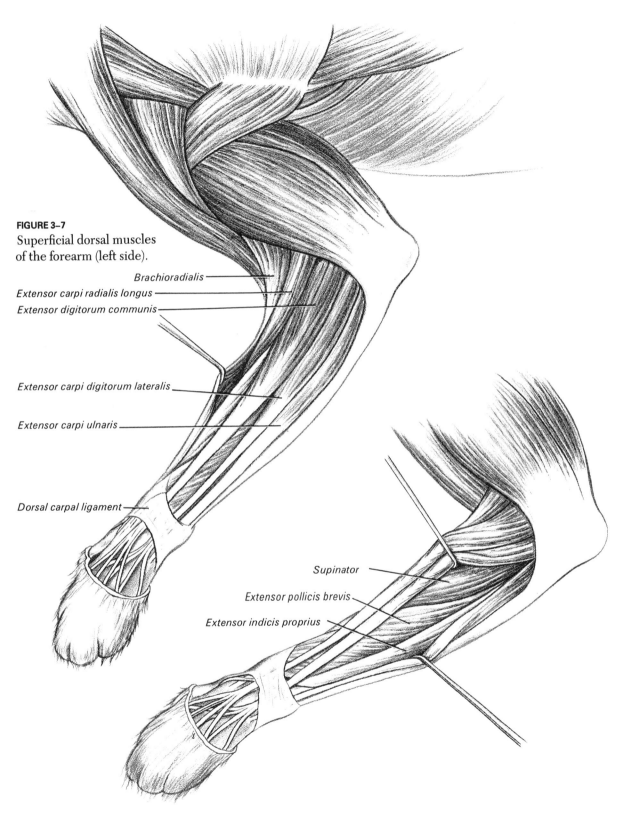

**FIGURE 3–7**
Superficial dorsal muscles
of the forearm (left side).

Brachioradialis

Extensor carpi radialis longus

Extensor digitorum communis

Extensor carpi digitorum lateralis

Extensor carpi ulnaris

Dorsal carpal ligament

Supinator

Extensor pollicis brevis

Extensor indicis proprius

**FIGURE 3–8**
Deep dorsal muscles of the forearm (left side).

from the ventral surface of the distal fleshy portion of the flexor digitorum profundus. Four tendons arise from the muscle and insert into the middle phalanx of the second through fifth digits.

**Flexor digitorum profundus**   This muscle has five heads of origin. The first is from the outer border of the ulnar; the second, third, and fourth are from the medial epicondyle; the fifth is from the middle third of the radius. The heads give rise to a common tendon that extends into the palm and there divides into five strong tendons. The second to fifth of these perforate the tendons of the flexor digitorum superficialis before they insert on the distal phalanges.

**Pronator quadratus**   A very deep, flat, quadrilateral muscle that can be seen by moving aside the tendon bundle of the flexor digitorum profundus proximal to the wrist.

There are several intrinsic muscles on the hand, which are quite small and difficult to dissect.

## MUSCLES OF THE THIGH

These muscles can be divided into three groups according to location. Those on the medial and lateral surfaces each have two superficial wide muscles that should be identified and then transected to expose the underlying members of the group. The third group, which is deeper and more difficult to dissect, can be ignored if time is limited.

### Medial Surface (Figs. 3-11, 3-12)

*Study the sartorius and gracilis, the two superficial muscles on the inner surface of the thigh. Then transect them across the middle and reflect their cut edges to expose the underlying muscles.*

**Sartorius**   This muscle occupies the anterior half of the inner surface of the thigh.

**Gracilis**   This is the large flat muscle covering most of the posterior inner surface of the thigh.

**Semimembranosus**   A very thick muscle lying deep to the gracilis. It is divided throughout most of its length into two portions.

**Adductor femoris**   This large triangular muscle lies deep to the gracilis and proximal to the semimembranosus.

**Adductor longus**   A relatively small muscle, not covered by the gracilis, located at the upper margin of the adductor femoris.

**Pectineus**   A very small muscle lying above the adductor longus, with which it is united.

**Iliopsoas**   A conical muscle emerging from the abdominal cavity onto the medial surface of the thigh adjacent to the pectineus.

**Rectus femoris**   Largely covered by the sartorius, this muscle is long and round. Together with the vastus medialis, lateralis, and intermedius, it makes up the **quadriceps femoris** muscle.

**Vastus medialis**   This large muscle lies entirely on the medial side of the thigh.

**Vastus intermedius**   This muscle becomes visible when the vastus lateralis and rectus femoris are separated.

### Lateral Surface (Figs. 3-13, 3-14)

*After studying the tensor fascia lata and biceps femoris, cut across their bellies and reflect their parts in order to reveal the underlying muscles.*

**Tensor fascia lata**   A large fan-shaped muscle occupying the anteroproximal part of the lateral surface of the thigh.

**Biceps femoris**   This large muscle covers the greater part of the lateral surface of the thigh.

**Semitendinosus**   This long cylindrical muscle lies on the ventral border of the thigh between the biceps femoris and the semimembranosus.

**Caudofemoralis**   A small muscle located near the base of the tail and just anterior to the proximal end of the biceps femoris.

**Gluteus maximus**   A short muscle just anterior to the caudofemoralis, which partly overlaps it.

**Gluteus medius**   This thick short muscle is almost entirely covered by the tensor fascia lata.

**Vastus lateralis**   Lies beneath the fascia lata along the anterior edge of the thigh.

**Tenuissimus**   A very long muscle band that lies beneath the biceps femoris. The thick white sciatic nerve runs parallel to it.

### Deeper Pelvic Muscles (Fig. 3-15)

*To expose these muscles, cut and reflect the caudofemoralis, gluteus maximus, and gluteus medius. Cut the first two near their origin and the gluteus medius near its insertion. After identifying the piriformis, cut it near its insertion to expose the gemellus superior and gluteus minimus.*

**Piriformis**   A small triangular muscle that lies beneath the gluteus maximus and gluteus medius. The sciatic nerve emerges from beneath the lower border of the piriformis. (Because this muscle is thin it may accidentally be cut and reflected with the gluteus medius.)

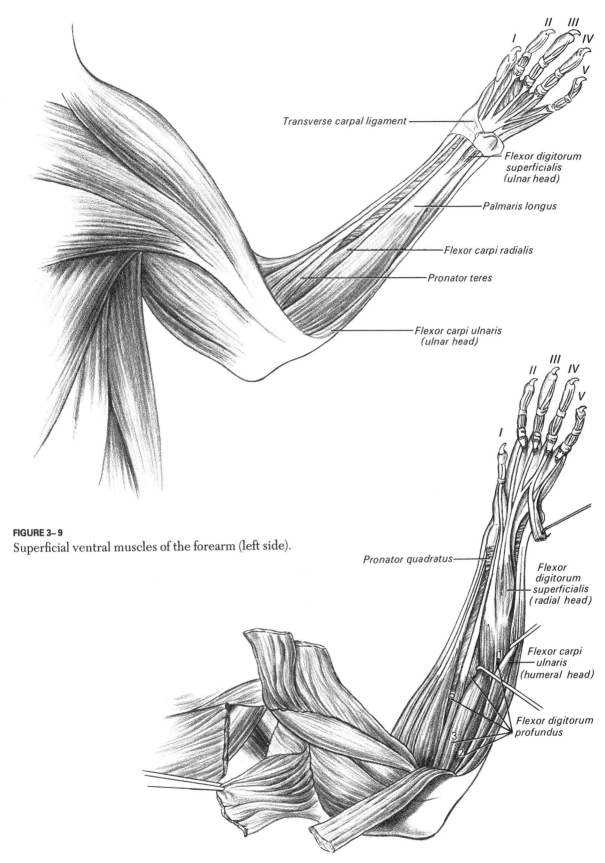

Transverse carpal ligament

Flexor digitorum superficialis (ulnar head)

Palmaris longus

Flexor carpi radialis

Pronator teres

Flexor carpi ulnaris (ulnar head)

**FIGURE 3–9**
Superficial ventral muscles of the forearm (left side).

Pronator quadratus

Flexor digitorum superficialis (radial head)

Flexor carpi ulnaris (humeral head)

Flexor digitorum profundus

**FIGURE 3–10**
Deep ventral muscles of the forearm (left side).

**Gluteus minimus**    A cylindrical muscle that lies beneath the gluteus medius. Its posterior border is in contact with the gemellus superior, from which it may be difficult to separate it. Its anteroventral border is characterized by conspicuous tendinous fibers.

**Gemellus superior**    This triangular muscle lies beneath the piriformis and gluteus maximus.

**Obturator internus**    A triangular muscle lying just caudal to the gemellus superior.

**Gemellus inferior**    A triangular muscle largely obscured by the overlying obturator internus (which should be kept intact). This muscle should be evident in the gap between the obturator internus and the quadratus femoris.

**Quadratus femoris**    A short thick muscle that lies deep to the origin of the biceps femoris and caudal to the obturator internus.

**Obturator externus**    A flat triangular muscle covering the lateral aspect of the obturator foramen. This muscle is difficult to see because it lies deep between the adductor femoris and quadratus femoris. In some specimens it may be seen by separating the adjacent borders of the gemellus inferior and the quadratus femoris. It may be exposed in toto by transection of the adductor femoris near its origin.

## MUSCLES OF THE LEG

Like the muscles of the forearm, those of the leg can be divided into extensors or flexors, but these groups are not so sharply delineated as those of the forearm. Thus, for purposes of description, the muscles will be divided into three groups, dorsal, lateral, and ventral, with the first being subdivided into superficial and deep muscles.

The muscles of the shank or leg are covered by a tough **fascia cruris**, which is partly united with the inserting tendons of the biceps femoris and gracilis.

### Superficial Dorsal Group (Fig. 3-15)

*Remove the remaining skin from the foot and clean the superficial fascia from the leg and foot.*

**Gastrocnemius**    This large calf muscle has two heads of origin that unite into a common muscle mass that ends with the strong **calcaneal tendon** of Achilles. This is also the tendon of insertion of the soleus.

**Soleus**    This flat spindle-shaped muscle lies deep to the gastrocnemius.

**Plantaris**    This fusiform muscle is enclosed almost completely by the lateral and medial heads of the gastrocnemius. The tendon of insertion of the plantaris extends through a tubular sheath formed by the tendons of the soleus and gastrocnemius, passes over the proximal end of the calcaneus, and ultimately inserts on an intrinsic foot muscle, the flexor digitorum brevis (see below).

### Deep Dorsal Group (Figs. 3-16, 3-19)

*Sever the calcaneal tendon, taking care to avoid severing the tibial nerve, which lies very close to the tendon. Reflect the superficial muscles cranially and identify the muscles belonging to this group.*

**Popliteus**    A triangular muscle located deep to the medial head of the gastrocnemius.

**Flexor digitorum longus**    A tapering muscle on the medial side of the leg. It lies close to the proximal half of the tibia.

**Flexor hallucis longus**    This long muscle lies adjacent to the flexor digitorum longus. Its tendon passes posterior to the medial malleolus to join with the tendon of the flexor digitorum longus and form a broad tendon plate on the sole (plantar surface) of the foot. The plate passes deep to the flexor digitorum brevis and, at the level of the toes, breaks up into four tendons that pass through perforations in the tendons of the flexor digitorum brevis to their insertions of the terminal phalanges.

**Flexor digitorum brevis**    The intrinsic foot muscle, which divides into four tendons that insert at the base of the middle phalanges.

**Tibialis posterior**    A slender fusiform muscle located between the flexor digitorum longus and flexor hallucis longus and deep to them.

### Lateral Group (Figs. 3-15, 3-18)

This group consists of three muscles that can be seen between the distal portions of the gastrocnemius and the extensor digitorum longus.

**Peroneus longus**    A slender fusiform muscle situated superficially on the lateral side of the leg just posterior to the extensor digitorum longus. Its long tendon passes through a groove on the surface of the lateral malleolus and then runs diagonally deep into the sole of the foot. There it breaks up into five tendons: two stout ones to the first and fifth metatarsals and three slender ones to the others.

**Peroneus tertius**    A slender tapering muscle lying just medial to the peroneus longus and covered by it. Its tendon passes through a groove on the posterior

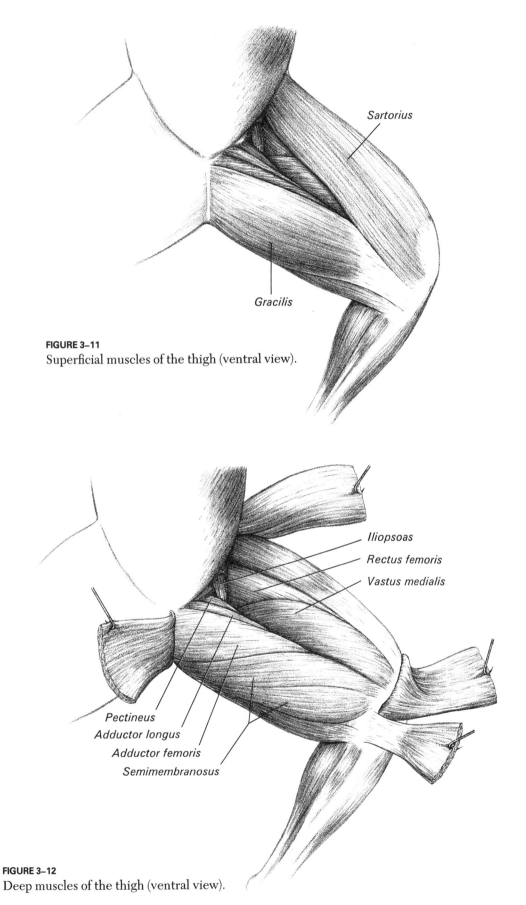

**FIGURE 3–11**
Superficial muscles of the thigh (ventral view).

Sartorius

Gracilis

Iliopsoas

Rectus femoris

Vastus medialis

Pectineus

Adductor longus

Adductor femoris

Semimembranosus

**FIGURE 3–12**
Deep muscles of the thigh (ventral view).

border of the lateral malleolus and down the dorsum of the fifth digit.

**Peroneus brevis**  A short muscle just medial to the peroneus tertius. Its tendon runs with the tendon of the peroneus tertius over the lateral malleolus to its insertion on the fifth metatarsal.

## Ventral Group (Figs. 3-17, 3-18)

**Tibialis anterior**  A superficial muscle located on the dorsolateral border of the tibia. Its long tendon crosses the anterior surface of the tibia, then passes beneath a transverse ligament near the ankle joint and over the dorsal surface of the foot to its insertion on the first metatarsal.

**Extensor digitorum longus**  A fusiform muscle located posterior to and partly covered by the tibialis anterior. After passing beneath the transverse ligament near the ankle joint, it divides into four tendons that pass down the dorsum of the foot to their insertion on the terminal phalanges. The tendons of this muscle join those of some of the intrinsic foot muscles, especially that of the extensor digitorum brevis, to form a common tendon.

The intrinsic muscles on the sole of the foot, which are arranged in five layers, are not discussed because they are quite small and difficult to dissect.

**TABLE 3-1**
Summary of the musculature of the cat.

| Muscle | Origin | Insertion | Action |
|---|---|---|---|
| *Cutaneous Muscles* | | | |
| Cutaneous maximus | Linea alba, latissimus dorsi | Dermis of skin | Moves skin of trunk |
| Platysma | Dorsal mid-cervical fascia | Dermis of skin | Moves skin of neck and face |
| *Shoulder Muscles* | | | |
| Superficial Group | | | |
| Acromiotrapezius | Spines of cervical vertebrae | Spine of scapula | Draws scapula dorsally |
| Spinotrapezius | Spines of thoracic vertebrae | Fascia of scapular muscles | Draws scapula dorsally |
| Clavotrapezius | Lambdoidal crest of skull | Clavicle | Draws scapula craniodorsally |
| Clavobrachialis | Clavicle | Ulna beneath semilunar notch | Flexes forearm |
| Levator scapulae ventralis | Atlas and occipital bone | Metacromion process | Draws scapula cranially |
| Acromiodeltoid | Acromion of scapula | Outer surface of spinodeltoid | Flexes and rotates humerus |
| Spinodeltoid | Spine of scapula | Deltoid ridge of humerus | Flexes and rotates humerus |
| Latissimus dorsi | Thoracic and lumbar vertebrae | Shaft of humerus | Pulls arm caudodorsally |
| Deep Group | | | |
| Rhomboideus | Upper thoracic vertebrae | Medial border of scapula | Draws scapula dorsally |
| Rhomboideus capitis | Lambdoidal ridge of skull | Angle of scapula | Draws scapula cranially |
| Splenius | Mid-dorsal fascial line | Lambdoidal ridge | Turns and elevates head |
| Supraspinatus | Supraspinous fossa | Greater tubercle of humerus | Extends arm |
| Infraspinatus | Infraspinous fossa | Greater tubercle of humerus | Rotates humerus outward |
| Teres major | Axillary border of scapula | Medial surface of humerus | Flexes and rotates humerus |
| Teres minor | Lateral border of scapula | Greater tubercle of humerus | Rotates humerus |
| Subscapularis | Subscapular fossa | Lesser tubercle of humerus | Draws humerus medially |

*(continued on page 186)*

185

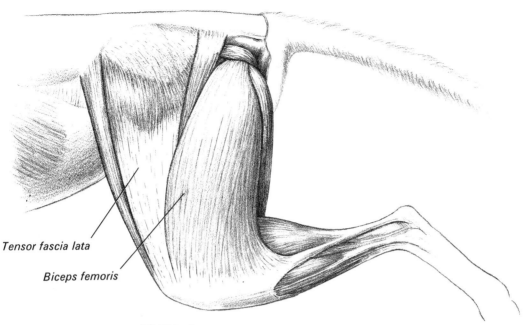

**FIGURE 3–13**
Superficial muscles of the thigh (lateral view).

Tensor fascia lata

Biceps femoris

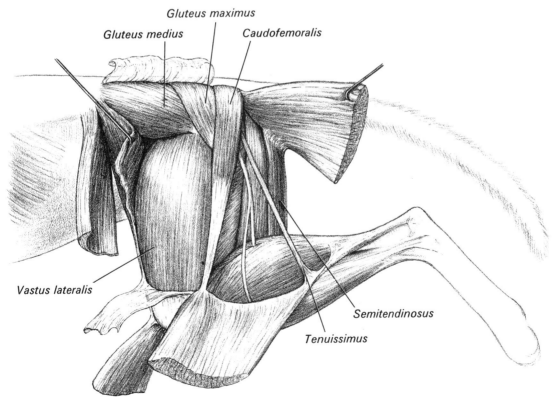

**FIGURE 3–14**
Deep muscles of the thigh (lateral view).

Gluteus maximus

Gluteus medius

Caudofemoralis

Vastus lateralis

Semitendinosus

Tenuissimus

**TABLE 3-1** (*continued*)

Summary of the musculature of the cat.

| Muscle | Origin | Insertion | Action |
|---|---|---|---|
| *Muscles of the Back* | | | |
| Serratus dorsalis superior | First nine ribs | Mid-dorsal raphe | Draws ribs cranially |
| Serratus dorsalis inferior | Last four ribs | Lumbar spinous processes | Draws ribs cranially |
| Spinalis dorsi | Last four thoracic vertebrae | Thoraco-cervical vertebrae | Extends vertebral column |
| Longissimus dorsi | Sacral and caudal vertebrae | Trunk and cervical vertebrae | Extends vertebral column |
| Iliocostalis | As separate muscle bundles from lower thoracic ribs | Three ribs craniad to the origin of each bundle | Draws ribs together |
| Multifidus spinae | As separate muscle bundles from lumbar transverse processes | On spinous process one vertebra craniad to origin of each bundle | Extends from vertebral column |
| *Thoracic Muscles* | | | |
| Pectoral Group | | | |
|   Pectoantebrachialis | Manubrium of sternum | Fascia of forearm | Adducts forelimb |
|   Pectoralis major | Cranial sternebrae | Pectoral ridge of humerus | Adducts forelimb |
|   Pectoralis minor | Body of sternum | Pectoral ridge of humerus | Adducts forelimb |
|   Xiphihumeralis | Xiphoid process of sternum | Proximal end of humerus | Adducts forelimb |
| Deep Thoracic Group | | | |
|   Serratus ventralis | First ten ribs | Medial surface of scapula | Draws scapula to thorax |
|   Levator scapulae | Last five cervical vertebrae | Medial surface of scapula | Draws scapula craniovent. |
|   Transversus costarum | Lateral border of sternum | First rib | Draws sternum cranially |
|   Scalenus | Ribs | Cervical transverse proc. | Flexes the neck |
|   External intercostal | Border of rib | Border of adjacent rib | Protracts the ribs |
|   Internal intercostal | Border of rib | Border of adjacent rib | Retracts the ribs |
| *Abdominal Muscles* | | | |
| External oblique | Lumbodorsal fascia and ribs | Linea alba and pubis | Constricts abdomen |
| Internal oblique | Lumbodorsal fascia | Linea alba | Compresses abdomen |
| Transversus abdominis | Costal cartilages of lower ribs | Linea alba | Constricts abdomen |
| Rectus abdominis | Pubis | Sternum and costal cartilages | Compresses abdomen |
| *Muscles of the Neck and Head* | | | |
| Superficial Group | | | |
|   Sternomastoid | Manubrium of sternum | Lambdoidal ridge of skull | Turns head |
|   Sternohyoid | First costal cartilage | Body of the hyoid | Draws hyoid posteriorly |
|   Digastric | Occipital bone of skull | Ventral border of mandible | Depresses lower jaw |
|   Mylohyoid | Inner surface of mandible | Median raphe | Raises floor of mouth |
|   Stylohyoid | Stylohyal bone of hyoid | Body of hyoid | Raises hyoid |
|   Masseter | Zygomatic arch | Mandible | Elevates mandible |
|   Temporalis | Surface of temporal fossa | Coronoid process of mandible | Elevates mandible |
| Deep Group | | | |
|   Sternothyroid | First costal cartilage | Thyroid cartilage | Draws larynx caudally |
|   Thyrohyoid | Thyroid cartilage | Posterior horn of hyoid | Raises larynx |
|   Cricothyroid | Cricoid cartilage | Thyroid cartilage | Tensor of true vocal cords |
|   Cleidomastoid | Mastoid process | Clavicle | Turns head |
|   Geniohyoid | Medial surface of mandible | Body of hyoid | Draws hyoid cranially |
|   Hyoglossus | Body of the hyoid bone | Dorsum of tongue | Retracts tongue |
|   Styloglossus | Mastoid process of skull | Apex of tongue | Retracts tongue |
|   Genioglossus | Medial surface of mandible | Root of tongue | Draws root of tongue forward |

(*continued on page* 188)

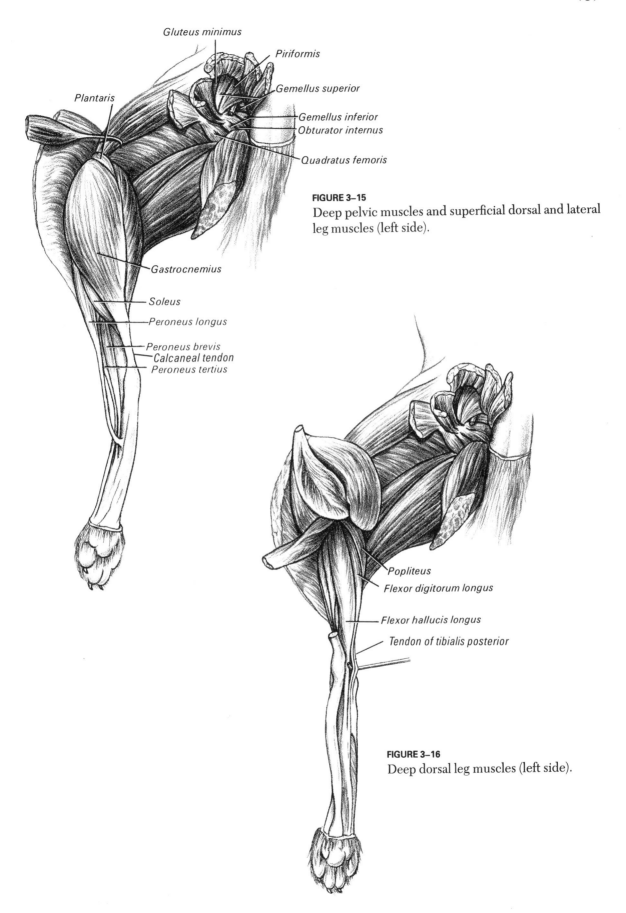

Gluteus minimus

Piriformis

Gemellus superior

Gemellus inferior
Obturator internus

Quadratus femoris

Plantaris

**FIGURE 3–15**
Deep pelvic muscles and superficial dorsal and lateral
leg muscles (left side).

Gastrocnemius

Soleus

Peroneus longus

Peroneus brevis
Calcaneal tendon
Peroneus tertius

Popliteus

Flexor digitorum longus

Flexor hallucis longus

Tendon of tibialis posterior

**FIGURE 3–16**
Deep dorsal leg muscles (left side).

**TABLE** 3-1 (*continued*)
Summary of the musculature of the cat.

| Muscle | Origin | Insertion | Action |
|---|---|---|---|
| *Muscles of the Upper Arm* | | | |
| Medial Surface | | | |
| Coracobrachialis | Coracoid process | Proximal end of humerus | Adducts humerus |
| Epitrochlearis | Ventral border of lat. dorsi | Olecranon process of ulna | Extends forearm |
| Biceps brachii | Scapula near glenoid fossa | Radial tuberosity | Flexes forearm |
| Lateral Surface | | | |
| Triceps brachii | Scapula and shaft of hum. | Olecranon process of ulna | Extends forearm |
| Anconeus | Distal end of humerus | Olecranon process of ulna | Tensor of elbow joint caps. |
| Brachialis | Lateral surface of humerus | Proximal end of ulna | Flexes forearm |
| *Muscles of the Forearm* | | | |
| Superficial Dorsal Group | | | |
| Brachioradialis | Middle fifth of humerus | Styloid process of radius | Supinator of hand |
| Extensor carpi radialis longus | Lateral supracondylar ridge | Base of second metacarpal | Extends hand |
| Extensor carpi radialis brevis | Lateral epicondyle of humerus | Base of third metacarpal | Extends hand |
| Extensor dig. communis | Lateral epicondyle of humerus | Prox. phalanges of dig. 2 to 5 | Extends digits 2 to 5 |
| Extensor dig. lateralis | Lateral supracondylar ridge | Tendons of ext. dig communis | Extends digits 2 to 5 |
| Extensor carpi ulnaris | Lateral epicondyle of humerus | Base of metacarpal 5 | Extends digit 5 |
| Deep Dorsal Group | | | |
| Extensor indicis proprius | Upper third of ulna | Middle phalanx of digit 2 | Extends digit 2 |
| Extensor pollicis brevis | Lateral ulnar surface and radius | Base of first metacarpal | Extends and abducts thumb |
| Supinator | Ligaments of elbow joint | Proximal third of radius | Supinator of hand |
| Superficial Ventral Group | | | |
| Pronator teres | Medial epicondyle of humerus | Middle third of radius | Pronation of hand |
| Flexor carpi radialis | Medial epicondyle of humerus | Base of metacarpals 2 and 3 | Flexor of digits 2 and 3 |
| Palmaris longus | Medial epicondyle of humerus | Prox. phalanges of dig. 2 to 5 | Flexes prox. phalanges |
| Flexor carpi ulnaris | Medial epicondyle and olecranon | Pisiform bone | Flexes hand |
| Deep Ventral Group | | | |
| Flexor dig. superficialis | Palmaris long., flexor dig. profundus | Middle phalanx of digits 2 to 5 | Flexes digits 2 to 5 |
| Flexor dig. profundus | Ulna, med. epicondyle, radius | Distal phalanx of all digits | Flexes all the digits |
| Pronator quadratus | Distal part of the ulna | Distal part of the radius | Pronator of hand |

(*continued on page* 190)

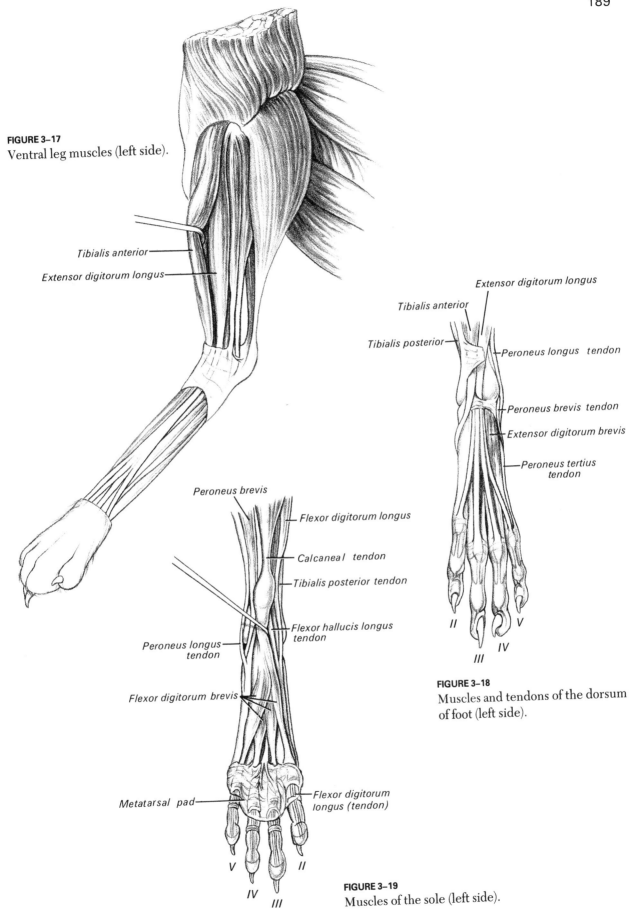

**FIGURE 3–17**
Ventral leg muscles (left side).

Tibialis anterior

Extensor digitorum longus

Extensor digitorum longus

Tibialis anterior

Tibialis posterior

Peroneus longus  tendon

Peroneus brevis  tendon

Extensor digitorum brevis

Peroneus tertius
tendon

II      V

IV

III

**FIGURE 3–18**
Muscles and tendons of the dorsum
of foot (left side).

Peroneus brevis

Flexor digitorum longus

Calcaneal  tendon

Tibialis posterior tendon

Flexor hallucis longus
tendon

Peroneus longus
tendon

Flexor digitorum brevis

Metatarsal  pad

Flexor digitorum
longus (tendon)

V      II

IV    III

**FIGURE 3–19**
Muscles of the sole (left side).

TABLE 3-1 (*continued*)

Summary of the musculature of the cat.

| Muscle | Origin | Insertion | Action |
|---|---|---|---|
| *Muscles of the Thigh* | | | |
| Medial Surface | | | |
| Sartorius | Ilium | Proximal end of tibia | Adducts and rotates thigh |
| Gracilis | Ischium and public symp. | Proximal end of tibia | Adducts thigh |
| Semimembranosus | Ischium | Distal end of femur | Extends thigh |
| Adductor femoris | Ischium and pubis | Shaft of femur | Extends thigh |
| Adductor longus | Pubis | Shaft of femur | Extends thigh |
| Pectineus | Pubis | Shaft of femur | Extends thigh |
| Iliopsoas | Lumbar vert. and ilium | Lesser trochanter of femur | Rotates and flexes thigh |
| Rectus femoris | Ilium | Patella | Adducts thigh |
| Vastus medialis | Shaft of femur | Patella | Extends leg |
| Vastus intermedius | Shaft of femur | Patella | Extends leg |
| Lateral Surface | | | |
| Tensor fascia lata | Ilium | Fascia lata of thigh | Tightens fascia lata |
| Biceps femoris | Ischium | Patella and tibia | Abducts thigh, flexes leg |
| Semitendinosus | Ischium | Proximal end of tibia | Flexes leg |
| Caudofemoralis | Caudal vertebrae | Patella | Abducts thigh, extends leg |
| Gluteus maximus | Sacral and caudal vertebrae | Greater trochanter of femur | Abducts thigh |
| Gluteus medius | Sacral and caudal vertebrae | Greater trochanter of femur | Abducts thigh |
| Vastus lateralis | Proximal part of femur | Patella | Extends leg |
| Tenuissimus | Second caudal vertebra | Fascia of biceps femoris | Abducts thigh |
| Deep Pelvic Muscles | | | |
| Piriformis | Sacral and caudal vertebrae | Greater trochanter of femur | Abducts thigh |
| Gluteus minimus | Ventral half of ilium | Greater trochanter of femur | Rotates thigh |
| Gemellus superior | Ilium and ischium | Greater trochanter of femur | Rotates and abducts thigh |
| Obturator internus | Ramus of ischium | Trochanteric fossa of femur | Abducts thigh |
| Gemellus inferior | Lateral surface of the ischium | Tendon of obturator internus | Abducts thigh |
| Quadratus femoris | Tuberosity of ischium | Greater and lesser trochanter | Extends and rotates thigh |
| Obturator externus | Ramus of pubis and ischium | Trochanteric fossa of femur | Rotates and flexes thigh |
| *Muscles of the Leg* | | | |
| Superficial Dorsal Group | | | |
| Gastrocnemius | Both epicondyles and patella | Proximal end of calcaneus | Extends foot |
| Soleus | Proximal third of fibula | Proximal end of calcaneus | Extends foot |
| Plantaris | Lateral epicondyle and patella | Flexor digitorum brevis | Flexes second phalanges |
| Deep Dorsal Group | | | |
| Popliteus | Lateral epicondyle of femur | Proximal end of tibia | Rotates thigh |
| Flexor digitorum longus | Head of fibula, shaft of tibia | Tendon of fl. hallucis longus | Flexes terminal phalanges |
| Flexor hallucis longus | Shaft of fibula and tibia | Tendon of fl. dig. longus | Flexes terminal phalanges |
| Tibialis posterior | Head of fibula, shaft of tibia | Medial cuneiform | Extends foot |
| Lateral Group | | | |
| Peroneus longus | Head and shaft of fibula | Bases of metatarsals | Flexes foot |
| Peroneus tertius | Middle of fibula | Tendon of ext. dig. longus | Extends and abducts digits |
| Peroneus brevis | Distal half of fibula | Base of metatarsal 5 | Extends foot |
| Ventral Group | | | |
| Tibialis anterior | Shaft and head of fibula and tibia | Medial surface of metatarsal 1 | Flexes foot |
| Extensor digitorum longus | Lateral epicondyle of femur | Term. phalanges of dig. 2 to 5 | Extends foot |

# Digestive and Respiratory Systems

# Anatomy of the Cat

The body cavity, or **coelom,** of the cat is divided into four separate compartments. The diaphragm, a muscular partition, divides the coelom into an anterior **thoracic cavity** and a posterior **abdominal cavity.** The thoracic cavity, in turn, is divided into three smaller cavities: two pleural cavities, each containing a lung, and the mediastinum, the midline region between the pleural cavities. The mediastinum contains the heart and the great blood vessels that enter and leave the heart (see Exercise 5, Circulatory System).

## DIGESTIVE SYSTEM

### Salivary Glands (Fig. 4-1)

The salivary glands are masses of gland tissue formed by outgrowths of the oral epithelium. The stalk of each outgrowth remains as the duct of the gland and serves to convey the secretion to the point at which it is deposited into the mouth. The glands are located among the muscles of the head and throat.

*Three main pairs of salivary glands can be uncovered on the undissected side of the head. The platysma and some small muscles of facial expression must be removed to expose the glands. Take special care in the region of the salivary ducts.*

**Parotid gland**  The largest of the glands, it lies in front of the ear and overlaps the masseter muscle. Its

duct passes over the external surface of the masseter to pierce the cheek opposite the last upper premolar tooth. Bits of glandular tissue are frequently found along the parotid duct. The duct should not be confused with either of the two branches of the facial nerve. These emerge from beneath the parotid gland and cross the masseter, dorsal and ventral to the duct.

**Submandibular gland**  This gland is located below the parotid. Its duct follows the ventral margin of the mandible to open onto the floor of the mouth just behind the lower incisor tooth. If the submandibular duct (and its companion sublingual duct) is traced, it will be found to extend lateral to the digastric muscle. Cut and reflect the digastric to observe the ducts extending medial to the posterior border of the mylohyoid. The latter may be cut and the ducts traced forward as far as possible.

**Sublingual gland**  This cone-shaped gland occupies a position anterior to the submandibular gland. The ducts of the submandibular and sublingual glands have a common opening into the mouth. The first part of the sublingual duct is overlapped by lymph nodes.

Two minor salivary glands are also in the head. The very small **molar gland** is near the angle of the mouth; the **infraorbital gland** is on the floor of the orbit near its lateral side. The molar gland need not be sought, and the infraorbital gland will be located when the eye is dissected (Exercise 7, Endocrine System and Sense Organs).

## Mouth (Fig. 4-2)

*Cut the muscles at each angle of the jaw. Press down on both sides of the mandible to expose the mouth. If exposure is not adequate, cut the angles of the mandible with bone shears. A more radical procedure can be used if maximal exposure is desired. In this case transect the masseter and temporalis muscles and remove them together with the posterior end of the zygomatic arch. Externally, cut through the floor of the mouth on each side as close to the inner surface of the mandible as possible. Then, using bone scissors, cut through the symphysis of the mandible and spread the halves apart.*

**Vestibule**   The part of the oral cavity or mouth between the lips and teeth.

**Oral cavity**   The part of the mouth behind the teeth.

**Lips**   Folds of skin lined internally by a mucous membrane. Each lip is attached to the mandible in the midline by a frenulum, or fold, of mucous membrane.

**Tongue**   Mobile muscular organ attached posteroventrally to the floor of the mouth by a frenulum. On the back part of the free surface, or **dorsum,** of the tongue is a V-shaped groove, the **terminal sulcus,** at the apex of which is located a small depression, the **foramen cecum.** Numerous small projections, or **papillae,** are located on the free surface of the tongue. They are classified into four types (Table 4-1).

**TABLE 4-1**
Papillae of the tongue of the cat.

| Papilla | Shape | Number | Location |
|---|---|---|---|
| Filiform | Conical | Most numerous | Middle of dorsum |
| Fungiform | Mushroom | Numerous | Behind filiforms |
| Vallate | Castle | About twelve | Near root |
| Foliate | Broad | Few | Root region |

**Teeth**   The cat is a diphyodont, i.e., it has both a temporary and a permanent set of teeth. The former, known also as deciduous, or milk, teeth appear within about three weeks after birth. There are 26 of them: 12 incisors, four canines, and 10 premolars (six in the upper jaw and four in the lower). When a cat is about four months old, the permanent teeth, 30 in all, begin to erupt and gradually push out and replace the deciduous teeth. Each side of the upper jaw has three incisors, one canine, three premolars, and one (small) molar. Each side of the lower has three incisors, one canine, two premolars, and one (large) molar. The number and kind of teeth can be conveniently expressed by a dental formula:

$$\frac{3\text{-}1\text{-}3\text{-}1}{3\text{-}1\text{-}2\text{-}1}$$

**Palate**   This partition separates the oral and nasal cavities. Its anterior portion, the **hard palate,** is supported by bone. On the mucosal surface of this anterior part, a number of transverse ridges, the **palatine rugae,** are present. The posterior portion of the palate, the **soft palate,** does not have a bony support.

**Palatine tonsils**   These are masses of lymphoid tissue located mainly on the lateral wall of the posterior part of the oral cavity near the end of the soft palate.

## Pharynx (Fig. 4-2)

The pharynx is the chamber that is common to both the digestive and respiratory systems.

*Slit the soft palate longitudinally in the midline to expose the pharynx somewhat. To obtain a better view, if the radical procedure for exposing the mouth has been followed, separate the hyoid from the tympanic bulla and make a longitudinal incision through the lateral wall of the pharynx to expose its cavity. If the radical procedure has not been followed, preparation of a sagittal section, according to the following description, will provide for very good visualization.*

*Mark out a line of incision in the midline of the head and carry this line back to the first few cervical vertebrae. By means of a bone (or hack) saw, cut through the cranium to split the skull anteriorly between the upper incisors and posteriorly in line with the spinous processes of the cervical vertebrae. Transect the tongue and mandibular symphysis and carry the sectioning back to about the fourth cervical vertebra. Then spread apart the two sides of the skull and wash them thoroughly. To determine whether the sagittal sections are off center, see if one section bears the median located mesoethmoid bone and whether the cerebrum has been cut between the hemispheres.*

**Nasopharynx**   The upper part of the pharynx that lies behind the internal openings of the nasal cavities, the **choanae** (internal nares). Near each side of the nasopharynx is a small dorsolateral slit, the pharyngeal **ostium of the auditory tube,** which communicates with the middle ear cavity.

**Oropharynx**   The middle part of the pharynx located behind the oral cavity.

**Laryngopharynx**   The lower part of the pharynx located behind the larynx.

**Glottis**   The opening into the larynx that is guarded by a small flap-like structure, the cartilaginous **epiglottis.**

193

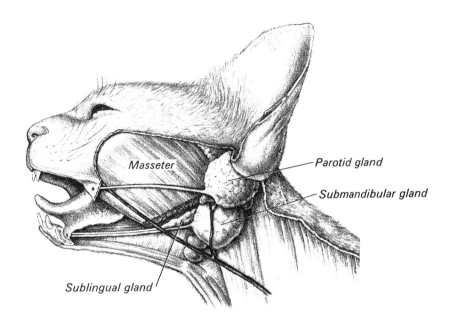

**FIGURE 4–1**
Salivary glands.

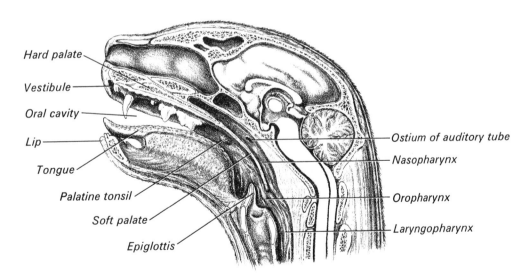

**FIGURE 4–2**
Mouth and pharynx.

## Digestive Organs (Fig. 4-3)*

*Make a cut slightly to one side of the mid-ventral line of the abdomen, continuing it forward from the pubic region to the thorax. Then cut transversely through the body wall at each end of the longitudinal incision and lay back the resultant flaps.*

**Esophagus**  This is the narrow tube that begins at the termination of the laryngopharynx, extends through the thoracic cavity, pierces the diaphragm, and terminates in the most cranial parts of the abdomen. Thus, the abdominal portion of the esophagus is very short and is continuous with a dilated tube, the stomach.

**Stomach**  This pear-shaped dilation of the digestive tract is divided into four parts. The **cardia** is the proximal part joined to the esophagus. The **fundus** is the bulge on the left side of the stomach. The **body** is the large section between the fundus and the distal part of the stomach, the **pylorus.** A constriction, the **pyloric sphincter,** is located at the terminal part of the stomach. The long convex surface on the left and posterior sides of the stomach is known as the **greater curvature;** the shorter concave border on the opposite side is the **lesser curvature.**

**Small intestine**  This tubular continuation from the stomach consists of three parts. The **duodenum** begins at the pyloric sphincter and exhibits a very sharp bend in its course. Its initial segment, which is caudally directed, lies along the right side of the pyloric part of the stomach. The duodenum then makes a U-shaped bend and continues, cranially and to the left, to join the next segment of the small intestine. The **jejunum** is a continuation of the duodenum. Its proximal and distal ends, like those of the other parts of the small intestine, are not identifiable by gross examination. The terminal segment of the small intestine is the **ileum,** which ends in the colon.

**Colon**  The beginning of the large intestine has a short blind pouch, the **cecum.** The first part of the colon, the **ascending colon,** lies on the right side of the abdominal cavity and extends in an anterior direction from its junction with the ileum. The colon then passes transversely to the left as the **transverse colon.** The transverse colon is continuous with the **descending colon,** which runs posteriorly toward the midline. The terminal part of the colon is the **rectum,** which opens to the outside at the anus. Two large **anal glands** are located lateral to the anus and their ducts empty into it.

*See also Fig. S-1, p. 287.

## Digestive Glands (Fig. 4-3)

**Liver**  This large organ is located just posterior to the diaphragm, to which it is attached. The liver is divided into **right** and **left lobes** by a ligament, and each lobe is subdivided into a large **lateral lobe** and a small **median lobe.** By lifting up the liver, a small triangular **caudate lobe** can be seen at the base of the right lateral lobe. On the ventral surface of the right median lobe, known also as the cystic lobe, is the pea-shaped **gall bladder.** The narrow end of the gall bladder opens into the **cystic duct. Hepatic ducts** from the liver, which may be difficult to see, join the cystic duct to form the **common bile duct.** The common bile duct passes posteriorly to empty into the duodenum near the pyloric sphincter.

**Pancreas**  This is a flat elongated gland that bends sharply near its middle. It consists of a **body** dorsal to the pyloric portion of the stomach and a **head** that lies in the concavity of the duodenum. Before the common bile duct penetrates the wall of the duodenum, it is joined by the main **pancreatic duct,** which may be exposed by carefully picking away pancreatic tissue. The **accessory pancreatic duct,** which may be absent in some specimens, opens independently into the duodenum slightly posterior to the point of entry of the common bile duct.

**Spleen**  This reddish-brown lymphatic organ lies on the left side of the abdominal cavity near the greater curvature of the stomach.

## Mesenteries (Fig. 4-3)

The arrangement of the peritoneum in the abdominal cavity is basically similar to that in the dogfish shark and the mud puppy. The following mesenteries should be identified.

**Falciform ligament**  The part of the original ventral mesentery extending between the ventral body wall and the liver, which it divides into right and left lobes.

**Coronary ligament**  Peritoneal duplication that extends between the caudal surface of the diaphragm and the cranial surface of the liver, attaching them to one another.

**Greater omentum**  This mesentery is suspended from the greater curvature of the stomach and covers the transverse colon and a large part of the small intestine. The greater omentum acts as a fat storage organ.

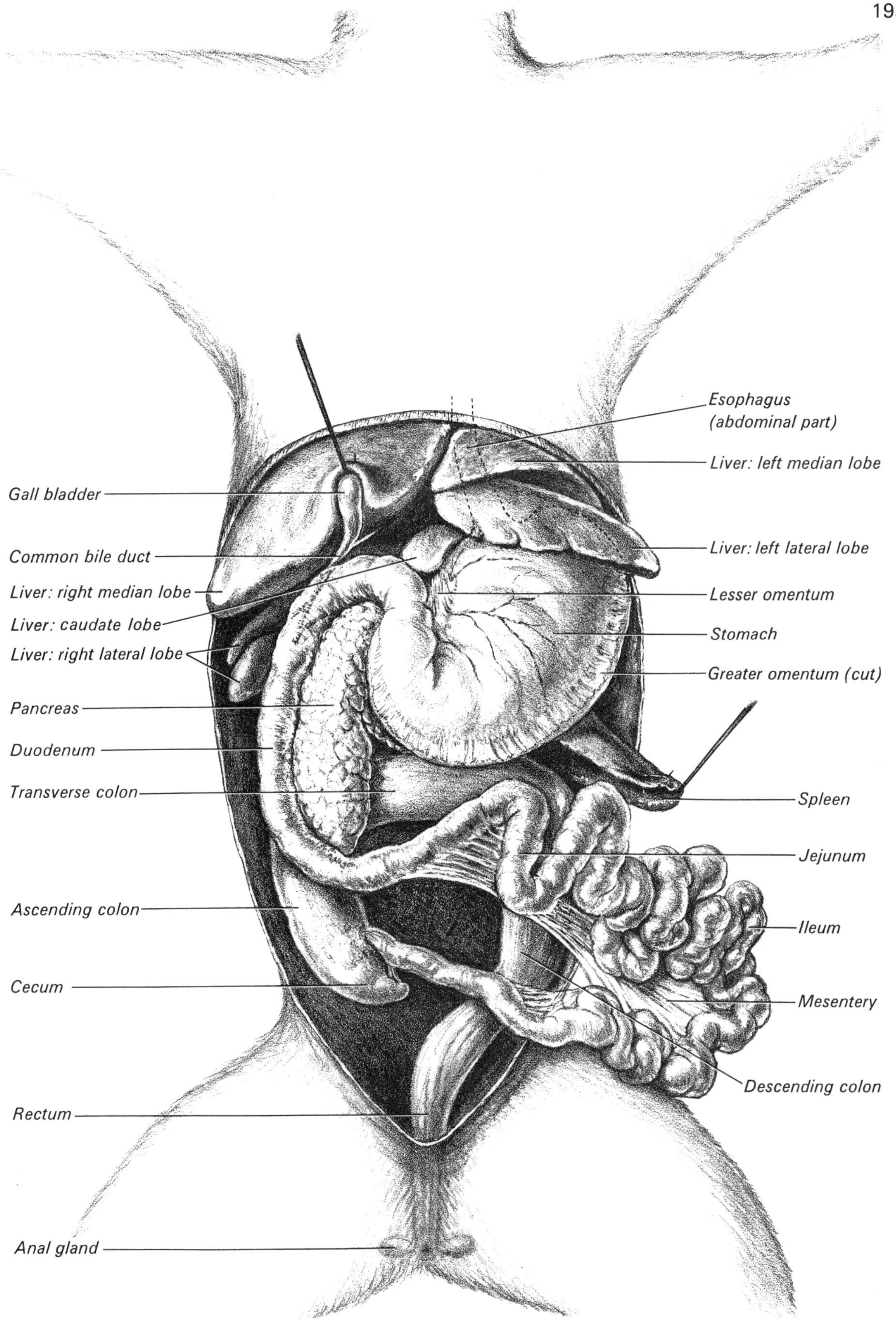

**FIGURE 4–3**
Digestive system.

**Lesser omentum** The part of the ventral mesentery extending from the liver to the lesser curvature of the stomach and first part of the duodenum. The part extending to the stomach is known as the **hepatogastric ligament;** that extending to the duodenum is the **hepatoduodenal ligament.**

**Mesentery** The part of the dorsal mesentery that suspends the small intestine from the mid-dorsal line is the mesentery proper.

**Mesocolon** Extends between the colon and the mid-dorsal line.

**Mesorectum** This mesentery suspends the rectum from the mid-dorsal line.

## RESPIRATORY SYSTEM

The respiratory system consists of the nasal passages, pharynx, larynx, trachea, and lungs. Air, taken in at the external nares, is warmed as it passes over the mucous membranes of the turbinate bones of the nasal cavities. The air leaves the nasal cavities through the choanae, passes through the nasopharynx, oropharynx, laryngopharynx, and the larynx.

### Larynx (Figs. 4-4, 4-5)

The larynx is the cartilaginous expansion of the cranial end of the trachea near the base of the tongue. It lies ventral to the esophagus and consists of five cartilages held together by ligaments.

*Cut through the base of the tongue a short distance in front of the epiglottis. Carefully remove the lower jaw and tongue. Then clear away the muscles and other soft tissues in front of the larynx. The hyoid apparatus may also be reviewed at this point in the dissection (see Exercise 3, Muscular System).*

**Thyroid cartilage** Large unpaired cartilage forming most of the ventral and lateral walls of the larynx.

**Cricoid cartilage** Smaller unpaired ring-shaped cartilage located directly caudal to the thyroid cartilage. It can be seen projecting anteriorly from the thyroid cartilage to which it is attached.

**Epiglottis** The small unpaired cartilage that guards the opening into the larynx. It projects anteriorly from the thyroid cartilage, to which it is attached.

**Arytenoid cartilages** A pair of very small pyramidal-shaped cartilages situated on the cranial border of the cricoid cartilage near the mid-dorsal line and slightly cranial to the cricoid cartilage. They can best be seen if the larynx is freed from the esophagus.

*Make a mid-ventral longitudinal incision through the larynx and the anterior part of the trachea and lay back the halves to expose the vocal cords.*

**False vocal cords** These are folds of mucous membrane that extend from the arytenoid cartilages to the base of the epiglottis.

**True vocal cords** These are comparable folds, supported by elastic ligaments, located slightly caudal to the false cords. They extend from the arytenoid cartilages to the mid-caudal border of the thyroid cartilage. The free margin of the true vocal cords bounds the glottis.

**Laryngeal ventricles** Lateral pockets on either side of the cavity of the larynx between each false and true vocal cord.

### Trachea (Figs. 4-4, 4-5)

The trachea is a long tube that extends caudally from the larynx to its point of bifurcation into the bronchi. It lies just ventral to the esophagus. Its walls are strengthened by a series of C-shaped **trachael cartilages** that are open dorsally. These cartilages are interconnected by ligaments and the dorsal gap is closed by muscle. The thyroid gland is located on either side of the trachea near its cranial end. This thyroid gland is part of the endocrine system.

### Lung (Fig. 4-6)

*Open the thoracic cavity by making a pair of longitudinal incisions about one-half inch on either side of the sternum and cutting through the ribs. Then make a transverse incision across the base of the sternum, extending it laterally and dorsally to include the most posterior ribs. When the incisions have been completed the sternum will remain intact anteriorly but can be elevated posteriorly so that the underlying blood vessels are visible. After the vessels have been noted, make a transverse incision above the most anterior ribs and extend it laterally and dorsally until it is close to the vertebral column. Now the lateral thoracic walls can be bent back. To expose the thoracic cavity fully, break each rib near its dorsal end.*

**Bronchi** These are the two short tubes formed by bifurcation of the trachea at the level of the sixth rib. Each bronchus is supported by a series of incomplete cartilages. The bronchi undergo many subdivisions within the substance of the lung.

**Lungs** The spongy organs in which respiratory ex-

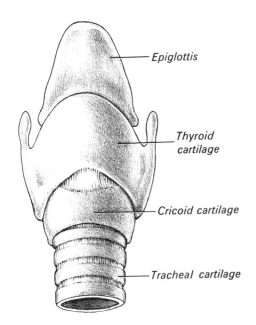

**FIGURE 4–4**
Larynx and trachea (ventral view).

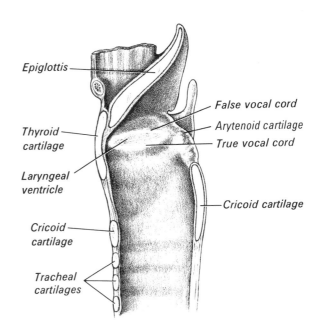

**FIGURE 4–5**
Larynx and trachea (longitudinal section).

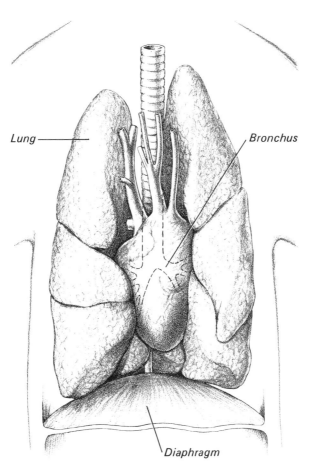

**FIGURE 4–6**
Respiratory system.

change takes place. They are enclosed in the **pleural cavities** and are covered by visceral pleura, a serous membrane. Parietal pleura line the pleural cavities. The right lung consists of four lobes, the left lung of three.

**Mediastinum**  This region in the median plane between the pleural cavities is filled with a number of structures. The ventral part of the mediastinum contains the large thymus gland and some small blood vessels. The middle part contains the pericardial cavity and the great blood vessels as well as some nerves. The dorsal part contains the trachea, esophagus, aorta, and large blood vessels. In the mid-ventral thoracic region the visceral and parietal pleura meet to form a ventral partition, the **mediastinal septum.**

**Diaphragm**  This dome-shaped sheet of muscle originates from the lower ribs and the bodies of the lumbar vertebrae and the sternum. The diaphragm inserts on a central tendon that lies at the base of the heart.

# Circulatory System

# Anatomy of the Cat

The circulatory system may be divided into three components: the heart, the arterial and venous vascular systems, and the lymphatic system. In the course of evolution these components have undergone changes correlated with increased physical activity and an increased rate of metabolism.

In fishes the heart has two chambers: a thin-walled dorsal atrium that receives blood from behind, and a thick-walled ventricle that propels it forward (see Anatomy of the Dogfish Shark, Exercise 5). In amphibians and reptiles the atria are paired, with systemic or venous blood entering the right atrium and pulmonary or arterial blood entering the left (see Anatomy of the Mud Puppy *Necturus*, Exercise 5), and the sinus venosus is markedly reduced in adults. Birds and mammals have the same type of four-chambered heart even though they evolved from a different reptilian stem. The sinus venosus is absent as a distinct part of the heart, having become incorporated into the right atrium.

The arterial system of mammals is considerably modified from that of reptiles. Among other things, a perfect double circulatory system has been completed by the bifurcation of the conus arteriosus into the pulmonary trunk and the base of the arch of the aorta, which spring from the right and left sides of the heart, respectively.

The venous system has undergone the fewest changes. The major alteration has been the fusion of the renal portal system with the posterior vena cava, which has completely eliminated portal circulation through the kidneys. Mammals thus have only one portal circulation—that through the liver—and as a result the hepatic portal vein is known simply as the portal vein. This system, as in all other vertebrates, conveys all the venous blood of the digestive tract into the liver sinusoids where it undergoes intensive biochemical alteration. The primitive lateral abdominal veins are represented in the embryo by the umbilical veins but are lost in the adult.

The blood of the circulatory system carries oxygen from the lungs, hormones from the endocrine glands, and nutritive substances absorbed from the digestive system to the various cells of the body. It also carries the end products of metabolism to the kidneys for excretion.

## CAT HEART

The heart of the cat, like that of all other mammals, consists of four completely separated chambers. Blood from all parts of the body empties into the right atrium and then passes to the right ventricle. The right ventricle pumps the blood into the pulmonary arteries for transport to the lungs, where oxygenation occurs. The aerated blood returns to the left atrium of the heart through the pulmonary veins and then passes into the left ventricle, which pumps it into the aorta for distribution to all parts of the body.

The heart is located in the middle of the mediastinum in the **pericardial cavity.** It is enclosed by a pericardial sac, which is lined on its internal surface by **parietal pericardium.** The parietal pericardium is reflected over the surface of the heart as the **visceral pericardium,** or epicardium.

## External Features (Fig. 5-1)

*Dissect away the mediastinal septum and remove the thymus gland and any fat deposits covering the heart and great vessels. Slit open the pericardial sac in the midline and expose the ventral surface of the heart. To expose the large vessels (pulmonary trunk and aorta) that leave the anterior end of the ventricles, the fat in this region should be picked away. Connective tissue dorsal to the superior vena cava and from the root of the lungs should be carefully removed in order to expose the pulmonary veins that come from the lungs and enter the left atrium.*

**Base**   The cranial end of the heart.

**Apex**   The caudal end, which is slightly to the left and is in contact with the diaphragm.

**Ventricles**   The large smooth posterior parts of the heart, which have a pronounced muscular appearance. The separation of the right and left ventricles is manifested on the surface by a faint groove extending obliquely across the ventral surface from the left of the base to the right of the apex.

**Atria**   Two thin-walled, irregularly shaped chambers at the anterior end of the heart. They are separated from the ventricles by a groove, the coronary sulcus. The ear-like appendage of each atrium is the **auricle.** Its border is scalloped.

**Pulmonary trunk**   Vessel arising from the right ventricle and passing toward the left. It divides into right and left pulmonary arteries.

**Conus arteriosus**   The part of the right ventricle that leads to the pulmonary trunk.

**Aorta**   The arterial arch arising from the left ventricle dorsal to the pulmonary trunk. Two major branches arise from the aorta, the brachiocephalic artery and, to the left of this artery, the left subclavian artery. A short fibrous band, the **ligamentum arteriosum,** connects the pulmonary trunk with the aorta. It represents the obliterated shunt, the ductus arteriosus, between these two vessels that existed in fetal life.

**Superior vena cava**   This vein empties into the anterodorsal side of the right atrium.

**Inferior vena cava**   This vein, which is caudal to the superior vena cava, also empties into the right atrium and can be seen if the apex of the heart is lifted.

**Pulmonary veins**   These are small vessels that empty into the left atrium.

**Coronary arteries**   Blood vessels on the surface of the heart, which they supply and drain. There are two coronary arteries, which originate at the base of the arch of the aorta. One may be seen deep between the pulmonary trunk and left auricle; the other, deep between the pulmonary trunk and right auricle.

## Internal Features (Fig. 5-2)*

The dissection of the internal structures of the heart is best deferred until after the study of the vascular system as a whole has been completed. It is discussed at this point only for the sake of continuity.

*Remove the heart from the pericardial cavity by severing its major vessels near their origin, leaving the large stumps attached to the heart. Divide the heart into dorsal and ventral halves by making a transverse slit in the lateral wall of the right atrium. Wash out any blood remaining in the right atrium. Then probe the ventricle through the opening made in the atrium. By inserting scissor tips through this incision, cut the lateral margin of the right ventricle. Make similar incisions through the atrium and ventricle on the left side of the heart. Finally, cut the septum, dividing the heart into right and left halves.*

**Right atrium**   The inner surface of the auricle is lined with parallel muscular ridges known as **musculi pectinati.** The rest of the atrial surface is smooth. The right atrium is separated from the left atrium by the **interatrial septum.** An oval depression, the **fossa ovalis,** is located in the wall of the septum. The opening of the superior vena cava has no valve, but that of the inferior vena cava is guarded by a rudimentary valve. A slit located posterior to the opening of the inferior vena cava is the opening of the coronary sinus, which receives blood drained by the **great cardiac vein.** This vein is formed by two trunks that drain most of the heart musculature, or myocardium. Several small anterior cardiac veins open independently into the right atrium.

**Left atrium**   The contours of the walls of this chamber are similar to those of the right atrium. The only openings present in these walls are those for the pulmonary veins.

**Right ventricle**   The internal surface of the wall of this chamber is corrugated by muscular cords called **trabeculae carneae.** An oval aperture, the **atrioventricular orifice,** provides communication with the right atrium and is surrounded by a **tricuspid valve.** The free border of each of the three cusps, or flaps, is connected by thin fibers, the **chordae tendineae,** to muscular pillars known as **papillary muscles.** These muscles are anchored to the wall of the ventricle. **Semilunar valves** are located at the base of the pulmonary trunk.

**Left ventricle**   This chamber has a **bicuspid valve** at its atrioventricular orifice, which communicates with the left atrium. The elements of the valve are similar to those of the valve in the right ventricle. Semilunar valves are at the opening of the aorta.

*See also Fig. S-2, p. 289.

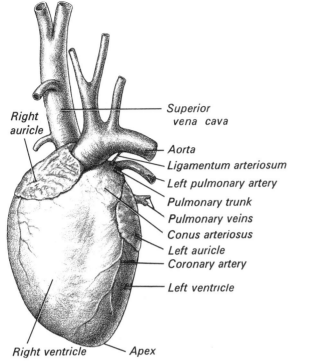

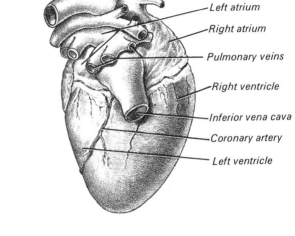

Right
auricle

Superior
vena cava

Aorta
Ligamentum arteriosum
Left pulmonary artery
Pulmonary trunk
Pulmonary veins
Conus arteriosus
Left auricle
Coronary artery

Left ventricle

Right ventricle — Apex

VENTRAL VIEW

Superior vena cava
Aorta
Left atrium
Right atrium
Pulmonary veins
Right ventricle
Inferior vena cava
Coronary artery
Left ventricle

DORSAL VIEW

**FIGURE 5–1**
Cat heart.

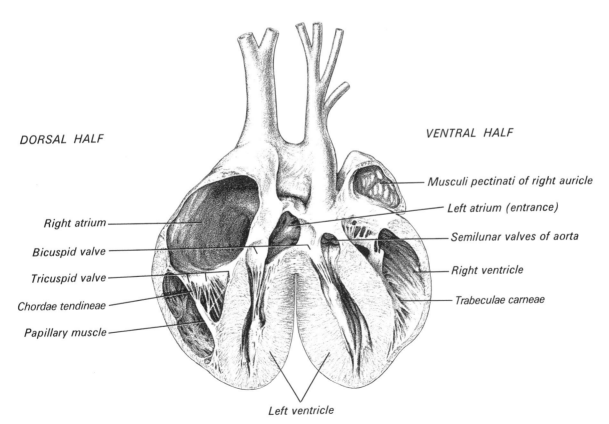

DORSAL HALF

VENTRAL HALF

Right atrium
Bicuspid valve
Tricuspid valve
Chordae tendineae
Papillary muscle

Musculi pectinati of right auricle
Left atrium (entrance)
Semilunar valves of aorta
Right ventricle
Trabeculae carneae

Left ventricle

**FIGURE 5–2**
Cat heart (internal view).

## SHEEP HEART

Dissection of the cat heart presents a difficult challenge because of its small size. A larger mammalian heart, such as that of a sheep, provides a better opportunity for studying the anatomy of this organ, especially its internal structure.

*In situ,* the sheep heart is enclosed within a fibroserous pericardial sac. Several fibrous bands attach the ventral surface of the sac to the sternum. The heart is an elongated, conical-shaped, muscular organ. Its long axis is oblique, and the heart in the animal's body lies obliquely across the median plane, with the apex on the left side projecting toward the diaphragm.

The external surface of the sheep heart should be examined first ventrally, then dorsally, and finally its internal components should be exposed. At the conclusion of the dissection, if time permits, the anatomy of the cat heart (see p. 188) should be reviewed.

*Examine your specimen and note that almost all the parietal pericardium, the fibrous sac enclosing the heart, has been removed. In some specimens remnants of it may be identified at the bases of the major blood vessels. Carefully remove the fat from the heart surface and especially from around the large blood vessels to facilitate their identification. Familiarize yourself with the external anatomy of the heart and its associated vessels. (If you probe into them, do so carefully to protect their valves.)*

### External Features: Ventral View (Fig. 5-3)

*Place the heart so that the interventricular sulcus is evident and the apex projects downward. (Remember that the directives "right" and "left" refer to the heart and the animal, not to the dissector.)*

**Base**    The wide upper or anterior end of the heart.

**Apex**    The somewhat blunt tip of the heart, composed entirely of the left ventricle.

**Atria**    The right and left atria are represented by their leaf-like atrial appendages or **auricles**.

**Pulmonary trunk**    A continuation of the conus arteriosus of the right ventricle. This prominent vessel passes anteriorly towards the left side by coursing dorsally then caudally. It subsequently divides into two branches, the right and left pulmonary arteries, which carry unoxygenated blood to the lungs. In sheep the left pulmonary artery is larger than the right.

**Aorta**    This very large vessel leaves the heart at the base of the left ventricle and curves to the left over the pulmonary trunk. It carries oxygenated blood from the left ventricle for distribution throughout the body.

**Ligamentum arteriosum**    This short band of tissue extends between the descending aorta and the pulmonary trunk. It represents the degenerated embryonic ductus arteriosus. The closure of this channel, shortly before birth, results in separation of the pulmonary from the aortic circulation. Consequently, the left ventricle becomes fully efficient, with all of its oxygenated blood pumped to the body.

**Brachiocephalic artery**    This large branch of the aorta ascends on the right side to divide into two major vessels. One passes to the right forelimb as the right subclavian artery, and the other passes to the neck and head as the right common carotid artery.

**Interventricular sulcus**    A longitudinal groove that extends obliquely from lower left to upper right across the surface of the heart. It contains both the coronary arteries that supply the heart muscle (myocardium) and the cardiac veins that drain the heart muscle. These vessels are usually obscured by the subepicardial fat.

**Right ventricle**    The chamber that lies above the interventricular sulcus.

**Left ventricle**    The chamber that lies beneath the interventricular sulcus. It occupies a larger portion of the ventral surface of the heart than does the right ventricle.

### External Features: Dorsal View (Fig. 5-4)

*Turn the heart over and identify the following structures.*

**Superior vena cava**    This major blood vessel drains all the venous blood from the head, the neck, both forelimbs, and the upper chest wall into the right atrium. The superior vena cava is formed at the lower end of the neck by the union of the two external jugular veins and two subclavian veins.

**Inferior vena cava**    This large blood vessel returns the venous blood to the right atrium from most of the trunk and both lower limbs. This vein begins in the abdomen and enters the thorax through a foramen in the diaphragm. It runs forward to pierce the pericardial sac and open into the right atrium.

**Pulmonary veins**    These veins carry oxygenated blood from the lungs back to the heart. They open into the left atrium.

**Aorta**    This major vessel is a wide, thick-walled structure located next to the brachiocephalic artery.

**Pulmonary artery**    This vessel is seen as divided into **right** and **left branches**. The branches carry venous blood from the right ventricle to the lungs, where it will be oxygenated.

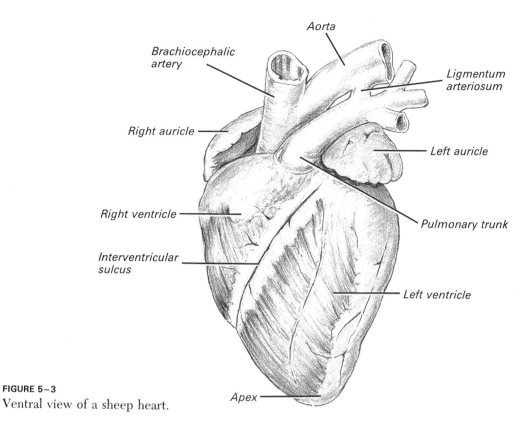

**FIGURE 5–3**
Ventral view of a sheep heart.

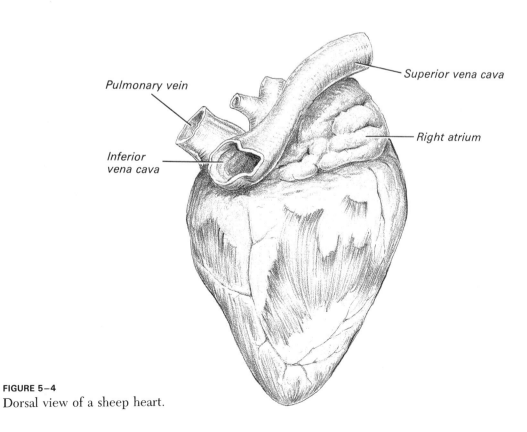

**FIGURE 5–4**
Dorsal view of a sheep heart.

### Internal Features: Right Atrium and Ventricle (Fig. 5-5)

To best investigate the interior of the sheep heart, the right atrium and ventricle should be exposed first and then both chambers on the left side should be studied.

*With the dorsal surface of the heart facing you and using sharp, pointed scissors, cut longitudinally through the length of the pulmonary artery. Then extend the cut through the wall of the right ventricle most of its length towards its apex. Next cut through the dorsal wall of the superior vena cava and continue through the wall of the right atrium until both cuts meet. Open up the chambers and identify the structures in the right atrium and ventricle. Note the thickness of the walls of these chambers.*

**Right atrium**    This chamber occupies the right base of the heart. It consists of two parts, the venous sinus and right auricle. The **venous sinus**, the main part of this atrium, is separated from the left atrium by the interatrial septum. The **right auricle** is lined by smooth endocardium. The inner surface is made irregular by the underlying muscular ridges, the **musculi pectinati**. Five orifices communicate with the right atrium: (1) the opening of the superior vena cava, (2) the opening of the inferior vena cava, (3) the right atrioventricular opening, (4) the coronary sinus opening, and (5) the openings of the small cardiac veins (located between the musculi pectinati, cranial to the coronary sinus opening on the ventral atrial wall). Identify also in the right atrium the following.

**Coronary sinus opening**    Cardiac veins drain blood from the myocardium and deposit it in the coronary sinus. This horizontal collecting vessel is located in a groove on the dorsal heart surface between the atria and ventricles. The collected venous blood is then deposited in the right atrium at the opening of the coronary sinus, located ventral to the opening of the inferior vena cava.

**Pulmonary semilunar valve**    Located at the base of the pulmonary trunk, this valve controls blood flow through its circular orifice. The valve consists of three interlocking cusp-shaped units made up of endocardium.

**Fossa ovalis**    An oval depression found only on the right side of the interatrial septum. It represents the site of the fetal foramen ovale that opened into the left atrium.

**Right ventricle**    Somewhat triangular in shape, this ventricle makes up half of the cranial border of the heart. Its base is largely connected with the right atrium. Its left part forms the **conus arteriosus**, from which the pulmonary trunk arises. Structures to be identified in the right ventricle follow.

**Chordae tendineae**    Tough, thin connective tissue strands that extend from the edges of the tricuspid valve flaps to node-like **papillary muscles** projecting up from the ventricle wall. Contraction of these muscles results in the chordae tendineae producing tension on the valve flaps, preventing them from overting into the atrium during ventricular contraction.

**Trabeculae carneae**    The lattice-like arrangement of muscle tissue on the inner wall of the ventricle(s).

**Right atrioventricular valve**    The **tricuspid valve**, located between the two chambers on the right side, also consists of three endocardial leaflets. They prevent blood from flowing back into the right atrium during ventricular contraction.

### Internal Features: Left Atrium and Ventricle (Fig. 5-6)

*With the ventral surface of the heart facing you, cut the aorta open along its length, extending the incision posteriorly through the myocardium of the left ventricle as far as the apex of the heart. Next open the left atrium along its lateral surface and then open up the chambers and identify the structures in the left atrium and ventricle, noting that they are similar to those on the right side. Also observe again the disparity in wall thickness between the chambers.*

**Left atrium**    This chamber lies behind the pulmonary trunk and the aorta and above the left ventricle. The interior wall of the atrium is smooth, while musculi pectinati are evident in the left auricular wall. Identify the following structures.

**Pulmonary veins**    These thin-walled blood vessels, which carry blood to the lungs, originate in the left atrium. There are five such veins opening into this atrium.

**Aortic semilunar valve**    Located at the base of the aorta, it consists of three cusps, which when they meet close the aorta, preventing blood from flowing back into the left ventricle.

**Left ventricle**    The caudal part of the ventricular mass. It communicates at its base with the left atrium through the atrioventricular opening, and its cranial part opens into the aorta. The chordae tendineae are larger but fewer than those of the right ventricle. The ventricular septum is concave and consists of small cranial membranous part and a larger caudal muscular part. Identify the following structure.

**Left atrioventricular valve**    This, the **bicuspid valve**, has two flaps, which are also connected to papillary muscles via chordae tendineae. This valve functions in an identical manner as the one on the right side.

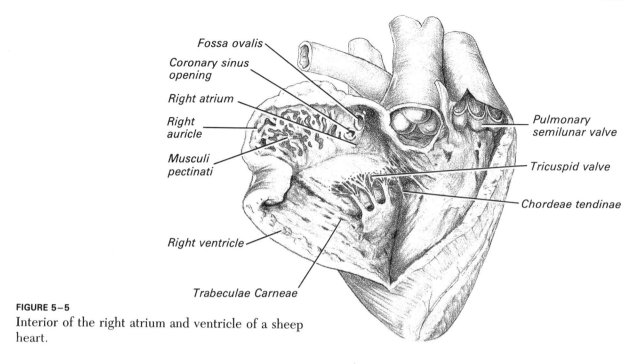

**FIGURE 5–5**
Interior of the right atrium and ventricle of a sheep
heart.

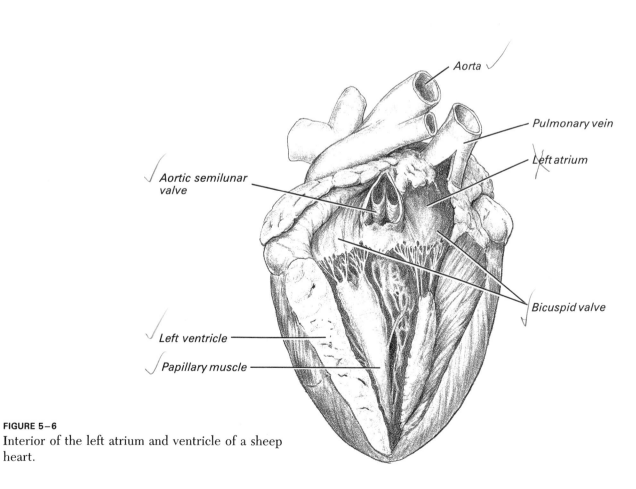

**FIGURE 5–6**
Interior of the left atrium and ventricle of a sheep
heart.

## ARTERIAL SYSTEM

### Anterior Arteries (Figs. 5-7, 5-12)*

The anterior arteries are those found in the neck, thorax, and forelimb. These arteries should all be traced back in order that their direct or indirect origin from the (dorsal) aorta may be observed.

**Aorta** This very large blood vessel begins at the base of the left ventricle and soon curves sharply to the left as the arch of the aorta (see definition). It continues posteriorly through the dorsal part of the thorax as the thoracic aorta (see definition). It leaves the thorax by piercing through the diaphragm and continues into the abdominal cavity as the abdominal aorta (see definition). The two **coronary arteries** arise from the stem of the aorta just cranial to its semilunar valves.

**Arch of the aorta** This is the site of origin of two major blood vessels, the brachiocephalic artery and, just to its left, the left subclavian artery. The **brachiocephalic artery** ascends directly cranially from the arch of the aorta. It ascends in an anterior direction, giving off first a small mediastinal artery to the mediastinum and then the large left common carotid artery. Finally, after curving laterally, it bifurcates into the right common carotid and right subclavian arteries (see definitions). The **left subclavian artery** ascends to the left of the brachiocephalic artery, to which it is parallel. It then curves sharply to the left and passes toward the left forelimb. The numerous branches of the brachiocephalic and left subclavian arteries are listed in Tables 5-1 and 5-2. In addition, the most important branches of the brachiocephalic are described below. To see them, remove the digastric and then cut and pull back the sternomastoid and cleidomastoid.

**Right and left common carotid arteries** These extend through the thorax into the neck in close relationship to the trachea. Each gives off the following six arteries.

   **Inferior thyroid artery** Runs cranially along the trachea on which it ramifies.

   **Superior thyroid artery** Arises at the level of the thyroid cartilage and passes caudomedially, giving off branches to the thyroid gland, the sternohyoid and the sternothyroid muscles, and the larynx.

   **Muscular branch** Arises at the same level as the superior thyroid artery and ramifies into three branches that supply the muscles on the dorsal part of the neck.

   **Occipital artery** Arises near the angle of the jaw

*See also Fig. S-3, p. 291.

and takes a circuitous course toward the base of the head, supplying deep neck muscles.

   **Internal carotid** This may be difficult to find because of its small size. It passes dorsally toward the skull.

   **External carotid artery** The name given to the continuation of the stem artery after origin of the internal carotid artery. It gives off branches to all parts of the head (see Table 5-1). These include the **lingual artery,** which runs craniomedially with the hypoglossal nerve; the **external maxillary artery,** which runs cranially to the mylohyoid, where it divides; the **posterior auricular artery,** which runs laterally to the region behind the ear; the **superficial temporal artery,** located between the auditory meatus and masseter; and the **internal maxillary artery,** the continuation of the stem artery into the interior of the skull.

**Right subclavian artery** Like the left subclavian artery, which arises independently from the aorta, this has five branches, as follows.

   **Internal mammary artery** Passes caudoventrally to the dorsal surface of the sternum and then continues down the thorax into the abdomen. This vessel may have been cut in exposing the thoracic cavity.

   **Vertebral artery** Passes dorsally to enter the foramen transversarium of the sixth cervical vertebra and ascends to the brain.

   **Costocervical trunk** Shortly after giving off a slender superior intercostal branch, it divides into two terminal vessels. One, the stout transverse cervical artery, courses laterally across the cranial surface of the first rib into an area between the scalenus and the serratus anterior muscles where it ramifies. The second, the slender deep cervical artery, ascends dorsally to leave the thorax and ramifies among the neck muscles.

   **Thyrocervical trunk** Proceeds cranially for a short distance and then divides into the ascending cervical artery, which ramifies among the neck muscles, and the transverse scapular artery, which runs laterally to pass into the scapular region.

   **Axillary artery** Runs through the axilla, giving off three branches in this region (see Table 5-2) and then continues in the forelimb, which it supplies as the **brachial artery.** The first branch of the axillary is the **ventral thoracic artery.** It arises slightly lateral to the first rib and extends to the medial surface of the pectoralis major, which it supplies. The **long thoracic artery** arises next, a few millimeters lateral to the first branch. It gives off branches to the pectoralis minor and terminates in the xiphihumeralis. The **subscapular artery** arises about a centimeter lateral to the pectoralis minor

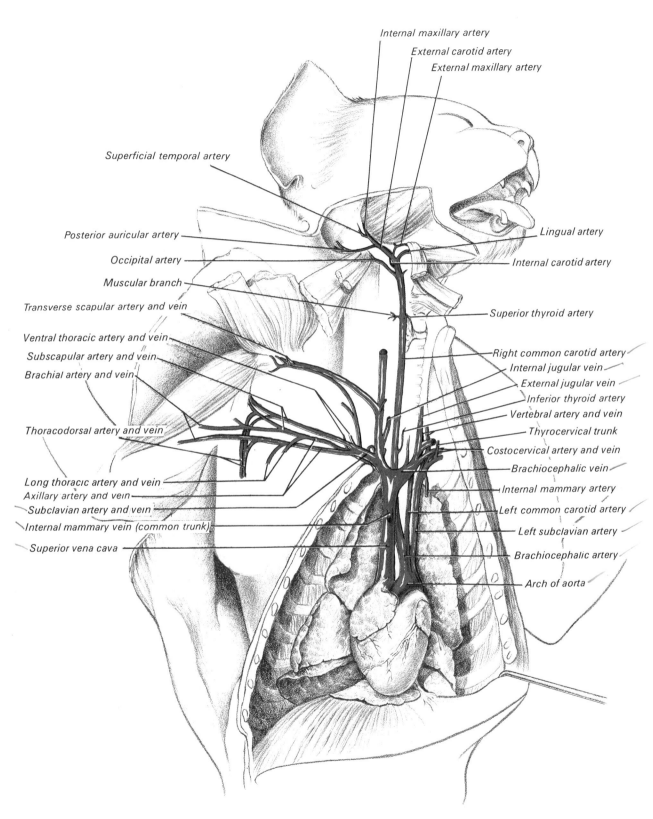

Internal maxillary artery
External carotid artery
External maxillary artery

Superficial temporal artery

Lingual artery

Posterior auricular artery

Internal carotid artery

Occipital artery

Muscular branch

Superior thyroid artery

Transverse scapular artery and vein

Ventral thoracic artery and vein

Right common carotid artery

Subscapular artery and vein

Internal jugular vein

Brachial artery and vein

External jugular vein

Inferior thyroid artery

Vertebral artery and vein

Thyrocervical trunk

Thoracodorsal artery and vein

Costocervical artery and vein

Brachiocephalic vein

Long thoracic artery and vein

Internal mammary artery

Axillary artery and vein

Left common carotid artery

Subclavian artery and vein

Left subclavian artery

Internal mammary vein (common trunk)

Brachiocephalic artery

Superior vena cava

Arch of aorta

**FIGURE 5–7**
Arteries and veins of the neck, thorax, and upper
arm.

and divides to give one branch that supplies the teres major, subscapularis, and latissimus dorsi. The other branch winds around the neck of the humerus to supply the triceps and deltoid muscles.

**Thoracic aorta** This section gives rise to a series of paired **intercostal arteries** that supply the intercostal and back muscles, a pair of **bronchial arteries** that go to the lungs, and several small **esophageal arteries** that supply blood to the musculature of the esophagus.

(The anterior veins and lymphatic vessels may be conveniently studied at this point.)

## Posterior Arteries (Figs. 5-8, 5-9, 5-13)*

The posterior arteries are those arising from the abdominal aorta. They can be divided into two groups: unpaired visceral and paired systemic arteries. The visceral arteries are the branches that supply blood to the digestive tract.

*After locating the abdominal aorta, trace its branches, starting at the anterior end of the abdomen. For initial exposure, the greater omentum, ileum, and jejunum can be removed (Fig. 5-4). Later, the colon and part of the stomach can also be cut out (Fig. 5-5).*

**Abdominal aorta** Extends from the diaphragm to the site from which the internal iliac arteries arise. It continues into the tail as the **caudal artery.** There are three major abdominal branches: the celiac trunk and the superior and inferior mesenteric arteries. Their branches are listed in Table 5-3. The most important branches are described below.

**Celiac trunk** A very short vessel that divides about a centimeter directly ventrad from its point of origin into three large branches: the hepatic, left gastric, and splenic arteries.

**Hepatic artery** Passes deep to the stomach and into the lesser omentum where, coursing along the left side of the hepatic portal vein, it gives rise to the **gastroduodenal artery.** The gastroduodenal artery gives off three branches (see Table 5-3), which ramify in this region to supply the pancreas and adjacent parts of the gastrointestinal tract. A **cystic artery** to the gall bladder arises from the hepatic near its termination.

**Left gastric artery** Runs directly ventral to the lesser curvature of the stomach where it divides into two branches that then ramify on the stomach wall.

**Splenic artery** This is the largest of the three branches of the celiac trunk. It divides very close to its origin

*See also Fig. S-4, p. 293.

into two rami that extend to the left to supply cranial and caudal ends of the spleen respectively.

**Superior mesenteric artery** Runs caudoventrally to the small intestine, passing beneath the transverse colon, and gives off the following branches: the posterior pancreaticoduodenal, ileocolic, middle colic, right colic, and intestinal arteries.

**Posterior pancreaticoduodenal artery** Arises from the stem artery after the latter has passed beneath the colon. It ascends along the duodenum to anastomose with a comparable branch from the gastroduodenal artery.

**Ileocolic artery** This artery runs to the right and ramifies into numerous slender branches as it approaches the cecum.

**Middle colic artery** This large vessel runs caudally to the left, sends branches to the transverse and descending colon, and finally anastomoses with the left colic artery, a branch of the inferior mesenteric.

**Right colic artery** A small branch that may be absent in some specimens.

**Intestinal arteries** The numerous terminal branches of the superior mesenteric that supply mainly the ileojejunal portion of the small intestine.

**Inferior mesenteric artery** Arises at the level of the last lumbar vertebra. It passes between the two layers of the mesocolon to the large intestine, near which it divides into an ascending and descending branch. The ascending branch, the **left colic artery,** ascends along the descending colon to anastomose with the middle colic artery. The descending branch, the **superior hemorrhoidal artery,** passes caudally along the descending colon and rectum to anastomose with the middle hemorrhoidal, a branch of the internal iliac artery.

*To see the deepest lying blood vessels, the pancreas and duodenum must be removed (Fig. 5-6). The stomach and the liver lobes should be pulled forward.*

The systemic arteries are paired retroperitoneal blood vessels. They are listed below and in Table 5-4.

**Adrenolumbar arteries** Each member of the pair runs directly laterally onto the dorsal body wall, giving off branches to the diaphragm (see below), the adrenal gland, and the adjacent musculature.

**Phrenic arteries** Each usually arises as a branch from the adrenolumbar artery and ascends to run along the abdominal surface of the diaphragm to unite with the corresponding artery of the opposite side.

**Renal arteries** Both of these vessels arise at the same level but because of the position of the kidneys the left passes caudolaterally and the right, craniolaterally. By subdivision, each artery may provide more than one branch to its kidney.

209

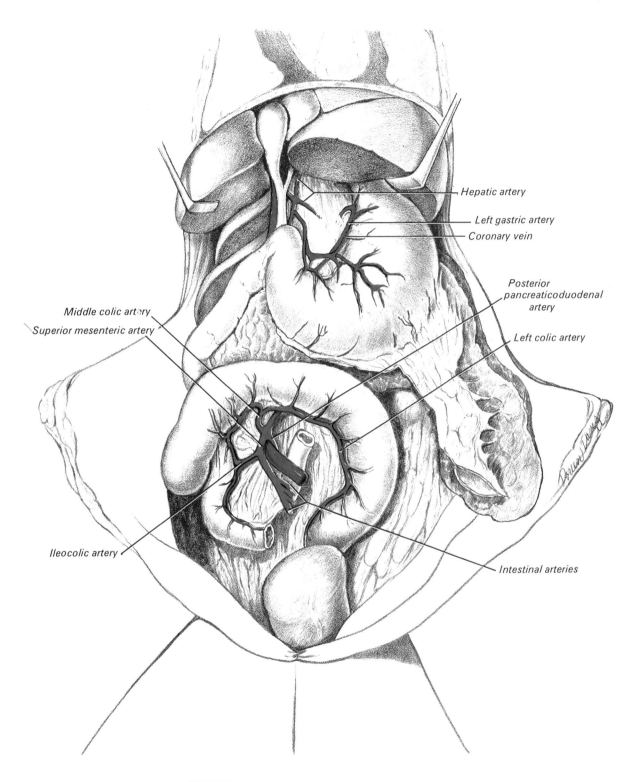

Hepatic artery

Left gastric artery

Coronary vein

Posterior pancreaticoduodenal artery

Left colic artery

Middle colic artery

Superior mesenteric artery

Ileocolic artery

Intestinal arteries

**FIGURE 5–8**
Arteries and veins of the abdomen (superficial view).

**Genital arteries** In the male these are the **internal spermatic arteries,** each of which arises from the aorta near the level of the caudal poles of the kidneys. Each vessel runs caudally to enter the inguinal canal and courses along the spermatic cord to the testis. In the female these are the **ovarian arteries,** each of which is large and convoluted and runs laterally to supply the ovary and the adjacent part of the uterus on its side.

**Iliolumbar arteries** These pass laterally. Each divides into two branches: a cranial branch that anastomoses with a branch of the adrenolumbar, and a caudal branch that runs laterally and supplies the adjacent abdominal musculature.

**Lumbar arteries** A series of about seven paired vessels. Each passes dorsally between the transverse processes of two adjacent lumbar vertebrae and divides into a spinal branch and muscular branch.

**External iliac arteries** These are the two terminal branches of the abdominal aorta at the level of the last lumbar vertebra. Each follows an obliquely caudal course and emerges from the abdomen to enter the thigh as the **femoral artery.** This artery descends along the medial surface of the thigh and soon fades into a tunnel formed by opposing borders of the adductor femoris and vastus medialis. It gives off muscular branches along its downward course and, in the lower third of the thigh, it perforates the adductor femoris to reach the undersurface and continues into the lower leg as the popliteal artery. Just before leaving the abdomen, each external iliac artery gives rise to a **deep femoral artery,** which in turn shortly divides into three branches. The first, the **vesicular artery,** passes ventromedially to the lateral surface of the bladder. The second, the **inferior epigastric artery,** runs to the dorsal surface of the rectus abdominis along which it ascends cranially to anastomose eventually with the internal mammary artery. The third branch, the **external spermatic artery,** is so called because a filament that it sends off as it crosses the adductor longus in the male passes down the spermatic cord to the testis. The major component of this branch passes to the ventromedial part of the thigh, where it terminates in the subcutaneous fat. After giving rise to the three branches, the deep femoral artery passes beneath the external iliac vein and then runs deeply between the adjacent surfaces of the pectineus and iliopsoas to provide branches to the deep muscles of the thigh.

**Internal iliac arteries** Each runs caudolaterally, giving off four vessels in the pelvis whose sites of origin are variable. Generally, the **umbilical artery** arises near the beginning of the stem artery; a short distance caudally, the **superior gluteal artery** arises.

Further caudally, the internal iliac divides into the **middle hemorrhoidal** and **inferior gluteal arteries.** The umbilical artery passes directly ventrally, close to the lateral surface of the descending colon and then lateral to the caudal end of the ureter, where it divides into branches that supply the lateral surface and neck of the urinary bladder. The superior gluteal artery passes down to the thigh muscles, many of which it supplies. The middle hemorrhoidal artery passes directly ventrally to the lateral surface of the rectum, which it supplies, and then divides. In the male the branches go to the penis and anal region; in the female the branches supply the uterus, urogenital sinus, and anal region.

## VENOUS SYSTEM

### Anterior Veins (Figs. 5-9, 5-14)

The anterior veins consist of the superior vena cava and its tributaries. The **superior vena cava** receives all the venous blood from the head, neck, thorax, and forelimbs. This large vein is formed by the union of the two large brachiocephalic veins, and in its course to the right atrium it receives the azygos, internal mammary, and right vertebral veins. The tributaries of the superior vena cava are listed below and in Table 5-5.

**Azygos vein** This unpaired vessel originates by the confluence of several small veins that drain the dorsal wall of the abdominal cavity. The azygos vein ascends along the right side of the thoracic vertebrae, receives the **intercostal, bronchial,** and **esophageal veins** along its course, and enters the right side of the superior vena cava near the root of the right lung.

**Internal mammary veins** These originate as the **superior epigastric veins** in the abdominal wall, ascend through the thoracic cavity, and unite into a common trunk that enters the superior vena cava opposite the third rib. A **sternal vein** is a tributary of the common trunk.

**Brachiocephalic veins** Formed by union of the subclavian and external jugular veins, each has as its tributary the **vertebral vein,** which tunnels through the intervertebral foramina with its corresponding artery. After it emerges from its bony canal, each vertebral is joined by a **costocervical vein.**

**Subclavian veins** Each subclavian vein is a continuation of the **axillary vein.** The axillary, the continuation of the **brachial vein,** which drains the forelimb, receives four tributaries from the chest and shoulder muscles (see Table 5-5). The tributaries

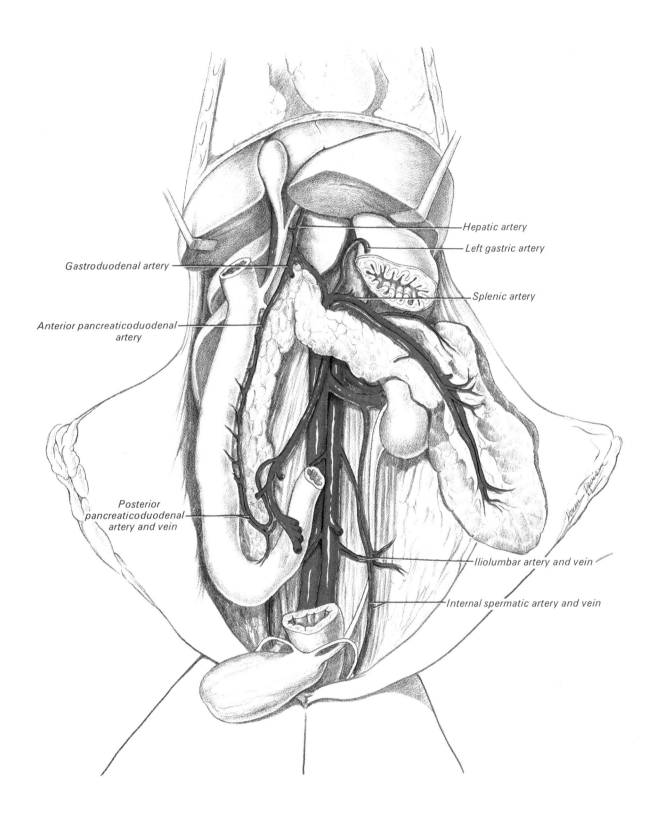

Hepatic artery

Left gastric artery

Gastroduodenal artery

Splenic artery

Anterior pancreaticoduodenal artery

Posterior pancreaticoduodenal artery and vein

Iliolumbar artery and vein

Internal spermatic artery and vein

**FIGURE 5–9**
Arteries and veins of the abdomen and pelvis (deeper view).

are the large **subscapular vein,** which arises from the underside of the scapula and joins the axillary near the base of the external jugular; the small **ventral thoracic vein,** which arises from the pectoral muscles and joins the axillary near the base of the subscapular; the **long thoracic vein,** which arises from the inner surface of the pectoralis minor; and the **thoracodorsal vein,** which arises from the latissimus dorsi. The last two veins join the axillary at variable sites.

**External jugular veins**    Each is formed by junction of the **anterior** and **posterior facial veins** near the ventral border of the sternomastoid muscles. Each passes obliquely across the sternomastoid to the triangular interval between it and the clavotrapezius where the **transverse scapular vein** is received. It continues to a site opposite the cranial end of the manubrium and receives the **internal jugular vein.** The external jugular continues on to unite with the subclavian, but before doing so it receives the thoracic duct, a major lymph-carrying channel (Fig. 5-7).

### Posterior Veins (Figs. 5-10, 5-15)*

The posterior veins consist of the hepatic portal system and the systemic veins.

*Unless the specimen is triply injected, the hepatic portal system is difficult to follow. The arrangement of this venous system may vary considerably in different specimens. The hepatic portal system may be studied by starting either with its most anterior or with its most posterior tributaries. Only if it is necessary should the mesenteries and omenta be torn and the fat and lymph glands removed to expose these tributaries. The greater omentum, with the exception of an anterior three-fourths inch segment, can be removed at this point.*

The **hepatic portal system** consists of the hepatic portal vein and its tributaries. This system collects venous blood from the digestive tract and thus conveys nutritional metabolic breakdown products to the liver. In the liver some of the metabolites may be converted for storage or undergo further change for utilization by the body elsewhere. The large **hepatic portal vein** drains the digestive organs and the spleen. It lies behind the bile duct in the margin of the lesser omentum. This vein ramifies in the liver into sinusoids that are drained by the hepatic veins into the inferior vena

*See also Fig. S-5, p. 293.

cava. The two major tributaries forming the hepatic portal vein are the gastrosplenic and the superior mesenteric veins. These and the other tributaries of the hepatic portal are listed below and in Table 5-6.

**Coronary vein**    After being formed, this small vessel passes dextrally through the lesser omentum to join the hepatic portal vein.

**Right gastroepiploic vein**    Runs dextrally along the pyloric portion of the greater curvature of the stomach. It then passes, in most specimens, behind the region of the pyloric sphincter to join the hepatic portal vein.

**Anterior pancreaticoduodenal vein**    A small vessel that may be seen best at the site where it joins the hepatic portal vein near the coronary vein.

**Gastrosplenic vein**    One of two major vessels forming the hepatic portal. It originates from four tributaries (see Table 5-6) near the lesser curvature of the stomach. It then ascends for a short distance to join the superior mesenteric vein, thereby forming the hepatic portal vein.

**Superior mesenteric vein**    This is the other major vessel forming the hepatic portal. It originates in the mesentery and receives several other tributaries (see Table 5-6) as it ascends to join the gastrosplenic near the pyloric portion of the stomach.

*Variation in the arrangement of the systemic veins, too, is found among the different specimens. In exposing these veins take care not to destroy the urogenital organs or the sympathetic ganglia. Many of the tributary veins bear the same names as the arteries that accompany them. To identify them, start from the most posterior or most anterior branch.*

The **systemic veins** consist of the inferior vena cava and its tributaries.

**Inferior vena cava**    This large vessel drains the hind limbs, pelvis, dorsal body wall, kidneys, liver, and the hepatic portal system. It is formed by the union of the common iliac veins, passes through the diaphragm, and empties into the right atrium. The tributaries of the inferior vena cava are the **phrenic, hepatic, right adrenolumbar, renal, right genital, lumbar, iliolumbar,** and **common iliac veins** (see Table 5-7). They all accompany the arteries of the same name.

**Common iliac veins**    The two major vessels that unite with the caudal vein at the level of the first sacral vertebra to form the inferior vena cava. Each is formed by the union of the large external iliac vein with the smaller internal iliac vein. There is considerable variation in both the formation of the common iliac veins and their union to form the vena cava. The two most typical variations in the latter

213

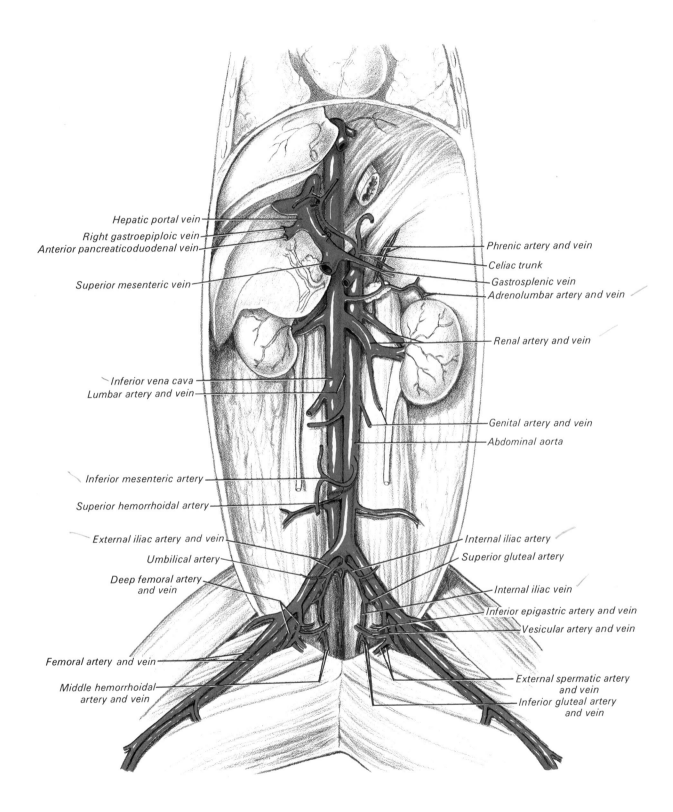

Hepatic portal vein
Right gastroepiploic vein
Anterior pancreaticoduodenal vein

Superior mesenteric vein

Phrenic artery and vein
Celiac trunk
Gastrosplenic vein
Adrenolumbar artery and vein

Renal artery and vein

Inferior vena cava
Lumbar artery and vein

Genital artery and vein
Abdominal aorta

Inferior mesenteric artery

Superior hemorrhoidal artery

External iliac artery and vein
Umbilical artery
Deep femoral artery
and vein

Internal iliac artery
Superior gluteal artery

Internal iliac vein
Inferior epigastric artery and vein
Vesicular artery and vein

Femoral artery and vein

Middle hemorrhoidal
artery and vein

External spermatic artery
and vein
Inferior gluteal artery
and vein

**FIGURE 5–10**
Arteries and veins of the abdomen, pelvis, and thigh
(deepest view).

are: (1) the two common iliacs may be longer than usual, so that the inferior vena cava is formed farther cranially; (2) there may be two separate (cardinal) veins in the abdomen instead of a single inferior vena cava, which unite in the region of the kidneys.

**External iliac vein** This vein follows the corresponding artery, collecting the blood from the posterior extremity. Its branches are named the same as those of the artery (e.g., deep femoral vein) and have in general the same distribution. The vein has, however, certain branches that the artery does not have, resulting in a somewhat different general arrangement of vessels.

**Internal iliac vein** Branches of this vein from the gluteal region follow those of the corresponding artery except that the umbilical vein from the bladder joins the middle hemorrhoidal vein. The latter runs forward from the anus alongside the rectum.

**Caudal vein** Known also as the median sacral vein, the caudal vein follows the course of the corresponding artery and in most cats enters the left common iliac. In some cats, however, it terminates in the right common iliac vein; in others it forks, one branch passing to the left common iliac, the other to the right one.

## LYMPHATIC SYSTEM (Fig. 5-11)

Besides the lymphatic vessels, other structures belonging to the lymphatic system include the thymus gland, spleen, tonsils, and bone marrow. The lymphatic vessels collect lymph from all parts of the body and carry it to the external jugular vein. This system of vessels originates as lymph capillaries that drain the tissue spaces. These capillaries combine into larger vessels that are interrupted by lymph nodes, which serve as filters of the lymph. The lymph vessels combine to form increasingly larger vessels and ultimately form lymph ducts.

### Anterior Lymph Ducts

*The main lymph ducts are difficult to find because their walls are thin. If they are located, trace them to their point of union with the external jugular veins.*

Two lymph ducts drain the anterior part of the body.

**Right lymphatic duct** This drains the lymph from the right forelimb and from the right sides of the thorax, neck, and head. It enters the right external jugular vein near that vein's junction with the subclavian vein.

**Thoracic duct** This duct drains both the posterior half of the body and the left side of the anterior half. It begins just beneath the diaphragm as a dilated channel, the **cisterna chyli,** and ascends in the thorax dorsal to the aorta. Beyond the aortic arch it passes along the left side of the esophagus to terminate in the left external jugular vein at that vein's junction with the subclavian vein. The thoracic duct, a brownish vessel, is sometimes divided into two channels.

### Posterior Lymphatic Vessels

Lymphatics in the mesentery may be traceable through the lymph nodes leading into larger vessels.

**Pancreas Aselli** This is the largest lymph node in the body of the cat. It is located in the center of the mesentery. Other mesentery lymph nodes are present in the mesentery and mesocolon.

**Lacteals** These are the tiny intestinal lymphatics in the fat deposits of the intestine. If a small section of mesentery is excised and held up to the light, the lacteals will be visible as thin streaks of fat.

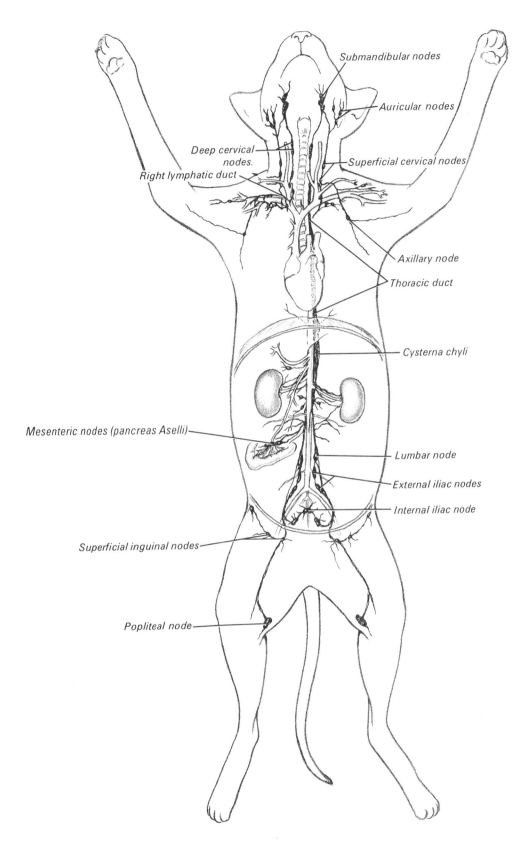

**FIGURE 5–11**
Lymphatic system.

**TABLE 5-1**
Branches of the brachiocephalic artery in the cat.

| Artery | Site of Origin | Area Supplied |
|---|---|---|
| Right common carotid | At level of third rib | |
| Inferior thyroid | Near origin of stem artery | Trachea and thymus |
| Superior thyroid | At level of thyroid cartilage | Thyroid gland |
| Muscular | At level of thyroid cartilage | Muscles of neck |
| Occipital | Near angle of jaw | Deep neck muscles |
| Internal carotid | Near angle of jaw | Brain |
| External carotid | Continuation of stem artery | |
| Lingual artery | Near origin of stem artery | Tongue |
| External maxillary | Middle of jaw | Superficial face |
| Submental | Near border of mandible | Lower jaw |
| Superior labial | Angle of mouth | Upper lip |
| Inferior labial | Angle of mouth | Lower lip |
| Posterior auricular | Angle of jaw | Muscle around ear |
| Superficial temporal | Angle of jaw | Temporal region |
| Internal maxillary | Continuation of stem artery | Deep face |
| Left common carotid (branches correspond to those of the right common carotid) | At level of third rib | |
| Right subclavian (branches correspond to those of the left subclavian; see Table 5-2) | At level of third rib | Right forelimb |

**TABLE 5-2**
Branches of the left subclavian artery in the cat

| Artery | Site of Origin or Course | Area Supplied |
|---|---|---|
| Internal mammary | Ventral surface opposite first rib | Pericardium, mediastinum, and diaphragm |
| Vertebral | Dorsal surface opposite first rib | Brain |
| Costocervical trunk | Dorsal surface opposite first rib | |
| Superior intercostal | Near origin of stem artery | Anterior intercostal muscles |
| Transverse cervical | Laterally | Superficial back muscles |
| Deep cervical | Dorsally | Deep neck muscles |
| Thyrocervical trunk | Near first rib | Neck and shoulder muscles |
| Ascending cervical | Terminal branch | Neck muscles |
| Transverse scapular | Terminal branch | Shoulder muscles |
| Axillary | Continuation of stem artery | |
| Ventral thoracic | Lateral to first rib | Pectoral muscles |
| Long thoracic | Lateral to ventral thoracic | Chest muscles |
| Subscapular | Lateral to long thoracic | Shoulder muscles |
| Brachial | Continuation of axillary | Arm and forearm muscles |

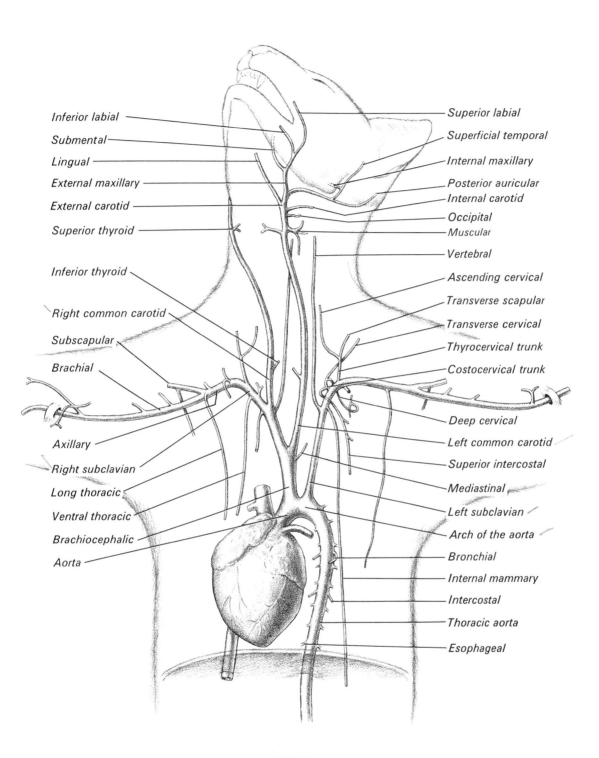

Inferior labial

Submental

Lingual

External maxillary

External carotid

Superior thyroid

Inferior thyroid

Right common carotid

Subscapular

Brachial

Axillary

Right subclavian

Long thoracic

Ventral thoracic

Brachiocephalic

Aorta

Superior labial

Superficial temporal

Internal maxillary

Posterior auricular

Internal carotid

Occipital

Muscular

Vertebral

Ascending cervical

Transverse scapular

Transverse cervical

Thyrocervical trunk

Costocervical trunk

Deep cervical

Left common carotid

Superior intercostal

Mediastinal

Left subclavian

Arch of the aorta

Bronchial

Internal mammary

Intercostal

Thoracic aorta

Esophageal

**FIGURE 5–12**
Anterior arteries.

**TABLE 5-3**
Branches of the abdominal aorta in the cat.

| Artery | Site of Origin or Course | Area Supplied |
|---|---|---|
| Celiac trunk | Just below diaphragm | |
|   Hepatic | Ascends | Liver |
|     Gastroduodenal | In lesser omentum | |
|       Pyloric | Descends | Pylorus |
|         Anterior pancreaticoduodenal | Descends | Duodenum and pancreas |
|         Right gastroepiploic | Along greater curvature | Stomach |
|   Cystic | Near liver | Gall bladder |
|   Left gastric | Along lesser curvature | Stomach |
|   Splenic | To left across abdominal cavity | Spleen, pancreas, and stomach |
| Superior mesenteric artery | Just posterior to celiac | |
|   Posterior pancreaticoduodenal | Ascends | Duodenum and pancreas |
|   Ileocolic | Descends | Cecum |
|   Middle colic | Descends | Transverse and descending colon |
|   Right colic | Descends | Ascending colon |
|   Intestinals | Through mesentery | Small intestine |
| Inferior mesenteric | Near pelvis | |
|   Left colic | Ascends | Descending colon |
|   Superior hemorrhoidal | Descends | Rectum |

**TABLE 5-4**
Branches of the systemic arteries in the cat.

| Artery | Site of Origin | Area Supplied |
|---|---|---|
| Adrenolumbar | Just posterior to superior mesenteric | Adrenal gland |
| Phrenic | Variable | Diaphragm |
| Renal | Posterior to adrenolumbar | Kidney |
| Genital | Posterior to renal | Testis or ovary |
| Iliolumbar | Near end of aorta | Adjacent muscles |
| Lumbar | At intervals along abdominal aorta | Adjacent muscles |
| External iliac | Near anterior end of pelvis | |
|   Deep femoral | Near origin of stem artery | Deep muscles of thigh |
|     Vesicular | Near origin of stem artery | Bladder |
|     External spermatic | Near origin of stem artery | Testis |
|     Inferior epigastric | Near origin of stem artery | Superior epigastric artery* |
|   Femoral | Continuation of stem artery | Hind limb |
| Internal iliac | Posterior to external iliac artery | |
|   Umbilical | Near origin of stem artery | Urinary bladder |
|   Superior gluteal | Near origin of stem artery | Thigh muscles |
|   Middle hemorrhoidal | One of the terminal branches | Rectum and genital organs |
|     Uterine | Near uterus | Uterus |
|   Inferior gluteal | Other terminal branch | Thigh muscles |

*Anastomoses with superior epigastric artery, an abdominal continuation of the internal mammary artery.

219

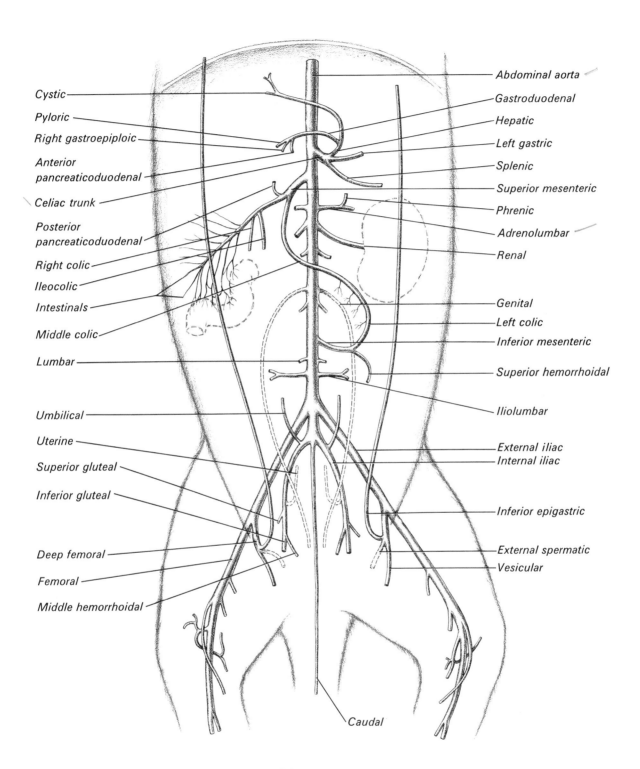

Cystic

Pyloric

Right gastroepiploic

Anterior
pancreaticoduodenal

Celiac trunk

Posterior
pancreaticoduodenal

Right colic

Ileocolic

Intestinals

Middle colic

Lumbar

Umbilical

Uterine

Superior gluteal

Inferior gluteal

Deep femoral

Femoral

Middle hemorrhoidal

Abdominal aorta

Gastroduodenal

Hepatic

Left gastric

Splenic

Superior mesenteric

Phrenic

Adrenolumbar

Renal

Genital

Left colic

Inferior mesenteric

Superior hemorrhoidal

Iliolumbar

External iliac

Internal iliac

Inferior epigastric

External spermatic

Vesicular

Caudal

**FIGURE 5–13**
Posterior arteries.

219
219

**TABLE 5-5**
Tributaries of the superior vena cava in the cat.

| Vein | Region Drained |
|---|---|
| Azygos | Thoracic and abdominal walls |
|   Intercostal | Rib interspaces |
|   Esophageal | Esophagus |
|   Bronchial | Lungs |
| Internal mammary | Ventral thoracic wall |
|   Sternal | Mediastinum |
|   Superior epigastric | Ventral abdominal wall |
| Brachiocephalic | Head, neck, and forelimb |
|   Vertebral | Brain |
|     Costocervical | Neck muscles |
|   Subclavian | Shoulder and forelimb |
|     Axillary | Axilla and forelimb |
|       Subscapular | Shoulder muscles |
|         Ventral thoracic | Pectoral muscles |
|       Thoracodorsal | Latissimus dorsi |
|       Long thoracic | Mammary region |
|       Brachial | Deep muscles of arm |
| External jugular | Face, neck, and forelimb |
|   Transverse scapular | Anterior shoulder |
|     Cephalic | Superficial arm |
|   Transverse jugular | Neck |
|   Anterior facial | Face and jaws |
|   Posterior facial | Ear region |
|     Posterior auricular | Postauricular |
|     Anterior auricular | Preauricular |
|     Superficial temporal | Temporal region |
| Internal jugular | Brain and deep neck |

221

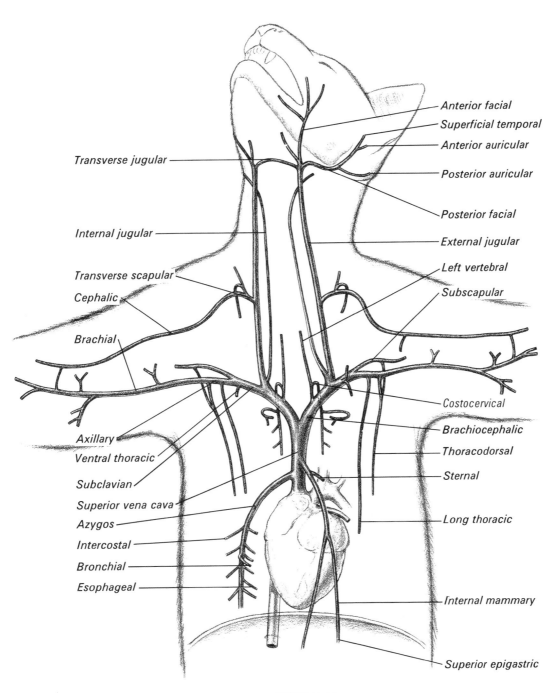

Anterior facial
Superficial temporal
Anterior auricular
Posterior auricular
Posterior facial
External jugular
Left vertebral
Subscapular
Costocervical
Brachiocephalic
Thoracodorsal
Sternal
Long thoracic
Internal mammary
Superior epigastric

Transverse jugular
Internal jugular
Transverse scapular
Cephalic
Brachial
Axillary
Ventral thoracic
Subclavian
Superior vena cava
Azygos
Intercostal
Bronchial
Esophageal

**FIGURE 5–14**
Anterior veins.

**TABLE 5-6**
Tributaries of the hepatic portal vein in the cat.

| Vein | Region Drained |
| --- | --- |
| Coronary | Lesser curvature of stomach |
| Right gastroepiploic | Pylorus and stomach |
| Anterior pancreatico- duodenal | Pancreas and duodenum |
| Gastrosplenic | Stomach and spleen |
|   Pancreatic | Pancreas |
|   Middle gastroepiploic | Stomach and greater omentum |
|     Posterior splenic | Caudal end of spleen |
|     Anterior splenic | Cranial end of spleen |
|       Left gastroepiploic | Stomach and greater omentum |
| Superior mesenteric | |
|   Posterior pancreatico- duodenal | Pancreas and duodenum |
|   Inferior mesenteric | Large intestine |
|   Ileocolic | Cecum, adjacent ileum, and colon |
|     Intestinals | Small intestine |

**TABLE 5-7**
Tributaries of the inferior vena cava in the cat.

| Vein | Region Drained |
| --- | --- |
| Phrenic | Diaphragm |
| Hepatic | Liver (and hepatic portal system) |
| Right adrenolumbar | Right adrenal and body wall |
| Right renal | Right kidney |
| Left renal | Left kidney |
|   Left adrenolumbar | Left adrenal and body wall |
|   Left genital | Ovary or testis |
| Right genital | Ovary or testis |
| Lumbar | Body wall |
| Iliolumbar | Body wall |
| Common iliac | Pelvis and hind limb |
|   External iliac | Hind limb |
|     Deep femoral | Deep part of thigh |
|     Femoral | Superficial part of thigh |
|     Inferior epigastric | Abdominal wall |
|   Internal iliac | Pelvis |
|     Umbilical | Bladder |
|     Middle hemorrhoidal | Rectum |
|   Caudal | Tail |

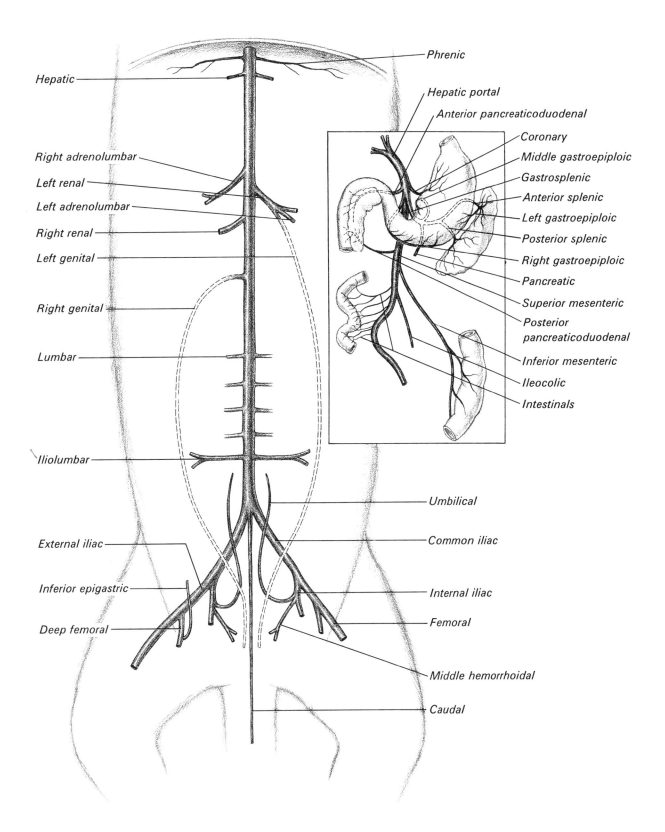

Hepatic

Phrenic

Hepatic portal

Anterior pancreaticoduodenal

Coronary

Middle gastroepiploic

Gastrosplenic

Right adrenolumbar

Anterior splenic

Left renal

Left gastroepiploic

Left adrenolumbar

Posterior splenic

Right renal

Right gastroepiploic

Left genital

Pancreatic

Superior mesenteric

Right genital

Posterior
pancreaticoduodenal

Inferior mesenteric

Lumbar

Ileocolic

Intestinals

Iliolumbar

Umbilical

Common iliac

External iliac

Inferior epigastric

Internal iliac

Deep femoral

Femoral

Middle hemorrhoidal

Caudal

**FIGURE 5–15**

Posterior veins and hepatic portal system (inset).

# Anatomy of the Cat

Since the excretory and reproductive systems in the cat are not as closely related anatomically as they are in lower forms, the two are described separately.

## EXCRETORY SYSTEM

This system consists of the paired kidneys and ureters and an unpaired bladder and urethra.

*Expose a kidney and ureter by removing the peritoneum and fat that envelop it. After studying the shape and adjacent structures of the kidney, cut it horizontally into dorsal and ventral halves and identify the prominent features listed below.*

### Kidneys (Figs. 6-1, 6-2, 6-3)

These are bean-shaped retroperitoneal organs located in the region of the fourth lumbar vertebra.

**Hilus**   The concavity on the medial side of the kidney. The renal artery and vein and the ureter join the kidney at the hilus.

**Renal cortex**   Outer layer of kidney tissue.

**Renal medulla**   Central part of the substance of the kidney.

**Renal papilla**   Cone-shaped projection of the medulla, whose apex is directed toward the hilus.

**Renal sinus**   The cavity in the hilus.

**Renal pelvis**   The expanded portion of the ureter located within the renal sinus.

**Adrenal glands**   These are small ovoid bodies that serve an endocrine function (see Exercise 7, Endocrine System and Sense Organs). They are located slightly craniomedially to each kidney.

### Ureters (Figs. 6-2, 6-3)

These are tubular retroperitoneal structures extending from the kidneys to the urinary bladder. Each ureter begins as the expanded renal pelvis, emerges at the hilus as a narrow duct, and terminates caudally by opening into the neck of the bladder.

### Urinary Bladder (Figs. 6-2, 6-3)

The urinary bladder is a musculomembranous, pear-shaped sac just anterior to the pubic symphysis. Note the following features and structures.

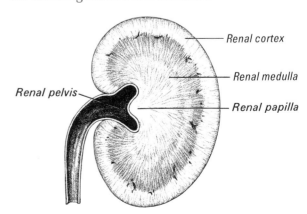

**FIGURE 6–1**
Kidney (vertically sectioned).

**Fundus**    Expanded larger part of the bladder.

**Neck**    Constricted caudal part of the bladder. It is continuous with the urethra (see the next section).

**Median suspensory ligament**    This remnant of the ventral mesentery extends from the bladder to the linea alba.

**Lateral ligaments**    Supporting structures going from each side of the bladder to the dorsal body wall.

## MALE REPRODUCTIVE SYSTEM (Fig. 6-2)

This system consists of the paired testes, epididymides, ductus deferentes, and bulbourethral glands as well as the prostate gland and penis.

*To expose one of the testes, incise the pouch in which it lies and reflect the skin from it. Then remove the connective tissue that surrounds the testis. Examine the other contents of the pouch in detail, identifying the structures noted below. Follow the duct leading from the pouch as it passes into the abdominal cavity.*

### Testes

These are the paired, oval, male gonads. At the time of birth the testes descend from the abdominal cavity into the scrotum. As they do so, a double layer of peritoneum is pushed down before them. It forms a flask-shaped sac whose round blind end lies within the scrotum, and the constricted portion forms a channel for the ductus deferens (see below), spermatic nerve, and blood vessels.

**Scrotum**    The external pouch covered by skin and lined internally by peritoneum that encloses each testis. It is divided into two chambers by a median septum.

**Tunica vaginalis**    The glistening peritoneal covering of the testis.

**Tunica albuginea**    The dense fibrous coat of the testis proper.

### Epididymides

Each epididymis lies on the dorsal surface of its testis. It consists of an extensive coiled tube that is connected to the testis by a number of delicate efferent ductules. The enlarged cranial end of the epididymis is the **head;** the narrow middle part is the **body;** and the expanded caudal portion is the **tail.**

### Ductus Deferentes

Each ductus deferens is a continuation of the tail of the epididymis. It ascends in coiled fashion in the **spermatic cord** from the scrotum to the peritoneal cavity through the **inguinal canal,** a short passageway through the abdominal musculature. In the abdomen the ductus deferens separates from the cord, turns medially to arch around the ureter, and doubles back, dorsal to the bladder, to converge caudally toward the duct from the opposite side. Both ducts then pierce through the prostate gland to open into the part of the urethra enclosed by this gland.

### Urethra

*Split the symphysis pubis and spread the hind limbs apart. For better exposure, it may be advisable to remove some of the muscles and parts of the pubis and ischium on one side.*

The urethra is a tube that extends from the neck of the bladder as far as the tip of the penis. The urethra has three parts.

**Prostatic urethra**    The part of the urethra that is surrounded by the prostate gland.

**Membranous urethra**    The short segment of the urethra between the prostate and the penis.

**Spongy urethra**    The part of the urethra passing through the penis.

### Prostate Gland

The prostate is a bilobed gland located at the junction of the ductus deferentes and the neck of the bladder. Tiny ducts arising within the glands open into the lumen of the prostatic urethra.

### Bulbourethral Glands

These two glands are located on either side of the membranous urethra, just dorsal to the root of the penis. Each opens into the urethra by means of a single fine duct.

### Penis

The penis is the cylindrical external copulatory organ.

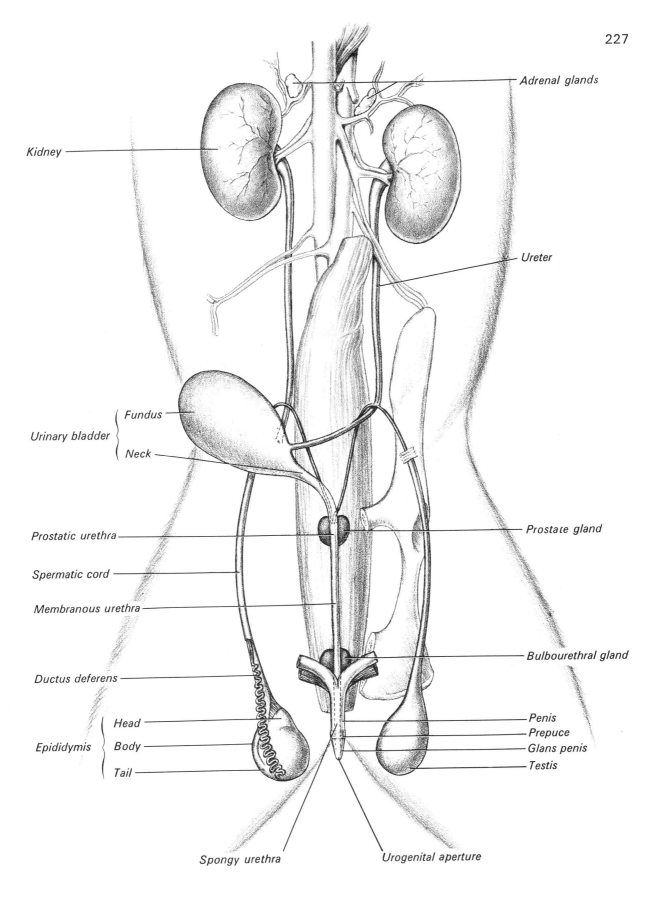

Adrenal glands

Kidney

Ureter

Urinary bladder {
Fundus

Neck

Prostatic urethra

Spermatic cord

Membranous urethra

Ductus deferens

Epididymis {
Head

Body

Tail

Prostate gland

Bulbourethral gland

Penis

Prepuce

Glans penis

Testis

Spongy urethra

Urogenital aperture

**FIGURE 6–2**
Male urogenital system.

*Remove the skin covering the penis and trace this organ back to its attachment to the pubic symphysis. Then make a cross section through the middle of the penis.*

**Glans penis**    The enlargement at the distal end of the penis.

**Prepuce**    The loose skin covering the glans penis.

**Urogenital aperture**    The opening at the end of the penis.

**Corpora cavernosum**    Two vascular bodies of erectile tissue lying side by side that form the dorsal part of the penis. The two corpora diverge laterally at their proximal ends. These basal parts are referred to as the **crura.**

**Corpus spongiosum**    Erectile tissue located on the ventral surface of the penis in the groove between the corpora cavernosa. The spongy urethra passes through this mass of vascular tissue.

**Os penis**    A small bone embedded in the posterior end of the penis. It helps stiffen this part of the penis.

## FEMALE REPRODUCTIVE SYSTEM (Fig. 6-3)

Cats are prolific animals, having from three to 12 young at a time; the length of a gestation period is about 63 days. They have three to four breeding seasons a year and are capable of reproducing from age 1 to 9. Their probable life span is 12 to 18 years. Like other eutherian mammals, cats are viviparous, that is, they give birth to live young. Their embryos develop within the uterus, surrounded by the various extra-embryonic membranes that are characteristic of amniotes (see discussion below).

The mother usually cares for the young during their early helpless condition and for a short time thereafter. At birth the mother's mammary glands are fully developed and functional. There are four or five nipples on each side, ensuring nursing for an average litter.

The female reproductive system consists of the ovaries, uterine tubes, uterus, and vagina.

### Ovaries

The ovaries are a pair of small oval organs located slightly posterior to the kidneys. Each is attached to the broad ligament of the uterus (see definition) by a mesentery, the **mesovarium.**

### Uterine Tubes

The uterine tubes, known also as the oviducts, are coiled ducts that convey the ovulated ova to the uterus. Each tube has an expanded free end, the ostium, which receives the ova. The opposite end of each tube leads into a horn of the uterus.

### Uterus

**Horns**    The enlarged posterior continuations of the uterine tubes.

*To study the body of the uterus it is necessary to expose the vagina and the more posterior parts of the system. In order to do this, first identify the external urogenital opening just ventral to the anus and trim away the hair around it. Then locate the position of the symphysis pubis beneath the muscles and with a scalpel split the entire symphysis. Spread the legs apart to expose the genital organs. After the various parts have been identified, the walls of the urogenital sinus, vagina, and uterus may be slit open.*

**Body**    The posterior median part of the uterus located between the bladder and rectum.

**Broad ligament**    Fold of peritoneum that attaches the horns and body of the uterus to the body wall.

**Round ligament**    The fibrous band that extends from the dorsal body wall to the middle of each uterine horn.

*If a specimen is pregnant, its uterine horns will be found to be enlarged and exhibit a series of swellings, each of which contains an embryo. The size of each swelling reflects the extent of development of the embryo it contains. Remove one of the swollen segments by sectioning a horn transversely at two sequential constricted zones. Open the excised segment by a longitudinal incision. In addition to the embryo identify the following.*

**Chorion**    The outermost membrane of the two that enclose the embryo.

**Amnion**    The thin, transparent membrane immediately around the embryo. The developing embryo lies suspended by the umbilical cord in this fluid-filled sac and is thus protected from pressure and the impact of physical shocks. The amniotic fluid also prevents adherence of the amnion to the fetus and thus permits changes in fetal position.

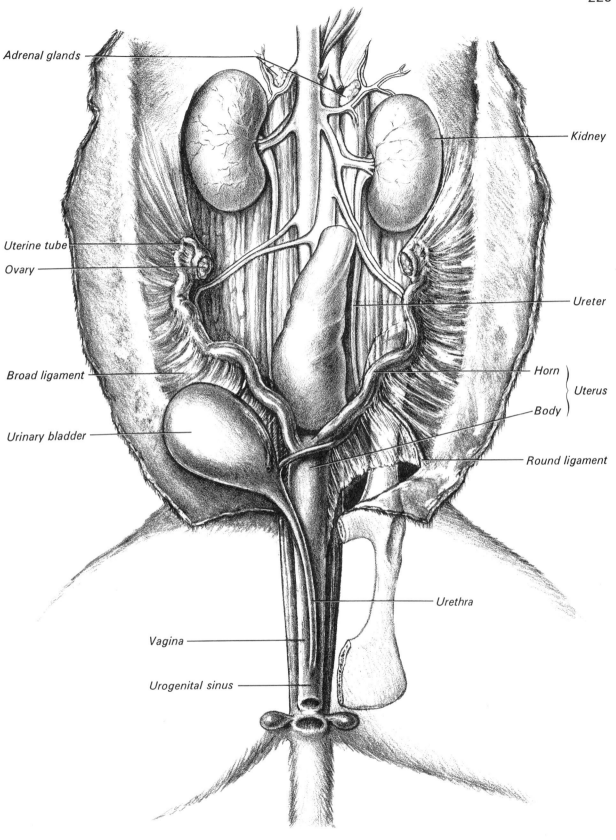

Adrenal glands

Kidney

Uterine tube

Ovary

Ureter

Broad ligament

Horn

Uterus

Body

Urinary bladder

Round ligament

Urethra

Vagina

Urogenital sinus

**FIGURE 6–3**
Female urogenital system.

**Placenta**  The broad, circular, thickened, vascular band of tissue encircling the chorion, about midway between the embryo's head and tail. The placenta serves as an organ for the transfer of nutrients, inorganic salts, and oxygen from the mother to the fetus and waste products from the fetus to the mother.

**Umbilical cord**  An elongated, cylindrical structure that extends from the placenta to the abdomen of the embryo. It carries the blood between mother and fetus through a pair of arteries and a single large vein. The umbilical arteries are branches of the internal iliacs of the embryo; the umbilical vein extends to the liver.

## Vagina

The vagina is the chamber extending from the body of the uterus to the urogenital sinus (see definition). The vagina lies in back of the urethra.

**Cervix**  The distal end of the body of the uterus that projects somewhat into the vagina.

**Os uteri**  The opening of the cervical portion of the uterus.

**Urogenital sinus**  The common chamber, known also as the vestibule, into which the vagina and urethra open. The external aperture of the sinus is the urogenital opening.

**Vulva**  The name given to all the external genital organs. These include the labia majora, the longitudinal folds that bound the urogenital opening, and the clitoris, the small prominence consisting of erectile tissue on the ventral wall of the urogenital sinus near its external aperture. The clitoris is the homolog of the penis.

# Endocrine System and Sense Organs

# Anatomy of the Cat

## ENDOCRINE SYSTEM

The endocrine system consists of a number of ductless glands whose secretions, hormones, are poured directly into the bloodstream. There are five known endocrine glands. Of these, only the thyroid and the adrenal glands need to be identified in the specimen.

### Thyroid Gland

The thyroid gland is located beneath the larynx on either side of the trachea. The gland consists of two **lobes,** which in some specimens are connected across the ventral surface of the trachea by a narrow isthmus. This gland secretes thyroxin, a hormone that accelerates the rate of cellular metabolism in the body.

### Parathyroid Glands

The parathyroids are four minute bodies embedded in the dorsal part of the thyroid gland. Because of their size, they are not evident during routine dissection. They secrete parathormone, a substance that regulates calcium metabolism in the body.

### Adrenal Gland

The adrenals consist of a pair of bean-shaped organs, which, because of their location just cranial to the kidneys, are known also as the suprarenal glands. Each gland consists of an outer cortex that secretes steroid

hormones and an inner medulla that secretes epinephrine and norepinephrine. The cortical steroids control water and electrolyte balance as well as carbohydrate metabolism. The medullary secretions supplement the action of the sympathetic division of the autonomic nervous system.

### Hypophysis (Pituitary Gland)

The hypophysis is a small organ suspended from the undersurface of the brain in the region of the diencephalon. It is divided into two lobes, each of which elaborates one or more hormones.

### Islets of the Pancreas

Although the pancreas is predominantly an exocrine gland, it contains isolated irregular nests of cells, the **islets of Langerhans,** which serve an endocrine function. These cells secrete insulin and glucagon, hormones that regulate the concentration of sugar in the blood.

### Other Hormone-Secreting Organs

Hormones are secreted by the testes and ovaries. They affect the functional activities of the accessory sex organs. Some of the organs of the gastrointestinal tract also produce hormones.

## SENSE ORGANS

The sense organs may be considered to be of two types. One is a group of widely distributed receptors of general sensation. These receptors are not routinely dissected in the student laboratory, but they are important because they receive stimuli for touch, pressure, temperature, and pain. The second group consists of the special sense organs. These specialized receptors are only in the head and consist of the olfactory and gustatory organs, the eye and the ear.

### Olfactory Organ

The sense organ responsible for smell reception consists of that part of the nasal mucous membrane, the **olfactory mucosa,** that lines the dorsocaudal part of each nasal cavity. The receptors for smell in this membrane are olfactory hair cells. Each hair cell is a bipolar neuron having a peripheral process that terminates as a series of fine hairs on the free surface of the epithelium. In addition, the hair cell has a central process, or axon, that emerges from the undersurface of the epithelium as a fiber of the olfactory nerve. This process then passes to the olfactory bulb where it terminates in a synapse.

### Gustatory Organ

The taste receptors are organized into specialized groups of cells, the **taste buds.** Most taste buds are located on the fungiform and circumvallate papillae of the tongue. From the front, or oral, part of the tongue, sensations of taste are transmitted via a branch of the facial nerve (N. VII); from the back, or pharyngeal, part, sensations are transmitted centrally via the glossopharyngeal nerve (N. IX).

### Cat Eye

The eye lies within its protective bony orbit. It is associated with several accessory structures.

**Eyelids**   These consist of an upper and lower extension of the integument. They protect the external surface of the eye.

**Conjunctiva**   The thin transparent mucous membrane that covers the underside of each eyelid and is reflected onto the surface of the eyeball.

**Nictitating membrane**   A fold located in the inner corner of the eye that can be expanded over the entire exposed surface of the eyeball. It can be seen clearly if the lateral corner of the eye is cut and the lids pulled apart.

**Harderian gland**   This small gland is located on the medial surface of the base of the nictitating membrane. The gland can be seen best after the membrane is removed. It supplements the secretions of the lacrimal gland.

*Free the eyeball from the bony orbital rim and push it anteriorly.*

**Lacrimal gland**   The tear gland is located at the outer angle of the eye. The fluid runs down toward the inner angle of the eye where it drains into lacrimal canals that carry the secretion down to the nasal passages.

*Push the eyeball dorsally and remove the fat and connective tissue from the floor of the orbit.*

**Infraorbital gland**   A small reddish salivary gland that lies on the floor of the orbit in line with, and just posterior to, the upper row of teeth.

*After identification of the aforementioned structures, remove with great care the skin of the eyelids and parts of the zygomatic arch below the eye until the eye is fully exposed within its orbit. As necessary, carefully remove fat and connective tissue to expose the structures being sought.*

**Extrinsic eye muscles** (Fig. 7-1)   Seven muscles control the movements of the eyeball. Four recti originate on the bone around the optic foramen and insert on the sclera just behind the midline, or equator, of the eyeball. They are known as the **superior, inferior, lateral,** and **medial recti** muscles, according to their position relative to the eyeball. Both the **superior oblique** and **inferior oblique** muscles are evident only if the lateral and dorsal walls of the orbit are removed. A **retractor** muscle also originates on the bony margin of the optic foramen. It divides into four heads that are inserted into the sclera around the entrance of the optic nerve. This muscle is partially hidden by the recti muscles.

*Remove the eyeball by pulling it forward and cutting the optic nerve at the back of the eyeball. Some of the muscles will also require severing. After the eye has been removed, cut it into two equal parts with an incision that passes through the equatorial plate. The internal structures of the cat eye are difficult to observe. They frequently are studied in larger specimens such as from a sheep or ox (see p. 234).*

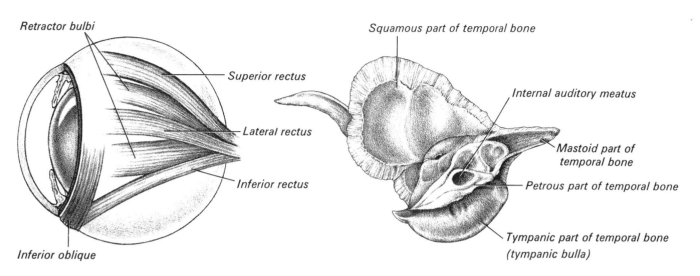

FIGURE 7–1
Extrinsic eye muscles.

FIGURE 7–2
Temporal bone (medial surface).

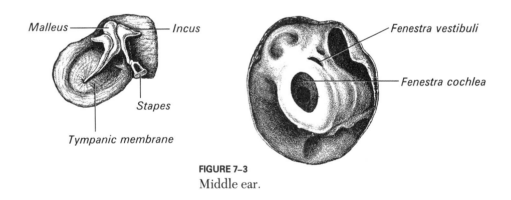

FIGURE 7–3
Middle ear.

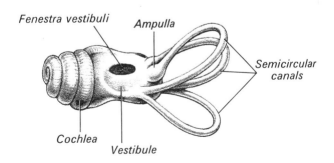

FIGURE 7–4
Bony labyrinth.

**Eyeball**    The structure of the eyeball is similar to that found in lower vertebrates. A discussion of the major details of the vertebrate eye is given in the series Anatomy of the Dogfish Shark (Exercise 7).

### Cat Ear (Figs. 7-2, 7-3, 7-4)

The ear contains the sense organs for both hearing and equilibrium. It consists of three parts: the external, middle, and internal ear.

**External ear**    This consists of the **pinna**, a sound-funneling structure, which is supported by cartilage and covered by skin. Its position is adjusted by extrinsic auricular muscles. Extending internally from the base of the pinna is the **external auditory meatus.** This canal terminates at the **tympanic membrane,** a delicate sheet that separates the external ear from the middle ear.

*Dissection of the middle ear is quite difficult and can more readily be carried out if that portion of the skull in which it lies is separated from the rest of the cranium. To do this, first remove all soft tissues from the area around the tympanic bulla (the tympanic part of the temporal bone). The cartilaginous external auditory meatus can also be removed. Then, with the aid of bone shears, separate the bulla and adjacent medial (petrous) portion of the temporal bone from the skull. Next, remove the medial wall of the tympanic bulla. The cavity within the bulla is separated into a larger lateral and a smaller medial chamber by a vertical bony wall. Carefully break away this vertical plate to expose the lateral chamber and reveal the inner surface of the tympanic membrane and the small middle-ear bone (malleus) attached to it. Then, still using the bone shears, fracture the thick bony connection between the caudal end of the petrous bone and the caudal part of the bony ring surrounding the external auditory meatus. These two bony parts may then be separated manually so that the tympanic membrane and two ossicles are on one side and the petrous bone and third ossicle (stapes) are on the other.*

**Middle ear**    The two-chamber cavity within the tympanic bulla is also known as the tympanic cavity. Three small bones, **or ossicles,** are located in the lateral of the two chambers and transmit sound from the external ear to the internal ear. The largest ossicle, the **malleus,** has a process attached to the inner surface of the tympanic membrane. The head of the malleus articulates with the body of the **incus,** and the incus, in turn, articulates with the **stapes.** The base plate of the stapes fits over an oval opening

in the lateral wall of the tympanic cavity, the **fenestra vestibuli,** which leads into the internal ear. A round opening, the **fenestra cochlea,** is located a little below the fenestra vestibuli.

*Because of its small size, the internal ear is extremely difficult to dissect. However, an attempt to do so can be made by carefully chipping away the bone medial to the tympanic cavity.*

**Internal ear**    This consists of two labyrinths, a membranous one and a surrounding **bony labyrinth.** The latter is made up of a small central chamber, the **vestibule** (containing the utriculus and sacculus), the three **semicircular canals,** and the **cochlea.** The coiled cochlea contains hair cells that are the actual sensory receptors for sound. The loop-shaped semicircular canals enclose ducts that at their expanded ends, or **ampullae,** contain sensory receptor cells responsible for the maintenance of equilibrium.

## SHEEP (OX) EYE

### External View (Figs. 7-5, 7-6)

*The description of the third eyelid is applicable only if the specimen being examined has eyelids. After identifying the third eyelid, cut it horizontally and observe its inner supporting cartilage. By identifying the third eyelid you will simultaneously determine if you have a right or left eye.*

**Third eyelid**    The thick, light-colored fold located in the medial corner of the eye between the upper and lower eyelids. It consists of the nictitating membrane with its embedded cartilaginous framework. In the ox, this membrane is very well developed and can sweep across the entire front of the eye, removing foreign materials from the surface of the cornea and lubricating it. Movement of the third eyelid is brought about when the true eyelids are squeezed, thus placing pressure on the orbital fat supporting the eyeball. The Y-shaped cartilaginous plate embedded in the third lid serves to shape the lid to the eyeball.

**Lacrimal gland**    The tear organ may be identified on the upper lateral aspect of the eyeball.

**Eye muscles**    The cut stumps of various extrinsic eye muscles are quite evident. The muscles fall into two categories: straight (recti) and oblique muscles.

**Optic nerve**    The circular, grayish stump emerging posteriorly from the eyeball.

**Eyeball**    Roughly spherical in shape, it consists of three concentric layers: fibrous, vascular, and ner-

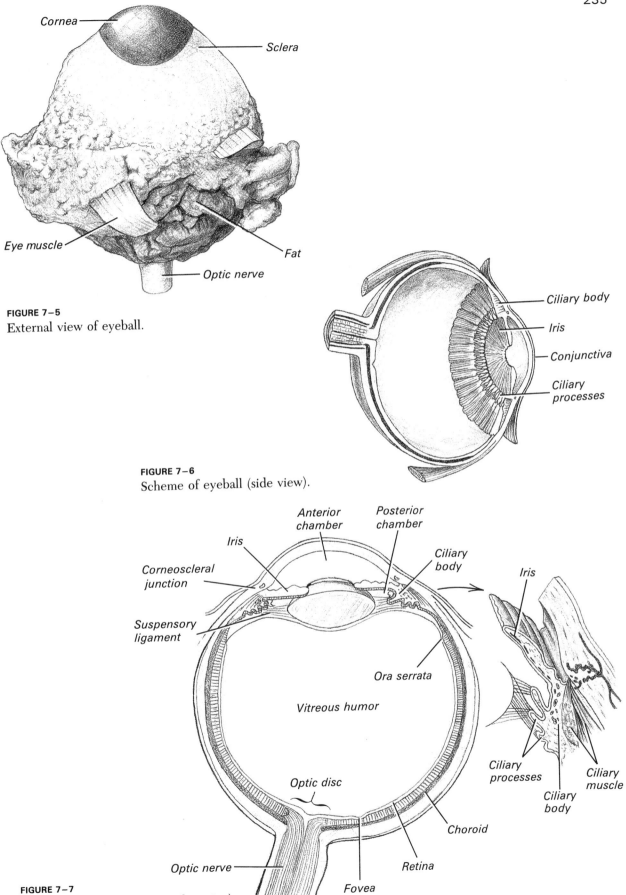

235

**FIGURE 7-5**
External view of eyeball.

**FIGURE 7-6**
Scheme of eyeball (side view).

**FIGURE 7-7**
Scheme of eyeball (horizontal section).

vous. Each of these layers is subdivided into a number of parts.

**Fibrous coat** This is the external tunic of the eyeball. It consists of two parts: the smaller cornea in front and the much larger sclera toward the back.

**Cornea** This is the avascular protruding anterior fifth of the eye. While in life this segment is transparent, after preservation with a fixative it becomes somewhat opaque. The cornea is continuous posteriorly with the bulk of the outer fibrous coat, the nontransparent sclera.

**Conjunctiva** Forms the lining of the eyelids (palpebral conjunctiva) and cornea (bulbar conjunctiva). On the inner lining of the eyelids it is evident only where remnants of the lids are present. This layer is reflected onto the surface of the cornea at the **corneoscleral junction**. The groove formed by the reflection of the conjunctiva from the eyelid to the cornea is called the **fornix**.

*Clean the exterior of the eyeball by holding the stump of the optic nerve with forceps while using sharp scissors to trim off adherent fat, muscles, and blood vessels. Retain the optic nerve stump, around whose periphery a muscle cone is evident, which should be removed after identification. Anteriorly the conjunctiva should be severed at the corneoscleral junction and discarded.*

**Retractor bulbi** The group of four distinct bundles of muscles that surround the optic nerve and insert on the sclera at the posterior aspect of the eyeball. In the orbit, these muscles interdigitate with the straight (recti) muscles. The presence of a retractor bulbi is characteristic of all mammals except man and monkeys and is especially well developed in the ox. The retractor bulbi acts to protect the eye by pulling it back into the orbit; hence its name.

**Sclera** This outer eyeball coat is exposed after removal of the retractor bulbi and its underlying connective tissue. The sclera is a grayish, shiny, fibrous layer that is continuous with the cornea and envelops the posterior four-fifths of the surface of the eyeball. The rostral oval end of the sclera is beveled and the cornea is attached to it in a manner similar to that of the crystal to a watch. The optic nerve penetrates the sclera at a point ventrolateral to the caudal pole. Bundles of nerve fibers pass through small holes forming a sieve-like area. The sclera is attached to the underlying choroid by a fine network of muscles as well as by various blood vessels and nerves.

## Horizontal Section (Fig. 7-7)

*Using a new scalpel blade, transect the entire eyeball across its equator into anterior and posterior halves. Remove the vitreous humor carefully with the scalpel handle.*

**Vitreous (body) humor** The jelly-like material that fills the posterior half of the eyeball.

**Nervous coat** This tunic is composed of three parts: optical part (retina), ciliary part (covering the inner surface of the ciliary body), and iridial part (covering the caudal surface of the iris). Only the retina is light sensitive. The ciliary and iridial parts lie rostral to the ora serrata.

**Retina** A brownish, delicate membrane that usually falls into folds because, with the removal of the vitreous humor, the retina separates from the underlying middle layer (choroid) except at the optic disc.

*With forceps, carefully lift off the retina, noting the site at which it is firmly attached, namely the optic disc. Then remove the remnants of the retina, except around the disc, to expose the choroid.*

**Optic disc** The site at which the optic nerve fibers leave the eyeball. It is the blind spot in the sensory receptive layer of the retina.

**Vascular coat** The middle layer of the eyeball, sandwiched between the outer fibrous and inner nervous tunics. The vascular coat is divisible into three parts: choroid, ciliary body, and iris.

**Choroid** The choroid is the largest part of the vascular coat. Thin and heavily pigmented, and thus dark brown or black in appearance, the choroid should be separated from the sclera, to which it is attached in a number of places, especially strongly around the optic disc. The attachment points are sites at which veins leave the sclera and bind it to the choroid. Rostrally the choroid extends as far as the light-sensitive layer of the retina, where it continues forward as the ciliary body.

**Tapetum lucidum** A wide oval or triangular iridescent area of the choroid, located above the optic disc. This area facilitates night vision in animals by reflecting rather than absorbing light and is thus responsible for the glow of the eyes at night. The tapetum lucidum is absent in humans.

## Interior View (Figs. 7-8, 7-9)

*Gently remove the vitreous humor from the inner aspect of the anterior half of the eyeball. This should be done, as before, carefully with the scalpel handle.*

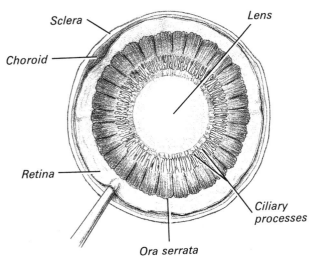

**FIGURE 7-8**
Interior view of the eyeball (back of lens and ciliary body).

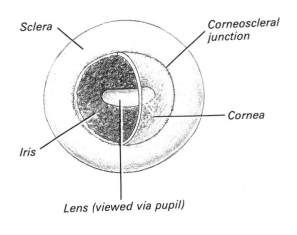

**FIGURE 7-9**
Front of eyeball—I (half of cornea removed).

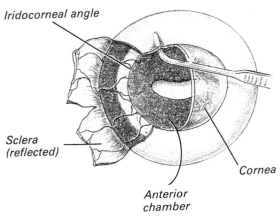

**FIGURE 7-10**
Front of eyeball—II (sclera partially reflected).

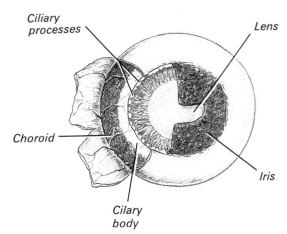

**FIGURE 7-11**
Front of eyeball—III.

**Ora serrata**   The scalloped circular margin where the dull grey, sensory layer of the retina ends (near the level of the corneoscleral junction).

**Ciliary body**   A rostral continuation of the choroid, connecting it from the level of the ora serrata to the circumference of the iris. The ciliary body is a scalloped, wedge-shaped circular structure that surrounds the lens. It consists of three parts: ciliary ring, ciliary processes, and ciliary muscles (see Fig. 7-7). The **ciliary ring** is a band-shaped zone about 3 mm wide directly anterior to the ora serrata. The **ciliary processes** are numerous black-pigmented finger-like meridional ridges that lie in a narrow circular zone anterior to the ciliary ring. Collectively, the processes form a crown-like structure (ciliary crown). The processes project towards the lens periphery. The **ciliary muscle** (see Fig. 7-7) lies between this crown and the sclera, in the external portion of the ciliary body. The muscle consists of radiating involuntary fibers. The muscle fibers extend primarily between the sclera and the ciliary processes and ring.

**Suspensory ligament**   Fine fibers that can be seen, in favorable specimens, as extending between the ciliary processes and the lens it suspends.

*Using a sharp scalpel, cut vertically across the entire diameter of the cornea. Next, with fine scissors, cut along the corneoscleral junction halfway around the circumference of the cornea, connecting up with the two ends of the first incision. Half of the cornea can now be removed, leaving the other half intact (Fig. 7-9).*

**Iris**   A rostral part of the choroid, which extends as a curtain in front of the lens. The central, elliptical opening at the edge of the iris is the **pupil**. The iris serves as a muscular diaphragm to regulate the amount of light passing through the lens to the retina.

**Anterior chamber**   The area, filled in life with the aqueous humor, located between the cornea and anterior surface of the iris.

**Iridocorneal angle**   The space at the site where the iris and cornea meet. (With half the cornea removed, it can now be probed.)

*Pass the tip of a probe through the pupil so that it reaches the space between the lens and the posterior surface of the iris, namely the posterior chamber.*

**Posterior chamber**   This chamber is also filled with aqueous humor and communicates with the anterior chamber through the pupil.

*For the next step in the dissection of the anterior chamber of the eye, place the end of a probe into the iridocorneal angle (Fig. 7-10) and then move it around the cornea-free perimeter of the iris forcefully enough to rupture the firm attachment of the anterior end of sclera to the ciliary body. Then make near-vertical incisions a short distance into the sclera close to both ends of the cornea-free section (Fig. 7-10). Reflect the sclera in the semicircle between the two initial incisions. The ciliary body and the adjacent choroid are now exposed on the side of the eyeball.*

**Ciliary body**   The narrow, white, circular band extending between the iris and choroid, previously seen from its posterior surface. This band marks the location of the **ciliary muscle**, which is responsible for accommodation, the act of focusing near objects on the retina. It does so by pulling the ciliary processes forward and inward upon contraction, thus relaxing the suspensory ligament and allowing the lens to become more convex.

**Choroid**   This layer was previously seen in a transverse section. It is continuous medially with the ciliary body. The choroid, ciliary body, and iris forms the continuum of the middle or vascular coat.

**Ciliary nerves**   In favorable specimens, these nerves may be seen as white streaks that extend from under the sclera into the ciliary body and iris.

*Finally, remove the remaining half of the cornea and with fine scissors cut out a section of the iris on the cornea-free side, thereby exposing the ciliary processes from in front (Fig. 7-11). The capsule enclosing the lens can now be incised by a horizontal anterior cut. The lens will either come out of its own accord or will emerge under gentle pressure.*

**Lens**   This transparent, biconvex structure refracts the light passing through it. It is curved more prominently on its caudal than on its rostral surface. The lens is positioned between the iris and the vitreous humor. An elastic, homogenous membrane, the capsule, covers the lens. The suspensory ligament is attached to the capsule.

# Anatomy of the Cat

The nervous system consists of the cerebrospinal division (brain and spinal cord), peripheral division (cranial and spinal nerves), and autonomic division.

## SPINAL CORD (Fig. 8-1)

*The spinal cord can be exposed partly or completely, depending on the time available. For partial exposure, remove the muscles of the cervical region on both sides of the neural spines. When the neural arches have been exposed, carefully cut away the arch from one of the last cervical vertebrae with bone shears. Without damaging the spinal nerves, proceed anteriorly and cut away, in succession, the remaining cervical arches. For complete exposure of the spinal cord, remove the thoracic and lumbar arches in a similar manner.*

**Spinal meninges**  Three membranes surround the cord. The outer membrane, the **dura mater,** which is tough and fibrous, lines the neural canal. The delicate **arachnoid** lies just beneath the dura. The inner membrane, the highly vascular **pia mater,** closely envelops the spinal cord and dips into its depressions.

**Spinal cord enlargements**  The spinal cord is slightly thicker in the cervical and lumbar regions (see Fig. 8-5). These enlargements are the sites of origin and termination of spinal nerves passing to the appendages.

**Spinal nerve**  Each is formed by union of **dorsal** and **ventral roots** that emerge from the cord. Each dorsal root exhibits a swelling, the **spinal ganglion.**

**Sulci**  Several longitudinal grooves dip into the surface of the cord. The **dorsal fissure** is a very narrow mid-dorsal groove; the **dorsolateral sulci** are less prominent grooves extending along the line of entrance of the dorsal roots; the **ventral fissure** extends along the mid-ventral line.

*With a sharp scalpel cut across the thoracic region of the cord, close to a pair of spinal nerves. Lift up one of the severed ends of the cord and study its appearance.*

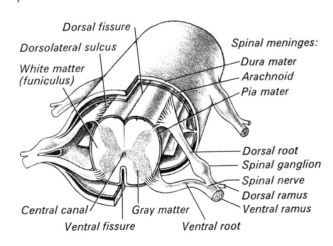

**FIGURE 8-1**

Spinal cord and meninges (cross section).

**Central canal**   Cavity in the center of the cord.

**Gray matter**   The H-shaped central area around the central canal.

**White matter**   The outer area around the gray matter. The sulci divide it into columns, or **funiculi.**

## CAT BRAIN (Figs. 8-2, 8-3, 8-4)

*If a prepared cat (or sheep) brain is unavailable, expose the brain by first removing all the muscles from the dorsal and lateral surfaces of the head. With bone shears or a bone saw, make an opening through the parietal bone. Taking particular care not to damage the brain, chip away the skull, using the opening as a starting point and proceeding anteriorly. Remove the posterior part of the skull in a similar manner, starting at the foramen magnum. Take special care when removing the transverse bony septum that lies between the cerebrum and cerebellum. Cut the spinal cord transversely just anterior to the atlas. Then beginning here, free the brain by carefully and progressively lifting it up. Note each cranial nerve as it is encountered and sever it as far away from the brain as possible. As the brain is being freed, take special care to minimize damage to the hypophysis and the olfactory bulbs. Then place the brain in a weak (5%) formalin solution and remove the dura mater.*

Three membranes surround the brain. These are continuous at the foramen magnum with those surrounding the spinal cord.

**Dura mater**   This layer, which is also the internal periosteum of the skull, protects the brain. A fold of dura, the **falx cerebri,** hangs down into the fissure between the cerebral hemispheres. Another dural fold, the **tentorium cerebelli,** lies transversely between the cerebrum and the cerebellum and has become ossified.

**Arachnoid**   The delicate membrane beneath the dura. It does not dip into the cleft-like depressions of the brain.

**Pia mater**   A thin vascular membrane directly covering the brain. It adheres closely to the cerebrum and dips into all of its clefts.

The *dorsal* surface of the brain (Fig. 8-2) reveals the following structures.

### Telencephalon

**Cerebrum**   The two **cerebral hemispheres** consist of a number of folds, or **gyri.** These folds are bounded by

grooves, or **sulci.** The hemispheres are separated by a deep **cerebral fissure.** In vertebrates the cerebrum has become expanded so that dorsally it overlies other parts of the brain. The hemispheres have a great deal of control over the movements of the body.

### Metencephalon

**Cerebellum**   This part of the brain lies behind the cerebrum and projects above the medulla oblongata (see definition). It consists of a median **vermis** and two lateral **cerebellar hemispheres.** The cerebellum influences muscle tone and helps to maintain body equilibrium.

### Myelencephalon

**Medulla oblongata**   This is the posterior part of the brain that is continuous with the spinal cord. It maintains many of the body's visceral reflexes, such as respiration.

The *ventral* surface of the brain (Fig. 8-3) reveals the following structures.

### Telencephalon

**Olfactory bulbs**   Small elongated structures that appear to project forward from the hemispheres. **Olfactory tracts,** which are bands of fibers, extend back from the bulbs and terminate in the anterior part of the brain.

### Diencephalon

**Optic chiasma**   The site at which the optic nerve fibers come together and partially cross.

**Tuber cinereum**   A small round elevation just posterior to the optic chiasma.

**Infundibulum**   A narrow stalk extending from the tuber cinereum.

**Hypophysis**   This structure, also known as the pituitary gland, is located at the end of the infundibulum.

**Mammillary bodies**   Two small elevations lying just posterior to the tuber cinereum.

### Mesencephalon

**Cerebral peduncles**   The large bundle of fibers run-

241

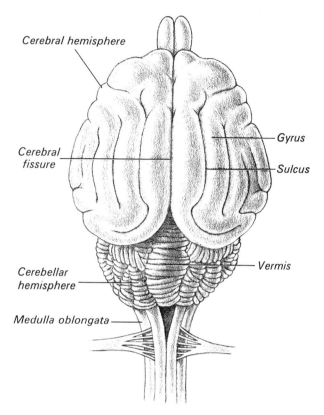

**FIGURE 8–2**
Brain (dorsal view).

Cerebral hemisphere

Gyrus

Cerebral fissure

Sulcus

Cerebellar hemisphere

Vermis

Medulla oblongata

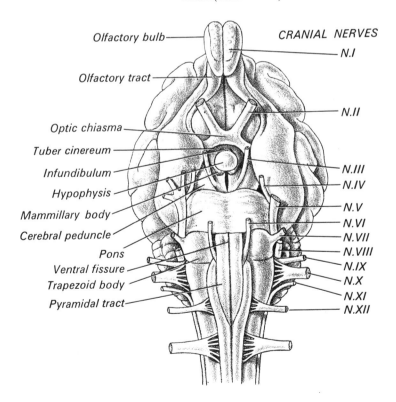

Olfactory bulb

CRANIAL NERVES

N.I

Olfactory tract

N.II

Optic chiasma

Tuber cinereum

Infundibulum

N.III

Hypophysis

N.IV

Mammillary body

N.V

Cerebral peduncle

N.VI

Pons

N.VII

Ventral fissure

N.VIII

Trapezoid body

N.IX

Pyramidal tract

N.X

N.XI

N.XII

**FIGURE 8–3**
Brain (ventral view) and cranial nerves.

ning obliquely forward on each side of the hypophysis. The peduncles relay impulses to and from the cerebral hemispheres.

## Metencephalon

**Pons** A broad structure at the level of the cerebellum, transversely striped by bands of fibers. Most of these fibers convey impulses to and from the cerebellum.

## Myelencephalon

**Ventral fissure** Median ventral groove along the medulla oblongata.

**Pyramidal tracts** A bundle of fibers on each side of the ventral fissure that narrow posteriorly. Known also as pyramids, they convey impulses posterior to the spinal cord that ultimately bring about contraction of the somatic musculature.

**Trapezoid bodies** Rounded masses located just lateral to the anterior end of the pyramidal tracts.

*With a long, sharp, wet knife cut the brain longitudinally into halves through the mid-sagittal plane.*

The *median sagittal* surface (Fig. 8-4) reveals the following structures.

**Corpus callosum** Longitudinal white band of fibers located at the bottom of the cerebral fissure. The fibers interconnect the cerebral hemispheres. The corpus callosum consists of an anterior **genu,** a middle **body,** and a posterior **splenium.** A fourth part, the **rostrum,** projects ventrally from the genu.

**Fornix** A white band extending ventrally from the corpus callosum.

**Septum lucidum** The thin membrane between the rostrum of the corpus callosum and the fornix.

**Anterior commissure** A small group of fibers just ventral to the fornix.

**Lamina terminalis** The membrane extending from the anterior commissure to the optic chiasma. It forms the anterior wall of the third ventricle.

**Thalamus** The thick lateral wall of the third ventricle (see definition). It consists of masses of gray matter that serve as relay stations for ascending fibers carrying impulses to the cerebral cortex.

**Third ventricle** This chamber is anteriorly bounded by the lamina terminalis and laterally by the thalami. It opens into the cerebral aqueduct (see definition) posteriorly.

**Middle commissure** Large round thalamic area, known also as the massa intermedia, in the central part of the third ventricle. It connects the two thalami.

**Epiphysis** The pineal body lies just beneath the splenium of the corpus callosum.

**Posterior commissure** The small round mass of fibers that crosses the midline just ventral to the epiphysis.

**Cerebral aqueduct** The longitudinal channel through the midbrain.

**Lamina quadrigemina** Forms the roof, or tectum, of the cerebral aqueduct.

**Fourth ventricle** Continuous with the aqueduct, it lies beneath the cerebellum.

**Arbor vitae** A white, branching structure especially evident in the sections through the vermis of the cerebellum. It consists of myelinated nerve fibers.

**Velum** The membrane just beneath the cerebellum that roofs the fourth ventricle.

## CRANIAL NERVES (Fig. 8-3)

There are 12 pairs of cranial nerves in the cat. All of these nerves pass through foramina at the base of the skull and all, except the tenth and eleventh, are distributed to structures of the head and neck.

*Identify the cranial nerves at the time the ventral brain surface is studied.*

**Olfactory nerve** (N. I) This nerve arises from an olfactory bulb, passes through foramina in the cribriform plate, and is distributed to the nasal epithelium.

**Optic nerve** (N. II) Originating at the eyeball, it extends back obliquely between the pyriform lobes of the brain. At the anterior end of the diencephalon the optic nerves cross to form the optic chiasma.

**Oculomotor nerve** (N. III) This motor nerve arises near the mid-ventral line between the hypophysis and the pons. It passes through the orbital fissure and ramifies to innervate the inferior oblique muscle and the superior, inferior, and medial recti muscles of the eye.

**Trochlear nerve** (N. IV) This motor nerve arises from the dorsal side of the midbrain and emerges ventrally along the outer edge of the pons. It passes through the orbital fissure to innervate the superior oblique muscle.

**Trigeminal nerve** (N. V) A large mixed nerve that arises on each side from two roots at the posterolateral border of the pons. It divides into three major trunks, the **ophthalmic, maxillary,** and **mandibular nerves,** which supply the teeth, eyelids, many head muscles, and the skin of the head. The major trunks of the trigeminal nerve are frequently identified as $V^1$, $V^2$, and $V^3$ (see Table 2-1).

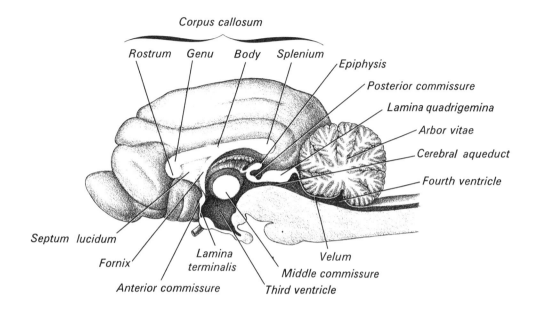

**FIGURE 8–4**
Brain (sagittal view).

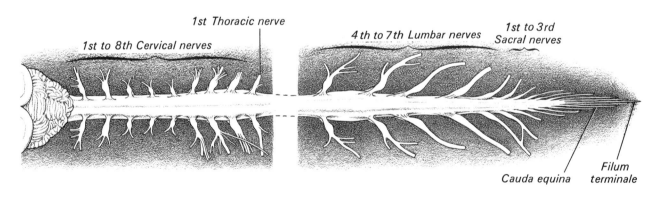

**FIGURE 8–5**
Spinal nerves.

**Abducens nerve (N. VI)** This motor nerve arises between the trapezoid body and pyramidal tract. Each passes through the orbital fissure to supply the lateral rectus.

**Facial nerve (N. VII)** This mixed nerve arises just behind the trigeminal nerve, lateral to each trapezoid body. It passes through the middle ear and emerges to supply many head and face muscles as well as the mucous membrane of the tongue.

**Vestibulocochlear nerve (N.VIII)** This sensory nerve arises just behind the facial. It passes into the inner ear through the internal auditory meatus to innervate the membranous labyrinth.

**Glossopharyngeal nerve (N. IX)** This mixed nerve arises as a series of rootlets just behind the auditory nerve. It emerges from the jugular foramen of the skull and supplies the muscles of the pharynx and the taste buds.

**Vagus nerve (N. X)** This mixed nerve also arises as a series of rootlets, posterior to the origin of the ninth. It emerges through the jugular foramen and passes down to most of the thoracic and abdominal viscera. One of the vagi should be traced down the side of the trachea, along the esophagus, and, if possible, as far as the curvature of the stomach.

**Spinal accessory nerve (N. XI)** This motor nerve arises as a series of rootlets from the sides of both the medulla and the spinal cord, passing out of the skull through the jugular foramen with the ninth and tenth nerves. It innervates the neck and shoulder muscles.

**Hypoglossal nerve (N. XII)** This motor nerve arises from the ventral surface of the medulla and leaves the skull through the hypoglossal foramen. It innervates the muscles of the tongue and neck.

## SPINAL NERVES (Fig. 8-5)

In the cat there are 38 or 39 pairs of these metameric structures: 8 cervical, 13 thoracic, 7 lumbar, 3 sacral, and 7 or 8 caudal nerves.

After each spinal nerve leaves the vertebral column it divides into two parts, the **dorsal** and **ventral rami** (see Fig. 8-1). Each ramus contains both motor and sensory fibers. The small dorsal ramus is distributed to the skin and muscles of the back. The larger ventral ramus supplies the ventral skin and musculature. The nerves described are ventral rami of spinal nerves.

### Cervical Nerves

The first through the fourth cervical nerves innervate the muscles of the anterolateral part of the neck.

The **phrenic nerve** is formed primarily from the fifth cervical nerve as it emerges from between the vertebrae, but it also receives a contribution from the sixth cervical. The phrenic passes down the neck to enter the thorax near the subclavian artery. It then descends to the diaphragm, which it innervates. The phrenic nerve is most readily found between the pericardial cavity and the point of entrance of the bronchus into the lung.

The fifth through the eighth cervical nerves and the first thoracic nerve anastomose to form a network known as the **brachial plexus** (Fig. 8-6). This network surrounds the large blood vessels that pass from the neck into the shoulder and forelimbs. The major nerves emerging from the brachial plexus are the following.

**Musculocutaneous nerve** It can be identified as it passes distally along the edge of the biceps brachii.

**Radial nerve** This is the largest nerve passing into the upper arm. It may be uncovered on the lateral surface of the arm between the lateral head of the triceps and the brachialis.

**Median nerve** This large nerve runs anterior to, but along with, the brachial artery.

**Ulnar nerve** This nerve runs posterior to the brachial artery.

### Thoracic Nerves

The first thoracic nerve joins the brachial plexus. The remaining 12 pass ventrolaterally as the **intercostal nerves,** supplying the intercostal muscles and the external oblique muscle.

### Lumbar Nerves

The first three pairs of lumbar nerves pass independently from the spinal cord to the abdominal wall, where they supply the muscles and skin. These nerves may be found in the back of the abdominal cavity, about an inch posterior to the last rib and beneath the peritoneum and muscles of this region.

The last four lumbar nerves and the three sacral nerves anastomose to form the **lumbosacral plexus** (Fig. 8-7). This network can be exposed by carefully picking away the muscles on either side of the bodies of the lumbar vertebrae. The nerves originating from the plexus pass into the hind limb. Two major nerves from the lumbosacral plexus should be identified.

**Sciatic nerve** The longest nerve in the body. It may be observed just under the biceps femoris (which may be removed). It may be traced to its emergence

245

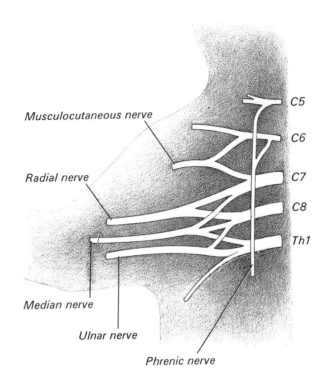

**FIGURE 8–6**
Major branches of the brachial plexus.

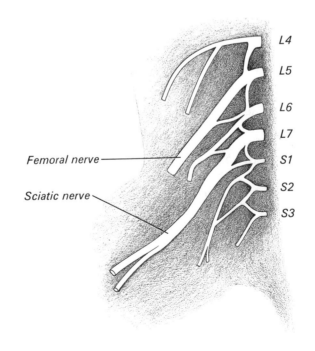

**FIGURE 8–7**
Major branches of the lumbosacral plexus.

from the pelvis and then distally to the site at which it divides into two large branches that continue into the leg.

**Femoral nerve** This is best observed on the inner surface of the thigh between the sartorius and gracilis muscles. It can be traced down into the leg.

## Sacral Nerves

The sacral nerves help to form the lumbosacral plexus but also give branches to the gluteal muscles, rectum, bladder, and external genitalia. These nerves are difficult to locate.

## Caudal Nerves

The caudal nerves innervate the musculature of the tail.

Only the more anterior spinal nerves leave the vertebral column at a point directly opposite their place of origin. The lower nerves assume an oblique course that gradually becomes more accentuated in the lowest regions, where they run parallel to the cord before emerging from the foramina. A fine thread, the **filum terminale,** extends from the tapered posterior end of the spinal cord. The sacral and caudal spinal nerves run parallel to the filum terminale and have a brush-like appearance that gives their configuration its name, the **cauda equina** (horse's tail).

## AUTONOMIC DIVISION

The autonomic division consists of the sympathetic and parasympathetic systems.

## Sympathetic System (Fig. 8-8)

The sympathetic system consists of two ganglionated chains, the **sympathetic trunks,** which run in parallel fashion along each side of the vertebral column, and a series of collateral ganglia and plexuses in the neck, thorax, and abdomen.

**Cervical portion of sympathetic trunk** Located at the side of the trachea adjacent to the vagus nerve. It forms a large **vagosympathetic trunk** just lateral to the common carotid artery. In this region the trunk has three sympathetic ganglia, one at the cranial end of this segment and the other two at the caudal end. These ganglia represent fusions of originally separate ganglia.

**Superior cervical ganglion** This represents the fused first four cervical nerve ganglia. It lies just medial

to the nodose or inferior ganglion of the vagus nerve. Fibers from this ganglion lead into the head following the path of the common carotid artery and its branches.

**Middle cervical ganglion** This very small ganglion represents the fused fifth and sixth cervical nerve ganglia. It lies above the first rib just cranial to the subclavian artery. From this ganglion the sympathetic trunk splits to loop around the subclavian artery and is reunited in the inferior cervical ganglion.

**Inferior cervical ganglion** Also known as the stellate ganglion because of its shape. It lies lateral to the vertebra between the heads of the first and second ribs. This large ganglion represents the fused seventh and eighth cervical and first three thoracic nerve ganglia. The middle and inferior cervical ganglia give off **cardiac nerves** to the heart as well as branches to other thoracic viscera.

**Thoracic portion of sympathetic trunk** Located on the lateral surfaces of the thoracic vertebrae, this part of the sympathetic trunk has 13 ganglia, corresponding to the 13 thoracic vertebrae. From each ganglion small branches, rami communicantes, are given off to the corresponding spinal nerves.

A little before the sympathetic trunk passes into the abdomen, preganglionic fibers arise from the lower thoracic sympathetic ganglia and unite to form the **greater** and **lesser splanchnic nerves.** They both pierce the diaphragm and terminate in the upper two abdominal collateral ganglia (see definition).

**Abdominal portion of sympathetic trunk** This portion consists of the lumbar and sacral parts of the trunk. It runs caudally, is more medially located than the thoracic portion of the trunk, and terminates in the pelvic cavity. Its branches extend along the major blood vessels to the three collateral sympathetic ganglia.

The upper two **abdominal collateral ganglia** are the closely joined **celiac** and **superior mesenteric ganglia** which lie near the base of the same named arteries, respectively. Postganglionic fibers follow the path of these arteries and their branches to the abdominal viscera, including the digestive glands, spleen, and kidneys. Preganglionic fibers from the abdominal sympathetic ganglia pass caudally to synapse in the third or **inferior mesenteric ganglion,** which lies above the base of the artery. Postganglionic fibers follow the inferior mesenteric artery and its branches to innervate the descending colon and rectum. Other fibers from this ganglion follow other aortic branches to innervate the urinary bladder and genital organs.

247

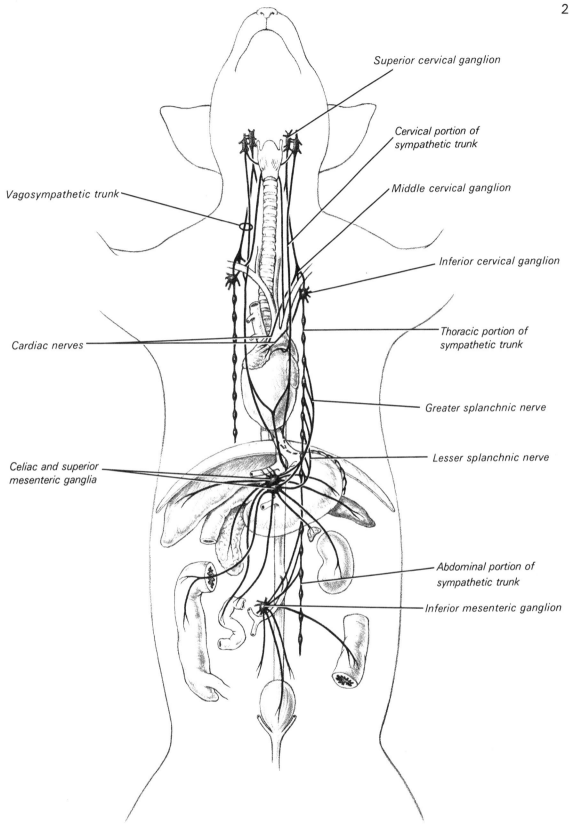

Superior cervical ganglion

Cervical portion of
sympathetic trunk

Middle cervical ganglion

Inferior cervical ganglion

Thoracic portion of
sympathetic trunk

Greater splanchnic nerve

Lesser splanchnic nerve

Abdominal portion of
sympathetic trunk

Inferior mesenteric ganglion

Vagosympathetic trunk

Cardiac nerves

Celiac and superior
mesenteric ganglia

**FIGURE 8–8**
Sympathetic nervous system.

As the sympathetic trunk proceeds caudally in the lumbar region, it swings medial to lie on the ventral surfaces of the vertebral bodies and is concealed by the iliopsoas muscles. The trunk extends into the sacral region, becoming gradually smaller as it approaches termination.

## Parasympathetic System

The fibers of this system are not organized into distinct nerves but rather pass to their effector organs by way of cranial nerves III, VII, IX, and X and sacral nerves two and three. These parasympathetic fibers terminate in secondary ganglia that lie close to their effector organs.

# Anatomy of the Cat

A study of the mammalian brain using a sheep brain in place of, or in conjunction with, that of the cat has been popular for a long time. This practice is common because sheep brains are large, available, and inexpensive. Moreover, gross sheep brain dissection can serve as a basis for acquiring knowledge about the location of the major clusters of neurons (nuclei) and bundles of interconnecting nerve fibers (tracts), which are a basic objective of neuroanatomical studies.

One should be aware, however, that gross brain dissection has its limitations. Some anatomically prominent structures, such as the hippocampus, are not fully understood in terms of function and, conversely, other structures, such as certain cranial nerve nuclei, whose functions are known precisely, are difficult to identify anatomically.

## MENINGES (Fig. 9-1)

*Examine carefully the brain membranes that enclose your specimen. If they are no longer present, which is usually the case for the dura and arachnoid, then attempt to obtain a sagitally sectioned demonstration specimen. Probe the depth of the fissures and observe how the pia and its vessels penetrate them.*

The brain is enclosed by three membranes that are known collectively as **cranial meninges.**

**Dura Mater** The tough, opaque membrane that is fused to the periosteum of the skull.

**Arachnoid** The delicate, semitransparent middle membrane that lies under the dura and is separated from it by **subdural space.** The arachnoid bridges the clefts of the brain and extends for only a short distance along all the cranial nerves except for the optic nerve, where it reaches the eyeball.

**Pia Mater** The very thin, transparent, vascular, innermost covering that intimately clothes the brain. The pia follows the surface convexities and clefts of the brain. The area between the arachnoid and pia is known as the **subarachnoid space.** Very delicate

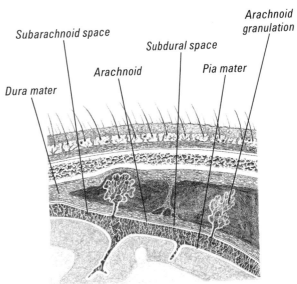

Subarachnoid space

Arachnoid

Dura mater

Subdural space

Arachnoid granulation

Pia mater

**FIGURE 9–1**
Sheep brain; cranial meninges.

strands connecting the two membranes extend across this fluid-filled zone.

## Cerebrospinal Fluid (Fig. 9-1)

The narrow central canal of the spinal cord extends into the brain and expands in four locations as ventricles. These chambers (and the central canal) contain the clear, colorless cerebrospinal fluid (CSF) that is continuously being secreted by the **choroid plexuses,** which extend into the ventricles (see Fig. 9-9). These plexuses are vascular tufts made up of two layers, the pia and ependyma (epithelium lining the ventricles). The CSF passes out of the brain through openings in the fourth ventricle into the subarachnoid space to bathe the brain (and spinal cord). The **arachnoid granulations** serve to bring the fluid back to the venous sinuses in the dura. The CSF thus serves to protect and support the soft, delicate nervous tissue.

## BRAIN

*For most commercial sheep brain preparations, the dura has already been removed and the arachnoid has been destroyed during its removal. The brain is thus ready for study. Examine first the dorsal and then the ventral surfaces.*

The *dorsal* surface of the brain (Fig. 9-2) reveals the following structures.

## Telencephalon

**Cerebrum**    The two **cerebral hemispheres** form the greater part of the brain. The anterior ends are known as the **frontal lobes,** the caudal ends as the **occipital lobes,** and the lateral and widest segments as the **temporal lobes.** The dorsal surface is characterized by a deep, median longitudinal **cerebral fissure** that separates the hemispheres. Small ridges, or **gyri,** and fissures, or **sulci,** cover the surface of the hemispheres. The gyri and sulci vary in terms of arrangement in different sheep and even between the hemispheres of the same animal.

**Corpus callosum**    A massive fiber tract located at the bottom of the longitudinal cerebral fissure. It is evident by separating the hemispheres. This tract serves to interconnect both hemispheres.

*To obtain a favorable dorsal view of the diencephalon (epithalamus) and mesencephalon, manually spread apart the lobes of the occipital hemisphere. If the visibility of the structures you seek is inadequate, slowly and carefully cut through the corpus callosum until the occipital lobes can be separated sufficiently to meet your needs.*

## Diencephalon (Fig. 9-2, insert)

**Pineal body**    A knob-like prominence that can be seen when the occipital lobes are parted carefully or lifted away from the cerebellum. This endocrine gland probably supresses puberty until the body is physiologically ready.

**Tela choroidea**    The vascular curtain that roofs a longitudinal slit, the **third ventricle.** The CSF-secreting anterior choroid plexus hands down from the tela into this chamber.

**Habenular trigone**    This consists of the **habenula,** a pair of short bands that forms the posterolateral V-shaped rim of the third ventricle, and the habenular commissure, which is an indistinct transverse band located between the pineal body and ventricle. The habenula is a significant olfactory center.

## Mesencephalon

**Corpora quadrigemina**    These four round elevations form the roof (tectum) of the midbrain. The larger anterior pair, the **superior colliculi,** cradle the pineal gland and are associated with eye reflexes. The posterior pair, the **inferior colliculi,** are concerned with auditory reflex activity.

## Metencephalon

**Cerebellum**    This segment lies behind the cerebral hemisphere and above most of the fourth ventricle. It consists of a median unpaired **vermis** and two lateral **cerebellar hemispheres.** The plate-like ridges, or folia, of the cerebellar cortex are smaller than those of the cerebral hemispheres and are disposed primarily in a transverse direction. The folia are separated by sulci.

## Myelencephalon

This last segment of the brain is known also as the **medulla oblongata.** It encloses the large diamond-shaped **fourth ventricle.**

**Anterior medullary velum**    This thin triangular layer of fibers forms the anterior part of the roof of the fourth ventricle. It can be seen best by pushing the cerebellum posteriorly.

**Tela choroidea**    This vascularized layer and its posterior choroid plexus forms the posterior part of the roof of the fourth ventricle.

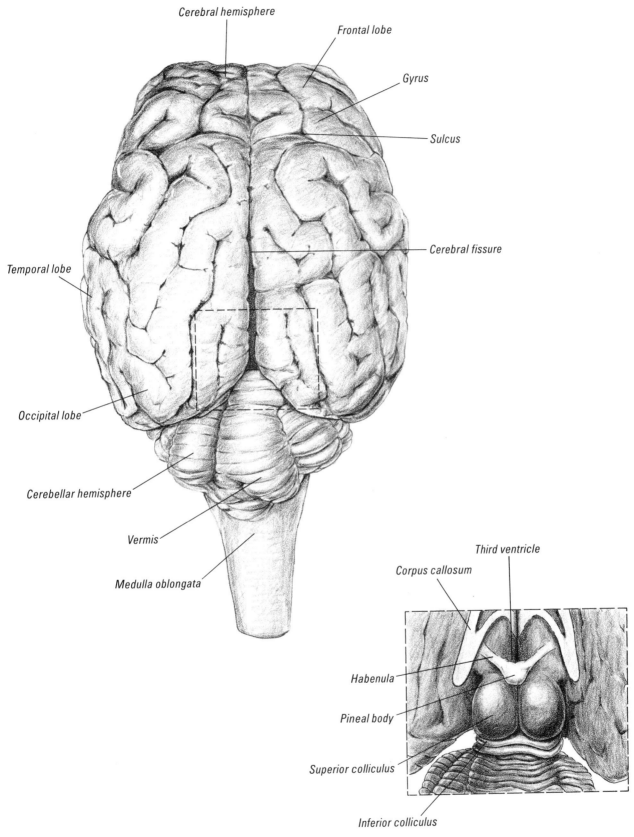

**FIGURE 9–2**
Sheep brain (dorsal view).
Inset: Diencephalon and Mesencephalon.

The *ventral* surface of the brain (Fig. 9-3) should reveal many of the following structures.

## Telencephalon

*Carefully removing the pia from this area will facilitate identification of the structures enumerated below.*

**Olfactory bulbs**   These are a pair of elongated oval bodies that lie in the ethmoid fossae of the interior of the skull. Numerous thin fibers, which constitute the olfactory nerves (N.I.), extend down from the bulbs, through the openings in the cribriform plate, and into the nasal olfactory mucosa (see Fig. 2-5).

**Olfactory tracts**   A pair of medial and lateral tracts extend at an angle from the posterior end of each of the bulbs. The **lateral olfactory tracts** are more distinct and are demarcated laterally by prominent **rhinal fissures.** Each lateral olfactory tract terminates caudally in the **pyriform lobe,** the most ventral part of the cerebral hemisphere, which is also bounded laterally by this fissure. The less prominent **medial olfactory tracts** pass obliquely toward the midline and onto the medial surface of the hemispheres. The pyriform lobe is homologous to the olfactory lobes of lower vertebrates, such as the shark.

## Diencephalon

The floor of the diencephalon is made up of the **hypothalamus.**

**Optic chiasma**   The X-shaped configuration that is present between the cranial ends of the pyriform lobes. It is formed by the (partial) decussation of the optic nerves. Extending posteriorly from the chiasma are the **optic tracts** which seem to disappear beneath the pyriform lobes.

**Band of Broca**   The area dorsolateral to the optic chiasma extending from the cranial end of the pyriform lobe to the median border of the cerebrum.

**Middle olfactory tract**   The cranially directed V-shaped area located between the medial and lateral olfactory tracts. Posterolaterally, this tract extends into the **anterior perforated substance** through which vessels pass into the substance of the hemispheres. Posteromedially, this tract terminates in the **olfactory tubercle.**

**Hypophysis**   The pituitary gland is attached to the brain by means of a thin stalk, the **infundibulum,** whose internal canal is an extension of the third ventricle. The site of attachment of the infundibulum to

the hypothalamus is a slight elevation, known as the **tuber cinereum.**

*The pituitary gland is frequently missing because the infundibular attachment to the brain is torn easily, and the pituitary is then removed accidentally. If still present, the gland can be removed after being identified to secure a bvetter view of the region.*

**Mammillary bodies**   A pair of rounded elevations that demarcates the caudal end of the hypothalamus; these bodies are integration centers for information received through the fornix. (see Fig. 9-10, inset A).

## Mesencephalon

**Cerebral peduncles**   A pair of columns that form the ventrolateral surface of the mesencephalon. They emerge from under the optic tracts and converge to disappear into the substance of the pons (see below). They represent massive bundles of nerve fibers from the telencephalon, including the major motor tracts, as well as from the diencephalon, which descend through the floor of the mesencephalon to more caudal centers. The oculomotor nerves emerge from the interpeduncular fossa on the medial sides and caudal ends of the cerebral peduncles.

**Posterior perforated substance**   The pitted area located between the cerebral peduncles just anterior to the pons. Blood vessels enter the brain through this region.

**Interpeduncular fossa**   This is a somewhat triangular depression that is bounded anteriorly by the mammillary bodies, laterally by the cerebral peduncles, and posteriorly by the pons.

## Metencephalon

**Pons**   This structure lies transversely across the ventral surface of the metencephalon. It consists chiefly of fibers connecting the cerebral cortex and the cerebellum, but it also includes motor fibers descending to more caudal levels.

## Myelencephalon

**Trapezoid body**   The transverse body located just posterior to the pons. It is covered centrally by the pyramids.

**Pyramids**   The two longitudinal fiber bands that emerge medially from the caudal border of the

253

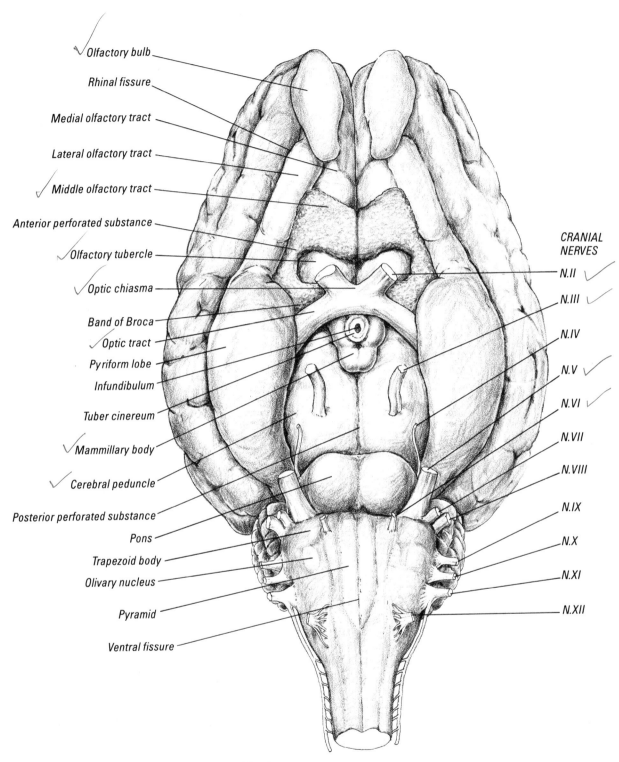

Olfactory bulb
Rhinal fissure
Medial olfactory tract
Lateral olfactory tract
Middle olfactory tract
Anterior perforated substance
Olfactory tubercle
Optic chiasma
Band of Broca
Optic tract
Pyriform lobe
Infundibulum
Tuber cinereum
Mammillary body
Cerebral peduncle
Posterior perforated substance
Pons
Trapezoid body
Olivary nucleus
Pyramid
Ventral fissure

CRANIAL NERVES
N.II
N.III
N.IV
N.V
N.VI
N.VII
N.VIII
N.IX
N.X
N.XI
N.XII

**FIGURE 9–3**
Sheep brain (ventral view).

pons and pass backward to converge and terminate in the distal third of the medulla oblongata. At mid-levels of the pyramids, the fibers of each pass over a small elevation, which is due to the underlying **olivary nucleus.**

**Ventral fissure**   The median ventral groove that extends the entire length of the medulla oblongata.

## CRANIAL NERVES

In the sheep brain there are also 12 numerically arranged cranial nerves, as in the cat. They emerge from the ventral surface of the brain and leave the cranium through foramina in the skull on their way to various structures innervated in the head and neck.

The cranial nerves, unlike the spinal nerves, are not distributed uniformly, nor are their nuclei necessarily arranged in continuous columns. The cranial nerves vary widely in terms of their composition and function; some are entirely motor or sensory, while others are mixed (like spinal nerves).

Identify the cranial nerves on the ventral brain surface. The olfactory nerve (N.I) cannot be seen grossly (see below).

**Olfactory nerve (N.I)**   This sensory nerve arises in the nasal olfactory mucosa and passes through the cribriform plate of the ethmoid as numerous small bundles of nerve fibers that terminate in the olfactory bulb. Only the stubs of these nerves remain attached to the brain.

**Optic nerve (N.II)**   This sensory nerve, in reality, is a fiber tract that extends from the eye to the optic chiasma. Each nerve originates in ganglion cells of the retina and passes through the optic foramen in the skull to terminate in the lateral geniculate body of the thalamus of the diencephalon. This body is an integrating and relay station from which some fibers continue on to the superior colliculus of the corpora quadrigemina, while most project to the occipital cortex. Most fibers from the lateral portion and all fibers from the medial part of the retina cross over at the optic chiasma.

**Oculomotor nerve (N. III)**   This motor nerve emerges from the interpeduncular fossa on the medial side of the cerebral peduncle toward its posterior end. It passes through the orbital fissure to innervate the levator palpebrae; superior, medial and inferior recti; and inferior oblique eye muscles.

**Trochlear nerve (N. IV)**   This motor nerve arises from the posterodorsal surface of the mesencephalon immediately in front of the base of the cerebellum. On emerging from the brain the nerve turns around the lateral border of the cerebral peduncle to reach the interpeduncular fossa region. It then passes through the orbital fissure to innervate the superior oblique muscle of the eye.

**Trigeminal nerve (N. V)**   This is the largest of the cranial nerves. It arises from the dorsocranial and lateral side of the pons by two roots, sensory and motor. The sensory root is the larger of the two, and it joins the **semilunar ganglion** a short distance from the pons adjacent to the foramen ovale. Three sensory branches arise from the ganglion.

**Ophthalmic nerve (N. V$^1$)**   This sensory nerve passes cranially in a groove in the floor of the cranial cavity and through the orbital fissure. Within the orbit, it ramifies to innervate the forehead, eyeball, and nose.

**Maxillary nerve (N. V$^2$)**   This sensory nerve follows the same course as the ophthalmic nerve, but it leaves the skull through the foramen rotundum. It ramifies to innervate the upper teeth, lip, palate, and part of the forehead and cheek.

**Mandibular nerve (N. V$^3$)**   This mixed nerve is formed by union of the third (sensory) branch from the semilunar ganglion with the motor root. It leaves the skull through the foramen ovale to provide motor innervation to the muscles of mastication and sensory innervation to the temporal region, ear, cheek, lower jaw, teeth, lip, and mouth, including the tongue.

**Abducens nerve (N. VI)**   This motor nerve arises from the region between the pons and the trapezoid body. It runs cranially in the groove along with the oculomotor and ophthalmic nerves to the orbital fissure to innervate the lateral rectus and retractor oculi muscles of the eye.

**Facial nerve (N. VII)**   This mixed nerve arises from the anterior surface of the trapezoid body. It enters the internal auditory meatus (along with N. VIII) and then passes through the facial canal to emerge at the stylomastoid foramen. Its sensory fibers innervate the taste buds of most of the tongue, while its motor fibers terminate in muscles of the head and face (except those of mastication).

**Vestibulocochlear nerve (N. VIII)**   This sensory nerve arises just caudal to N. VII. It enters the internal auditory meatus and divides into its two components: (1) the cochlear nerve, which reaches the spiral organ of Corti of the cochlea and transmits impulses to the brain, initiated by sound stimuli, and (2) the vestibular nerve, which carries sensations from the semicircular canals, utricle, and saccule.

**Glossopharyngeal nerve (N. IX)**   This mixed nerve originates by union of a number of rootlets that leave

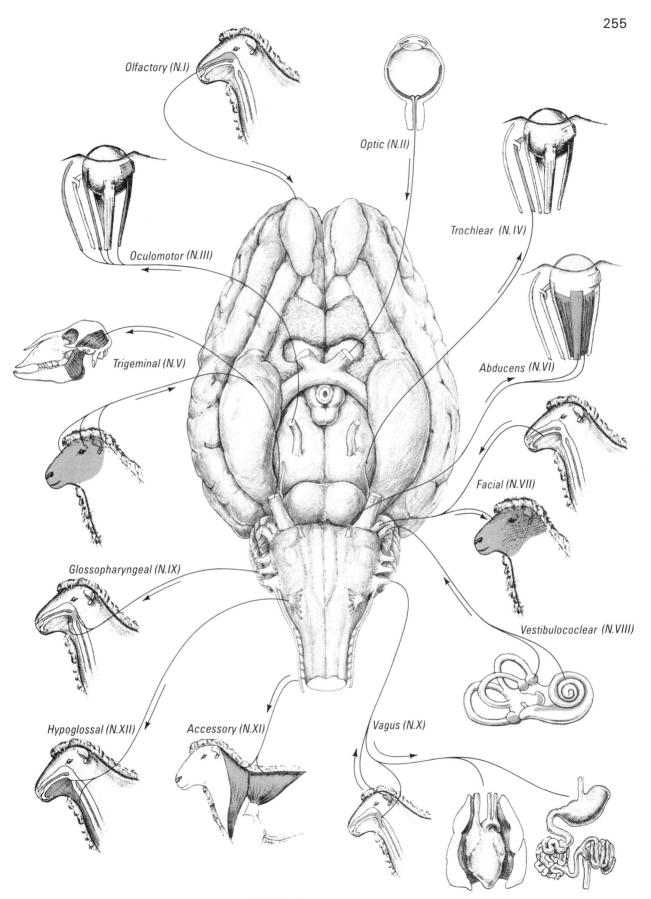

Olfactory (N.I)

Optic (N.II)

Oculomotor (N.III)

Trochlear (N.IV)

Trigeminal (N.V)

Abducens (N.VI)

Facial (N.VII)

Glossopharyngeal (N.IX)

Vestibulococlear (N.VIII)

Hypoglossal (N.XII)

Accessory (N.XI)

Vagus (N.X)

**FIGURE 9–4**
Sheep brain; cranial nerves.

the lateral border of the medulla oblongata caudal to N. VIII. It passes through the jugular foramen to divide into two branches; one travels to the tongue and the other passes to the mucosa and muscles of the pharynx. It serves a sensory function for the tongue and pharynx and a motor function for the pharyngeal muscles.

**Vagus nerve (N. X)**   This mixed nerve is the longest of the cranial nerves. It arises by a series of rootlets that are also derived from the lateral surface of the medulla, just posterior to N. IX. The vagus nerve emerges from the skull via the jugular foramen. It provides sensory innervation to the external ear and laryngeal mucosa and provides motor fibers to the muscles of the pharynx and larynx. Parasympathetic fibers of the vagus nerve are carried to most of the thoracic and abdominal viscera.

**Spinal accessory nerve (N. XI)**   This motor nerve is formed by rootlets from the medulla (in series with N. IX and N. X) as well as the first five cervical segments of the spinal cord. The accessory nerve leaves the skull through the jugular foramen and joins nerve fibers from N.X, N. XI, and the sympathetic system to contribute to the formation of the pharyngeal plexus that terminates in the musculature of the pharynx and larynx. Other fibers pass on to divide into branches that supply the sternocephalic and trapezius muscles.

**Hypoglossal nerve (N. XII)**   This motor nerve also originates by means of rootlets from the ventrolateral surface of the medulla. It passes out through the hypoglossal foramen to the muscles of the tongue and some ventral neck muscles.

A *sagittal section*, which is comparable in appearance to Figure 9-6, should reveal the following structures:

*Place your brain specimen on the dissection pan with its dorsal surface down. With a long, sharp, grease-free knife, make a single slow, continuous cut in the median sagittal plane along the entire length of the brain, i.e., from the frontal hemisphere lobes to the spinal cord. Successful cutting will be enhanced if the blade is coated with a 50% aqueous solution of glycerin containing a 1% wetting agent. The blade should be wiped and coated again if another cut is needed.*

## Telencephalon

**Corpus callosum**   The prominent curved band in the middle of the cerebral hemispheres. It is narrowest centrally at its **body** and expanded at its cranial end, the **genu,** and at its caudal end, the **splenium.** The corpus callosum is a major commissure that connects right and left cerebral hemispheres.

**Fornix**   The longitudinally directed, white fiber band that lies beneath the body of the corpus callosum and arches over the thalamus and third ventricle. It constitutes one of a pair of such bands that, viewed dorsally, have a )(-shape because they connect up on both sides of the median plane while coming together only in part of their course. Thus the fornix can be subdivided into a united **body** and diverging **anterior** and **posterior columns** (see Fig. 9-10, inset A). Only a portion of the fornix is seen in a midsagittal section. That visible portion extends from near the splenium in a curved, cranioventral direction to near the anterior commissure (see below). The fornix represents a major efferent pathway from the hippocampus to lower brain-stem centers.

**Anterior commissure**   A small, round body that is located near the ventral end of the part of the fornix which is seen in the sagittal section. This structure represents the transected bundle of fibers that passes in front of the anterior columns of the fornix in the anterior wall of the third ventricle and extends into the piriform lobes of the cerebral hemispheres and olfactory bulbs.

**Lamina terminalis**   A thin layer of gray matter that extends between the anterior commissure and the optic chiasma. It represents the anterior end of the forebrain and is attached laterally to the anterior columns of the fornix.

**Septum pellucidum**   The vertical sheet of fibrous tissue that extends between the corpus callosum and the fornix. This partition originated by fusion of the medial walls of both cerebral hemispheres and forms the wall between the two lateral ventricles. (Because it is delicate, special care should be taken to avoid damaging this septum during dissection.)

## Diencephalon

**Epithalamus**   This portion of the diencephalon is represented by two structures. The **pineal body,** which extends caudally between the superior colliculi of the corpora quadrigemina, and the **habenular trigone** which lies slightly in front of the pineal body.

**Thalami**   These paired structures form the sides of the diencephalon and thus the lateral walls of the third ventricle. A round elevation, the **massa intermedia,** represents a median section through the site of fusion of the two bulging thalami. The thalami have enlarged to enhance the relay of many fiber

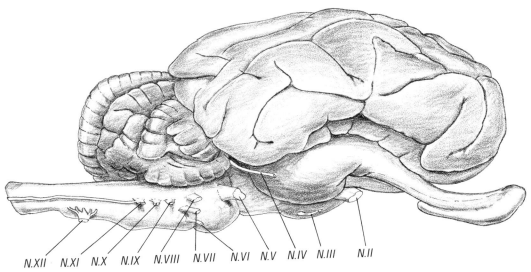

**FIGURE 9–5**
Sheep brain (lateral view).

N.XII  N.XI  N.X  N.IX  N.VIII  N.VII  N.VI  N.V  N.IV  N.III  N.II

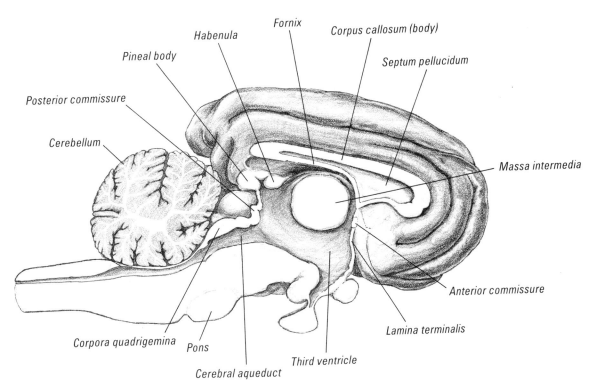

Pineal body
Habenula
Fornix
Corpus callosum (body)
Septum pellucidum
Posterior commissure
Cerebellum
Massa intermedia
Anterior commissure
Lamina terminalis
Corpora quadrigemina
Pons
Third ventricle
Cerebral aqueduct

**FIGURE 9–6**
Sheep brain (sagittal view).

systems, including sensory information, to the cerebral cortex.

**Posterior commissure**    A small, rounded mass that is located slightly beneath the pineal body. It represents a section through a bundle of transverse fibers that pass at the junction of the diencephalon and mesencephalon, which interconnect centers of these two regions.

**Third ventricle**    The cavity of the diencephalon is narrow as a result of thalamic expansion and is essentially restricted to the space around the massa intermedia. Anteriorly, it is bounded by the lamina terminalis. Its floor consists of the hypothalamic structures, namely, the optic chiasma, tuber cinereum, infundibulum, and mammillary bodies. (see ventral brain, p. 233). The lateral walls are the thalami, and the roof is formed by a thin epithelial layer that is covered by pia to form the tela choroidea.

## Mesencephalon

**Cerebral aqueduct**    The posterior extension of the third ventricle that extends through the mid-brain to the fourth ventricle.

**Corpora quadrigemina**    The paired sets of colliculi that form the roof of the cerebral aqueduct.

## Metencephalon

**Cerebellum**    The gray matter forms the cortical substance of the gyri, or cerebellar folia, while the white matter forms the core. This white matter represents tracts of fibers that extend between the cerebellum and other brain segments.

**Pons**    This structure lies beneath the floor of the anterior part of the fourth ventricle.

## Myelencephalon

**Medulla oblongata**    This portion of the brain lies under the floor of most of the fourth ventricle.

## INTERNAL ANATOMY

To gain an appreciation of both the complexity and integrated organization of the brain, it is necessary to review its internal anatomy. Since a comprehensive dissection is quite time consuming, the focus of this section is, of necessity, primarily limited to a group of readily identifiable structures and pathways that are linked together to form part of what is known as the **limbic system** and, especially, its **Papez circuit** (see Fig. 9-12), which plays a role in emotional behavior and other activities.

The neocortex (or neopallium) evolved during the evolutionary transition of reptiles to mammals. Great expansion of the neocortex in mammals accounts for the prominence of their cerebral hemispheres. To these structures sensory impulses are projected and from them motor impulses originate.

The hemispheres consist of gray matter, the cortex (or pallium), which covers the white matter.

## Cortical Tracts (Fig. 9-7)

*While keeping the half-brain moist, remove the gray matter from one or more gyri. This activity should be done with the handle of a forceps, a (bladeless) scalpel, or, preferably, with an orangewood manicure stick, and it should be executed using a gentle stroking motion. Note the substantial amount of gray matter that is removed and the shape of the core of white matter, which is arranged in the form of short fiber bundles.*

The white matter in the cerebral hemispheres consists of functionally heterogeneous fiber systems that can be classified into three categories: association fibers, commissural fibers, and projection fibers.

**Association fibers**    These fibers serve to connect near and distant gyri with one another and are of two types. The short **arcuate fibers** link adjacent gyri, while long association fibers serve to integrate more distant ones. The **cingulum** is an example of a long association tract. Its fibers run above and at right angles to those of the corpus callosum. The cingulum receives fibers coming from the rostral thalamic nuclei. (The cingulum may be exposed by gentle horizontal scraping of the cingulate gyrus, in which it lies, just dorsal to the body of the corpus callosum.)

**Commissural fibers**    These are fiber tract crossconnections between the two halves (hemispheres) of the brain. These fibers are exemplified by the anterior, posterior, and habenular commissures and the corpus callosum, which already have been identified in the sagittal brain section (see Fig. 9-6).

**Projection fibers**    These are fibers that ascend or descend to the neocortex from other brain or spinal cord structures (e.g., thalamocortical fibers and corticospinal fibers). They are not easily dissected as distinct fiber bundles and frequently need to be traced through stained serial sections.

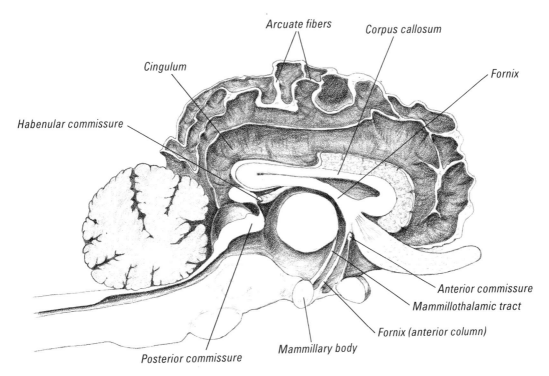

FIGURE 9–7
Sheep brain; cortical tracts.

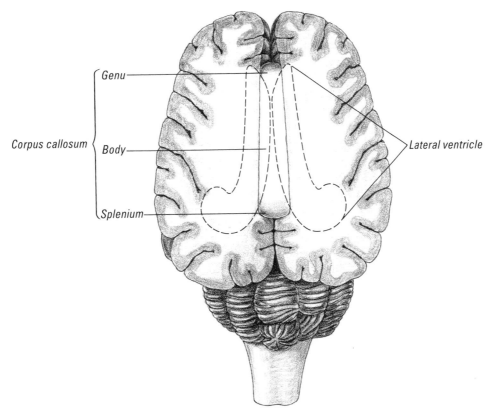

FIGURE 9–8
Sheep brain; corpus callosum.

## Other Tracts (Fig. 9-7)

*Carefully scrape away the tissue lying just above the anterior commissure and caudally between the commissure and the mammillary body.*

**Fornix**    A part of this tract, where its two components unite, was identified in the mid-sagittal section. A rostral part of one of the anterior columns of the fornix, which lies lateral to the median plane, is now exposed. It extends in a recurrent arc to the mammillary body in the hypothalamus.

*Gently scrape away some of the thalamic tissue that forms the lateral wall of the third ventricle just above the mammillary body.*

**Mammillothalamic tract**    This tract extends between the mammillary bodies and the cranial part (anterior nuclei) of the thalamus. It runs parallel to those fibers of the fornix that lie caudal to the anterior commissure.

## Dissection of a Cerebral Hemisphere (Fig. 9-8)

Using a new brain is desirable but, if unavailable, the (left) sagittal half-brain may be used in exposing a **lateral ventricle**, the longitudinal cavity in each cerebral hemisphere. As you slice more deeply into the cerebral hemisphere, note the significant depth of the sulci and the increasing amount of white matter that comes into view.

*Carefully cut away a succession of thin slices, in planes parallel with the corpus callosum, from a cerebral hemisphere. When the corpus callosum becomes visible, remove the remainder of the hemisphere with blunt forceps to avoid damaging the lateral ventricle, which projects above the medial part of the commissure, and to demonstrate the full extent of the corpus callosum.*

**Corpus callosum**    This transverse commissure forms a part of the roof of each lateral ventricle. The corpus should reveal transverse striations, reflecting the crossing fibers, but it may also show some faint longitudinal striations. Its lateral boundaries are indistinct. The **genu** forms the rostral wall of the roof of the anterior horn of the ventricle. The **splenium** overhangs the caudal ends of the thalami, pineal body, and superior colliculi.

## Exploring the Lateral Ventricle (Figs. 9-9, 9-10)

*The lateral ventricle should be exposed by making a longitudinal incision about 2 mm deep and about 4 mm lateral to the midline, extending from the genu to the splenium. The initial cut can be made with a scalpel and then, if desired, extended with a fine scissors. The flaps can be separated to see whether the ventricle has been fully exposed, and then they can be removed with the scissors or by blunt dissection.*

**Lateral ventricle**    This ependyma-lined chamber is relatively small compared to the size of the hemisphere in which it is located. It communicates with the third ventricle at the interventricular foramen, which is located near the anterior commissure. This chamber has a **central part** and **anterior, posterior, and inferior horns,** or extensions. The anterior horn extends forward, laterally, and downward. The posterior horn extends for a short distance toward the occipital lobe, while the inferior horn curves downward and forward into the pyriform lobe of the cerebral hemisphere. The central part of the ventricle is roofed over by the corpus callosum, while its medial wall is the septum pellucidum, which separates it from the ventricle on the other side. The lateral side is merely represented by the linear region of convergence of the roof and floor of the ventricle. The floor is formed, essentially, by three structures: the caudate nucleus anteriorly, the choroid plexus in the middle, and the hippocampus posteriorly.

**Caudate nucleus**    This nucleus is seen as a prominent grayish mass. Over its medial margin lies the conspicuous vascular network, the **choroid plexus.** The caudate nucleus is bent downward on itself like a horseshoe and is intimately associated with the lateral ventricle throughout its entire arc. The swollen end, or **head,** is pear-shaped and bulges into the anterior horn. The head tapers gradually into the **body** of the nucleus, which extends backward on the floor of the central part of the ventricle (along the lateral border of the upper surface of the thalamus). The balance of the caudate nucleus is drawn out into the long slender **tail.** The tail arches sharply to run downward with the inferior horn of the ventricle and then extends forward in the roof of the inferior horn to terminate in the amygdaloid nucleus, which is located at the tip of the horn (see Fig. 9-10, inset B).

**Choroid plexus**    This ependyma-covered vascular network extends along the medial border of the caudate nucleus and bulges into the lateral ventricle. The plexus contributes to the floor of the ventricle along with the caudate nucleus.

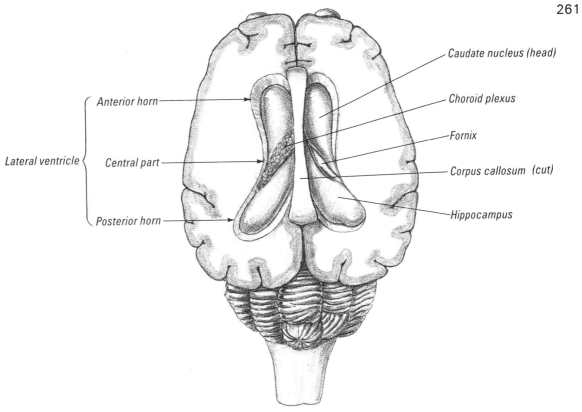

**FIGURE 9–9**
Sheep brain; lateral ventricles exposed.

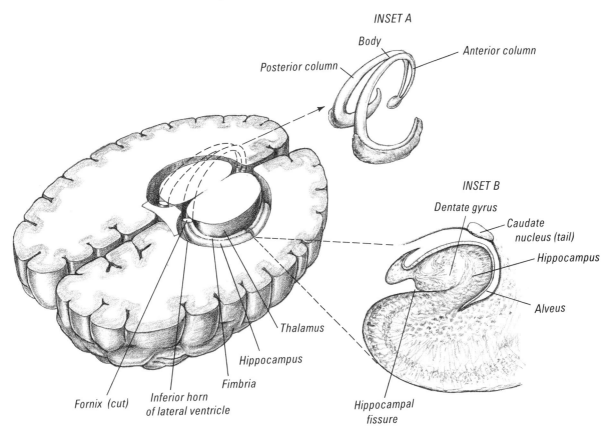

**FIGURE 9–10**
Sheep brain; inferior horn of the lateral ventricle
Inset A: Fornix; Inset B: Section through hippocampus.

**Hippocampus**   A portion of this structure is evident as a convex, elongated eminence that occupies the caudal end of the floor of the central part of the ventricle and continues into the inferior horn. The hippocampus will be seen when the inferior horn is exposed (see below).

## Exposing the Fornix (Figs. 9-9, 9-10)

*Using forceps carefully remove the choroid plexus from the floor of the lateral ventricle. Next, cut across the genu of the corpus callosum and remove the remainder of its body as far back as the splenium by cutting its connection with the septum pellucidum.*

**Fornix**   Part of the body and posterior column of this tract contributes indirectly to forming the floor of the lateral ventricle since they are covered by the choroid plexus. The dorsal surface of the body serves to anchor the septum pellucidum.

## Exposing the Hippocampus (Fig. 9-10)

*To fully expose the hippocampus, the inferior horn of the lateral ventricle must be uncovered by removing portions of the cerebral hemispheres associated with its posterolateral wall. This should be done by cutting out a ventrally tapering, wedge-shaped brain segment that overhangs the ventricle and hippocampus. The cranial surface of the inferior horn of the ventricle should be kept intact to avoid damaging the end of the tail of the caudate nucleus. Press the hippocampus caudomedially and observe the continuity of the tip of the hippocampus with the pyriform lobe.*

*A suitably stained, demonstration, transverse section should be used to identify the tail of the caudate nucleus, the dentate gyrus, and the hippocampal fissure (see Fig. 9-10 inset B).*

**Fimbria**   This band of fibers lies along the curved medial border of the hippocampus. It originates in the hippocampus from axons which coalesce there. The fimbria then continues along a course that arches dorsally toward the median plane, widening as it proceeds. As it nears the splenium of the corpus callosum, it continues as the posterior column (crus) of the fornix.

**Hippocampus**   This structure is a major visceral–somatic coordination center of the limbic system. It can be considered a cortical convolution that has been incorporated into the cerebral hemisphere and is embedded in the floor of the inferior horn of the lateral ventricle. The hippocampus, like the fornix, is a C-shaped structure. It hooks around from the floor of the ventricle lateroventrally and terminates in the pyriform lobe close to the external brain surface. The ventricular surface of the hippocampus is covered by a thin layer of white fibers, the **alveus,** which connects medially with the fornix.

**Tail of the caudate nucleus**   The horseshoe-shaped caudate nucleus ends in a long, slender tail that arches in the roof of the inferior horn of the lateral ventricle. The tail terminates in the **amygdala,** a ball-shaped structure that represents a group of nuclei, which protrude into the roof of the inferior horn at its anterior end (see Fig. 9-12).

**Dentate gyrus**   This slender, scalloped gyrus begins rostral to the hippocampal fissure and extends along its entire length. It lies along the medial margin of the hippocampus.

**Hippocampal fissure**   This groove extends from the pyriform lobe to near the splenium of the corpus callosum. It is formed by the involution of the hippocampus into the lateral ventricle wall.

## Exposing the Thalamus (Figs. 9-10, 9-11)

*Rock the hippocampus back and forth to see how the fimbria fits like a cap over the pulvinar and lateral geniculate body of the thalamus. Next cut across the fornix near the site at which the posterior column joins the body. Carefully remove the hippocampus, fimbria, and adjacent part of the wall of the inferior horn of the lateral ventricle as far as the pyriform lobe. This procedure will reveal the choroid plexus of the third ventricle as well as the structures on the medial side of the hippocampus. Grasp the tela choroidea near the interventricular foramen with a forceps and gently peel it away from the thalamus.*

**Thalamus**   This structure is the largest component of the diencephalon. The two thalami, representing complexes of nuclei, are oval masses of gray matter that fuse anteromedially at the massa intermedia and diverge caudally. The dorsal surface of each thalamus is convex and may show a slight rostral elevation, the **anterior tubercle.** (The lateral and ventral surfaces cannot be seen at this stage.) Laterally, an oblique groove that contains part of a long band of white fibers, the **stria terminalis,** separates the thalamus from the caudate nucleus. Medially, the thalamus is bounded by a longitudinal ridge of white fibers, the **stria medullaris.** The caudal end of the thalamus has a large projection, the **pulvinar.** On the ventrolateral surface of the pulvinar, there is a

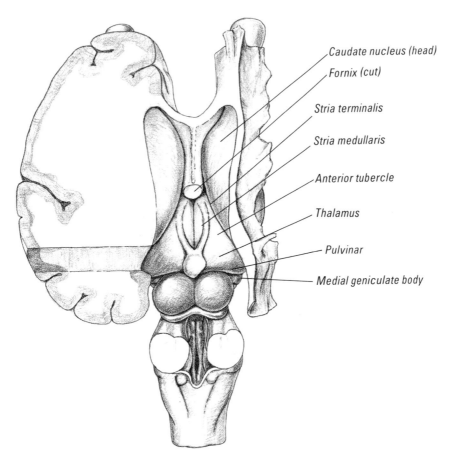

*Caudate nucleus (head)*

*Fornix (cut)*

*Stria terminalis*

*Stria medullaris*

*Anterior tubercle*

*Thalamus*

*Pulvinar*

*Medial geniculate body*

**FIGURE 9–11**
Sheep brain stem; thalamus exposed (cerebellum removed).

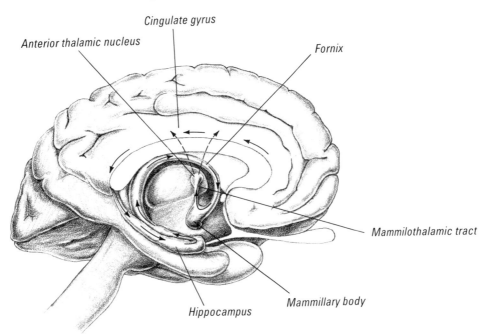

*Cingulate gyrus*

*Anterior thalamic nucleus*

*Fornix*

*Mammilothalamic tract*

*Mammillary body*

*Hippocampus*

**FIGURE 9–12**
Sheep brain; Papez circuit of the limbic system.

small, ill-defined eminence, the **lateral geniculate body,** which serves as a relay nucleus to the neocortex for optic tract fibers (visual pathway). The **medial geniculate body** is part of the thalamus overhung by the pulvinar and, in a similar fashion, serves to relay fibers to the auditory neocortex, including some fibers from the inferior colliculus. This distinct elevation lies in the region where the thalamus and midbrain join.

At the beginning of this section, it was noted that many of the internal structures are interrelated. The Papez circuit of the limbic system, a continuum of brain tissue, links many of these structures together (Fig. 9-12). Thus a linkage extends from the hippocampus to the fornix and then to the mammillary body. From there the mammillothalamic tract extends to the anterior nucleus of the thalamus, from which a connection is established with the cingulate gyrus, which, in turn, gives rise to the cingulum and thence back to the hippocampus.

The aforementioned circuit is part of the larger limbic system that includes the hypothalamus, thalamus, and amygdala. This system is responsible for emotional responses, instincts, and complex neuroendocrine regulatory functions. It is thus vital to maintain the body's physical and emotional equilibrium and, thereby, its viability.

# Glossary of Anatomic Terms

The early anatomists used Latin terms to describe parts of the body, and many of these have survived to the present time, e.g., *vena* vein. Some of the Latin names were derived from the Greek, e.g., *pterygium*, from Greek *pterygion* little wing. Most scientific names for later identified structures have been devised by combining Latin or Greek roots to form a new, compound term, e.g., adrenal, from Latin *ad* upon + *ren* kidney. Thus, knowledge of the root words that make up compound terms will not only aid in understanding new terms but will also help in retaining the meaning of anatomic terms in general.

This glossary includes only anatomic terms used in the manual. Specific structures defined in the glossary are of course described in greater detail in the text itself. A standard biological or medical dictionary is recommended for a wider vocabulary.

If the Latin or Greek root of the first portion of a compound term is defined in a preceding entry, its meaning is not repeated:

Epididymis (G *epi* upon + *didymos* testis)
Epiglottis (G *epi* + *glottis* tongue)

If the term being defined is identical with the Latin or Greek, only the meaning is given in parentheses:

Ampulla (L flask)

Some basic combining forms of terms, e.g., *chondro-*, *epi-*, and *gastro-*, are defined here. Consequently, some basic anatomic terms that are obvious compounds of two roots defined here, e.g., epidermis, have not been included in the glossary.

Only the singular form of a term will be found in the glossary, even though the plural form may be used in the text, e.g., Crus, not Crura.

Abbreviations used in the glossary are:

**dim** diminutive
**D** Dutch
**F** French
**G** Greek
**L** Latin
**super** Superlative

**Abducens** (L *ab* away + *ducere* to lead). Refers to the sixth cranial nerve.
**Abductor** (L *ab* + *ducere*). Drawing away from the median body axis.
**Accessory** (L *accessum* to be added). Supplementary.
**Acetabulum** (L vinegar cup). Socket of the hip bone.
**Acromion** (G *akros* top + *omos* shoulder). Lateral process extending from the scapular spine.
**Adductor** (L *ad* to + *duco* to lead). Drawing toward the median body axis.
**Adrenal** (L *ad* upon + *renes* kidneys). The endocrine gland near the kidney.
**Afferent** (L *ad* to + *ferre* carry). Conducting a fluid or nerve impulse to or into a structure (cf. Efferent).
**Ali-** (L *ala* wing). Prefix meaning "wing-like."
**Amnion** (G *amnion* fetal membrane). Inner embryonic membrane.
**Ampulla** (L flask). A sac-like dilation.
**Anconeus** (L *ancon* bend of arm). A muscle adjacent to the elbow.
**Angular** (L *angulus* angle). Occupying a corner.
**Annular** (L *annulus* a ring). Circular.
**Ante-** (L before). Prefix meaning "before" or "prior to."
**Anterior** (L foremost). Situated in front; ventral (cf. Posterior).
**Anus** (L fundament). The rear opening of the digestive tract.
**Aorta** (G *aeiro* to raise). The major arterial trunk.
**Aperture** (L *aperire* to open). An opening.
**Aponeurosis** (G *apo* from + *neuron* tendon). Flattened tendon.
**Appendicular** (L *appendere* to hang upon). Relating to the limbs.

**Aqueduct** (L *aqua* water + *ductus* leading). Any canal or passage in an organ.

**Arachnoid** (G *arachnes* spider + *eidos* shape). Middle membrane covering the brain and spinal cord.

**Arbor vitae** (L tree of life). Appearance of the white matter in a median section of the cerebellum.

**Arc-** (L *arcus* a bow). Prefix meaning "curved."

**Articular** (L *articulus* joint). Relating to a joint.

**Arytenoid** (G *arytaina* pitcher + *eidos* form). A laryngeal cartilage.

**Atlas** (G a Titan, G. *tlao* to bear). First cervical vertebra.

**Atrium** (L hallway). A cavity or passage; usually the main chamber of an auricle of the heart or the entire auricle.

**Auricle** (L dim of *auris* ear). An angular or ear-shaped lobe or process.

**Axial** (L *axis* axle of a wheel). Relating to the trunk.

**Axillary** (L *axilla* little axis). Relating to the axilla (armpit).

**Azygos** (G *a* without + *zygon* yoke). Unpaired structure.

**Basi-** (L *basis* base). Prefix pertaining to the base.

**Biceps** (L *bis* two + *caput* head). Two-headed muscle.

**Bicuspid** (L *bis* + *cuspis* points). Having two prongs.

**Brachial** (L *brachium* arm). Pertaining to the arm.

**Branchial** (G *branchion* gill). Pertaining to the gills.

**Brevis** (L short). Short.

**Bronchus** (G *bronchos* windpipe). One of the two branches of the trachea.

**Buccal** (L *bucca* cheek). Relating to the mouth.

**Bulbus** (G *bolbos* swollen root). Relating to a structure of globular form.

**Calcaneus** (L *calcaneum* heel). Heel bone.

**Canine** (L *canis* dog). Tooth type.

**Capitate** (L *caput* head). A wrist bone.

**Carina** (L keel). Any keel-like structure.

**Carotid** (G *karos* sleep). Relating to the artery supplying blood to the head.

**Carpus** (G *karpos* wrist). Wrist bones.

**Cauda equina** (L *cauda* tail + *equinus* horse). The collective roots of the sacral and coccygeal nerves.

**Caudal** (L *cauda* tail). Relating to the tail or posterior direction (cf. Cranial).

**Cecum** (L blind). An outpocketing closed at one end.

**Celiac** (G *koilia* belly). Relating to the abdominal cavity.

**Centrum** (G *kentron* center of a circle). Body of a vertebra.

**Cephalic** (G *kephale* head). Relating to the head.

**Cerata-** (G *keras* horn). Prefix denoting horny substance or horn shape.

**Cerebellum** (L dim of *cerebrum* brain). A brain segment in back of the cerebrum.

**Cerebral** (L *cerebrum* brain). Pertaining to the brain.

**Cervical** (L *cervix* neck). Pertaining to the neck.

**Chevron** (F rafter). V-shaped bones of the vertebral column.

**Choana** (G funnel). Internal opening of the nasal cavity.

**Chondro-** (G *chondros* cartilage). Combining form denoting cartilage.

**Chordae tendinae** (G *chordae* string + *tendo* stretch). Tendons of the papillary muscles of the heart.

**Chorion** (G skin). Outer embryonic membrane.

**Choroid** (G *chorion* skin + *edios* form). Vascular layer of the eye.

**Chromatophores** (G *chroma* color + *phorus* bearer). A pigment-containing cell.

**Ciliary** (L *cilium* eyelash). Relating to cilia or hair-like processes.

**Cirrus** (L lock of hair). Hair-like appendage.

**Cisterna chyli** (L *cistern* + G *chylos* juice). A reservoir of fatty lymph.

**Clav-** (L *clavis* key). Combining form denoting the clavicle (cf. Cleido-).

**Cleido-** (G *kleis* key). Combining form denoting the clavicle (cf. Clav-).

**Cloaca** (L sewer). Combined urogenital and rectal receptacle.

**Cochlea** (L snail shell). Spiral canal in the inner ear.

**Coelom** (G *koiloma* a hollow). Mesoderm-lined cavity.

**Colon** (G *kolon* gut). Large intestine.

**Columella** (*columna* little column). Bone in the middle ear.

**Commissure** (L *commissura* joining together). Strands of nerve fibers uniting like structures in the two sides of the brain.

**Communis** (L common). Belonging to many.

**Condyloid** (G *condylos* knuckle + *eidos* form). Relating to or resembling a condyle.

**Conjunctiva** (L *conjunctivus* connected). The mucous membrane covering the eyeball and lining the eyelid.

**Conus** (G *konos* cone). Cone.

**Coprodeum** (G *kopros* excrement + *hodos* way). A compartment within the cloaca.

**Corac-** (G *korax* crow). Combining form relating to the coracoid process of the scapula.

**Corn-** (L. *cornu* horn). **Combining form denoting a horny** consistency or shape.

**Coronary** (L *coronarius* crown). Encircling; denoting an anatomical structure.

**Coronoid** (L *korone* crow + *eidos* form). Shaped like a crow's beak; denoting processes and other parts of bones.

**Corpus** (L body). The main part (body) of an organ or other anatomical structure.

**Costo-** (L *costa* rib). Combining form denoting a rib.

**Cranial** (G *kranion* skull). Relating to the head or anterior direction (cf. Caudal).

**Cranio-** (G *kranion*). Combining form denoting head.

**Cribriform** (L *cribrum* sieve + *forma* shape). Sieve-like appearance.

**Crico-** (G *krikos* ring). Combining form denoting ring.

**Cricoid** (G *krikos*). Ring-shaped.

**Crista** (L crest). A projecting ridge.

**Crus** (L leg). Resembling a leg or root; any leg-like organ.

**Cuboid** (G *kybos* cube + *eidos* form). Cube-shaped.

**Cucullaris** (L *cucullus* hood). Hood-shaped muscle.

**Cuneiform** (L *cuneus* wedge + *forma* form). Wedge-shaped.

**Cutaneous** (L *cutis* skin). Relating to the skin.

**Cystic** (G *kystis* bladder). Relating to a bladder or cyst.

**Deltoid** (G Δ, delta + *eidos* form). Triangular in shape.

**Dens** (L tooth). Tooth-like process on the body of the axis.

**Dentary** (L *dens*). Lower jaw bone.

**Dermis** (G *derma* skin). The lower skin layer.

**Diaphragm** (G *diaphragma* a partition). The musculotendinous wall between the thoracic and abdominal cavities.

**Diapophysis** (G *dia* through + *apophysis* offshoot). The upper articular surface of the transverse process of a vertebra.

**Diastema** (G interval). A gap between teeth.

**Diencephalon** (G *dia* through + *enkephalos* brain). The posterior part of the forebrain.

**Digastric** (G *di-* two + *gaster* belly). Having two muscular parts.

**Digitorum** (L *digitus* finger). Finger.

**Dorsal** (L *dorsum* back). Relating to the back; posterior (cf. Ventral).

**Dorso-** (L *dorsum*). Combining form denoting the dorsal direction (cf. Ventro-).

**Ductus** (L to lead). Duct; any tubular structure.

**Duodenum** (L. *duodeni* twelve). The first segment of the small intestine.

**Dura mater** (L hard mother). Fibrous outer membrane covering the brain and spinal cord.

**Efferent** (L *ex* out + *ferre* carry). Conducting (a fluid or nerve impulse) from a structure outward (cf. Afferent).

**Endo-** (G *endon* within). Prefix signifying within.

**Endocrine** (G *endon* + *krinein* to separate). Secreting internally; relating to such secretion.

**Epi-** (G upon). Prefix denoting above or on top.

**Epididymis** (G *epi* + *didymos* testis). The first, convoluted portion of the excretory duct of the testis, located on its posterior border.

**Epiglottis** (G *epi* + *glottis* tongue). The cartilage that extends over the opening of the larynx.

**Epiphysis** (G *epi* + *physis* growth). Pineal gland.

**Extensor** (L *extendere* to stretch out). A muscle that straightens a limb or part by contraction. (cf. Flexor).

**Ethmoid** (G *ethmos* sieve + *eidos* form). A perforated bone of the base of the skull; resembling a sieve; cribriform.

**Falciform** (L *falx* sickle + *forma* shape). Sickle-shaped.

**Fascia** (L band). Fibrous connective tissues enveloping an organ or area.

**Fenestra** (L window). An aperture.

**Filiform** (L *filum* thread + *forma* form). Thread-like.

**Fissure** (L *fissus* split). A cleft.

**Flexor** (L *flectere* bend). A muscle that bends a limb or part by contraction (cf. Extensor).

**Foliate** (L *foliatus* leafy). Leaf-like.

**Foramen** (L opening). A perforation.

**Fornix** (L arch). An arched space or recess.

**Fossa** (L ditch). A depression.

**Fovea** (L small pit). A small depression.

**Fundus** (L bottom). The base of an organ.

**Fungiform** (L *fungus* mushroom + *forma* form). Mushroom-shaped.

**Funiculus** (L dim of *funis* cord). Column of white matter of the spinal cord.

**Gastro-** (G *gaster* belly). Combining form denoting the stomach.

**Geniculate** (L *geniculare* bend the knee). Bent like a knee.

**Genio-** (G *geneion*). Combining form denoting the chin.

**Glans** (L acorn). Conical-shaped body.

**Glenoid** (G *glene* socket + *eidos* form). Resembling a socket.

**Gloss-** (G *glossa* tongue). Combining form denoting the tongue.

**Glottis** (G opening of windpipe). Aperture to the larynx.

**Gluteus** (G *gloutos* rump). Pertaining to the buttock.

**Gular** (D *gula* throat). Pertaining to the throat.

**Gyrus** (G *gyros* circle). A convolution on the brain surface.

**Hemal** (G *haima* blood). Relating to blood.

**Hepat-** (G *hepar* liver). Combining form denoting the liver.

**Heterocercal** (G *heteros* different + *kerkos* tail). A tail having unequal lobes.

**Hilus** (L *hilum* little thing). A recess in an organ, usually where nerves, vessels, etc., enter.

**Holo-** (G *holos* whole). Combining form denoting entirety or relationship to a whole.

**Hyo-** (G *hyoeides* U-shaped). Combining form referring to the hyoid bone.

**Hyper-** (G above). Prefix denoting a position above.

**Hypo-** (G under). Prefix denoting a position below.

**Hypophysis** (G *hypo* + *physis* growth). Pituitary gland.

**Ileum** (G *eileo* twist). The lower portion of the small intestine.

**Ilio-** (L *ilium* groin). Combining form denoting the ilium.

**Incisor** (L *incidere* to cut into). Cutting tooth.

**Incus** (L anvil). The anvil-shaped middle-ear bone.

**Indicis** (L index). The index or pointing finger.

**Infra-** (L below). Prefix denoting a position below a designated part.

**Infundibulum** (L funnel). A funnel-shaped structure.

**Inguinal** (L *inguen* groin). Relating to the groin.

**Innominate** (L *innominatus* unnamed). Nameless; applied to various arteries and veins.

**Integument** (L *integumentum* covering). Skin.

**Inter-** (L between). Prefix signifying between or among.

**Intercalary** (L *inter* between + *calare* insert). Occurring in between.

**Intra-** (L within). Prefix signifying within or into.

**Ischi-** (G *ischion* hip). Combining form denoting the ischium or the hip.

**Jejunum** (L *jejunus* empty). Second part of the small intestine.

**Jugal** (L *jugum* yoke). Relating to the malar bone; connecting.

**Jugular** (L *jugulum* collarbone). Pertaining to the throat.

**Labial** (L *labium* lip). Pertaining to the lip.

**Labyrinth** (G *labyrinthos* labyrinth). An intricate system of connecting passageways.

**Lacrimal** (L *lacrima* tear). Relating to the secretion of tears.

**Lacteal** (L *lac* milk). Intestinal lymphatic vessel; relating to or resembling milk.

**Lagena** (L flask). An extension of the sacculus of the inner ear.

**Lambdoidal** (G the letter *lambda* + *eidos* form). Resembling the Greek letter lambda, Λ.

**Lamina** (G thin plate). Flat layer.

**Larva** (L mask). Immature, free-living stage of an animal's life cycle.

**Larynx** (G windpipe). Upper end of trachea.

**Lens** (L lentil). Refractive element of the eye.

**Lieno-** (L spleen). Combining form denoting the spleen.

**Ligament** (L *ligamentum* band). A band of tough, dense connective tissue.

**Lingual** (L *lingua* tongue). Pertaining to the tongue.

**Lobe** (G *lobos* lobe). Rounded part or projection of an organ.

**Lumb-** (L *lumbus* loin). Combining form denoting the loin or lumbar.

**Lymphatic** (L *lympha* clear water). A vessel conveying lymph.

**Macula** (L spot). A spot or patch of color; small spot differing in color from surrounding tissue.

**Malar** (L *mala* cheek). Pertaining to the cheek.

**Malleus** (L hammer). The hammer-shaped middle-ear bone.

**Mammillary** (L *mammilla* nipple). Relating to or shaped like a nipple.

**Manubrium** (L handle). Upper segment of the sternum.

**Mastoid** (G *mastos* breast + *eidos* form). Breast-shaped.

**Medial** (L *medius* middle). Relating to the middle or center; nearest the sagittal plane.

**Mediastinum** (L *mediastinus* being in the middle). The space between the pleura.

**Medulla** (L marrow). Any soft marrow-like structure, especially in the center of a structure.

**Melanophore** (G *melanos* black + *phorus* bearer). A black pigment-containing cell.

**Membranous** (L *membrana* skin). Resembling or consisting of membrane.

**Meninx** (G membrane). One of the coverings of the brain or spinal cord.

**Mental** (L *mentum* chin). Pertaining to the chin.

**Mes-** (G *mesas* middle). Prefix meaning the middle.

**Mesencephalon** (G *mesos* middle + *enkephalos* brain). The middle of the three primitive brain divisions.

**Mesentery** (G *mesentarium* intestine). The double layer of peritoneum suspending the (small) intestine.

**Mesonephros** (G *mesos* + *nephros* kidney). The functional kidney of fishes and amphibians.

**Meta-** (G after). Prefix meaning after or beyond; often used to denote change, transformation, or occurrence after something else in a series.

**Metencephalon** (G *meta* + *enkephalos* brain). The front part of the hindbrain that gives rise to the cerebellum and pons.

**Molar** (L *molaris* grinder). A grinding tooth.

**Mucus** (L *mucus* slime). A clear viscid liquid of the mucous membranes. (NB. The adjectival form is "mucous.")

**Multifidus** (L *multus* many parts). A multisegmented back-muscle.

**Myelencephalon** (G *myelos* marrow + *enkephalos* brain). The posterior of the two subdivisions of the primitive hindbrain.

**Mylo-** (G *myle* mill). Combining form denoting the molar teeth.

**Myo-** (G *mus* muscle). Combining form denoting muscle.

**Myomere** (G *mus* muscle + *meros* part). A muscle segment.

**Naso-** (L *nasus* nose). Combining form relating to the nose.

**Navicular** (L *navicula* small boat). Boat-shaped hand or foot bone.

**Neuro-** (G *neuron* nerve). Combining form denoting nerve.

**Neuromast** (L *neuron* nerve + *mastos* hillock). A sensory papilla.

**Nictitating** (L *nicto* to wink). Winking.

**Notochord** (G *noton* back + *chorde* string). The fibrocellular, primitive skeletal axis.

**Obturator** (L *obturare* to shut). Any structure that occludes an opening.

**Occipital** (L *occiput* back of head). Pertaining to the rear of the skull.

**Ocellus** (L little eye). Pigment spot.

**Oculo-** (L *oculus* eye). Combining form denoting the eye.

**Odontoid** (G *odous* tooth + *eidos* form). Tooth-shaped.

**Olecranon** (G *olea* elbow + *kranion* head). The large process at the upper end of the ulna.

**Olfactory** (L *olfacere* to smell). Pertaining to smelling.

**Omentum** (L fat-skin). Peritoneum connecting the stomach to another organ.

**Omo-** (G *omos* shoulder). Combining form denoting the shoulder.

**Ophthalmo-** (G *ophthalmos* eye). Combining form denoting the eye.

**Opisth-** (G behind). Prefix meaning behind.

**Oro-** (L *os* mouth). Combining form denoting the mouth.

**Os** (L bone). A bone (plural, ossa).

**Os** (L mouth). A mouth. (plural, ora).

**Ossicle** (L *ossiculum* dim of *os*). Any small bone.

**Ostium** (L door). Entrance to a hollow organ.

**Otic** (G *otos* ear). Pertaining to the ear.

**Otolith** (G *otos* ear + *lithos* stove). Calcified body in the sacculus.

**Palato-** (L *palatum* palate). Combining form denoting the palate.

**Pancreas** (G *pan* all + *kreas* flesh). Abdominal digestive and endocrine organ.

**Papilla** (L nipple). An elevated nipple-like structure.

**Para-** (G beside). Prefix meaning near.

**Parietal** (L *paries* wall). Relating to the wall of a cavity.

**Parotid** (G *para* + *otos* ear). Structure near the ear.

**Patella** (L little dish). The sesamoid bone in front of the knee.

**Pectoral** (L *pectus* breast). Relating to the chest.

**Peduncle** (L *pedunculus* little foot). A narrow supporting stem or stem of a structure.

**Peri-** (G around). Prefix meaning surrounding.

**Peritoneum** (G *peri* + *teino* stretch). The membrane lining the abdominal cavity and viscera.

**Phalanx** (G fingerbone). A finger or toe bone.

**Pharynx** (G throat). Throat.

**Phrenic** (G *phren* diaphragm). Pertaining to the diaphragm.

**Pia** (L *pius* delicate). The thin membrane enveloping the brain or spinal cord.

**Pineal gland or body** (L *pinea* pine cone). Median outgrowth from the roof of the brain.

**Pinna** (L feather). The protecting part of the external ear.

**Piriform** (L *pirum* pear + *forma* form). Pear-shaped; a gluteal muscle.

**Pisiform** (L *pisum* pea + *forma* form). Pea-shaped; a small wrist bone.

**Pituitary** (L *pituita* mucus). Pertaining to an endocrine-secreting gland, the hypophysis.

**Placenta** (L cake). The organ of metabolic interchange between fetus and mother.

**Placoid** (G *plax* plate). Plate-like shape.

**Planum** (L *plana*). Plane or level surface.

**Pleur-** (G *pleura* rib). Combining form denoting relationship to the rib, the side, or the pleura.

**Plexus** (L network). A network of interlacing structures.

**Pons** (L bridge). A bridge-like connection.

**Popliteal** (L *poples* ham). Pertaining to the region behind the knee.

**Posterior** (L latter). Situated behind or in back; dorsal (cf. Posterior).

**Pre-, pro-** (L before). Prefixes meaning before.

**Proboscis** (G *proboskis* trunk). The snout projecting from the head of an animal.

**Process** (L *processus* process). A projection or outgrowth.

**Profundus, -a** (L deep). Deep-seated; applied to arteries.

**Proprius** (L one's own). Special.

**Pter-** (G *pteron* wing). Combining form denoting wings or relationship to wings.

**Pubis** (L *pubes* mature). The anteroventral segment of the pelvis girdle.

**Pulmonary** (L *pulmo* lung). Relating to the lungs.

**Pulp** (L *pulpa* solid flesh). The soft interior part of an organ.

**Pylorus** (G *pyloros* gate-keeper). The terminal segment of the stomach.

**Pyramidal** (L *pyramis* pyramid). Conical.

**Quadratus** (L square). Name applied to muscles that are more or less square in shape.

**Quadri-** (L *quattuor* four). Prefix meaning four.

**Radix** (L root). Root; point of origin of a structure, as of an aorta or nerve.

**Ramus** (L branch). A branch or division; applied to nerves and blood vessels.

**Raphe** (G *rhaphe* seam). A line of junction of two similar structures.

**Renal** (L *ren* kidney). Pertaining to the kidneys.

**Retina** (L *rete* net). The light-sensitive layer of the eye.

**Rostrum** (L beak). Any beak-shaped structure.

**Ruga** (L wrinkle). A fold or crease, as of skin or of mucous membrane.

**Sacculus** (L little sac). The smaller sac of the inner ear.

**Sacrum** (L sacred). Fused sacral vertebrae.

**Sagittal** (L *sagitta* arrow). Section or division in median longitudinal plane; in an anteroposterior direction.

**Scaphoid** (G *skaphe* boat + *eidos* form). Shaped like a boat; applied to carpal and tarsal bones.

**Sciatic** (G *ishion* hip). Pertaining to the hip.

**Sclera** (G *skleros* hard). The tough outer coat of the eye.

**Sella** (L saddle). Saddle-shaped.

**Seminal** (L *semen* seed). Pertaining to the semen.

**Septum** (L *sepes* fence). A thin partition.

**Serratus** (L *serra* saw). Serrate (notched on edge like a saw).

**Sino-** (L *sinus* curve). Combining form denoting the sinus.

**Siphon** (G a hollow tube). A tube conveying water to or from the gills.

**Spermatheca** (L *sperma* sperm + *theke* case). A small sac that stores sperm.

**Spermatophore** (L *sperma* + *phorus* bearer). A sperm-containing packet.

**Spheno-** (G *sphen* wedge). Combining form meaning wedge or wedge-shaped, or relating to sphenoid bone.

**Spino-** (L *spina* thorn). Combining form indicating a sharp thorn-like process.

**Spiracle** (L *spiraculum* breathing hole). A respiratory vent.

**Splenium** (G *splenion* bandage). The round posterior end of the corpus callosum.

**Squamosal** (G *squama* scale). Relating to the squama of the temporal bone.

**Stapes** (L stirrup). The smallest of the three middle-ear bones.

**Stato-** (G *statos* standing). Combining form denoting equilibrium of the body.

**Stern-** (G *sternon* breast). Combining form denoting the sternum.

**Stylo-** (G *stylos* pillar). Combining form denoting connection with the styloid process of the temporal bone.

**Sub-** (L under). Prefix meaning under.

**Sulcus** (L furrow). A groove.

**Supinator** (L *supinus* bend back). A muscle (in the forearm) that produces supination.

**Supra-** (L above). Prefix meaning above.

**Sympathetic nervous system** (G *syn* with + *pathos* suffering). System of nerves supplying viscera and blood vessels.

**Synapticula** (G *syn* together + *hapto-* to fasten). One of the small rods connecting septa.

**Talus** (L heel). Heel bone.

**Tarsus** (G *tarsos* broad flat surface). Ankle.

**Tela** (L web). A web-like tissue.

**Telencephalon** (G *tele* far + *enkephalos* brain). The anterior division of the forebrain.

**Temporal** (L *tempus* temple). Pertaining to the temple.

**Tensor** (L *tendere* to stretch). A muscle that stretches a structure.

**Tentorium** (L tent). A partition of dura mater.

**Teres** (L round). Round and long (cylindrical shape); denoting certain muscles and ligaments.

**Thalamus** (G *thalamos* chamber). The side walls of the diencephalon.

**Thoraco-** (G *thorax* chest). Combining form denoting the thorax.

**Thyro-** (G *thyreos* shield). Combining form denoting the thyroid.

**Torus** (L bulge). A curved surface.

**Trabecula** (L little beam). A bundle of fibers extending into an organ.

**Trachea** (G *trachys* rough). The windpipe.

**Transversus** (L *trans* across + *verto* to turn). Transverse to an organ or part.

**Trapezium** (G *trapezion* table). A carpal bone.

**Trapezoid** (G *trapezion* table + *eidos* form). Shaped like a trapezium (a four-sided figure with no two sides parallel) except that two of its opposite sides are parallel.

**Tri-** (L three). Prefix meaning three.

**Trich-** (G *thrix* hair). Combining form meaning hair.

**Trigeminal** (L *trigeminis* triplet). Pertaining to the fifth cranial nerve.

**Trochanter** (G runner). Any of several projections at the upper end of the femur.

**Trochlea** (L pulley). A pulley-like structure.

**Tuber** (L swelling). Rounded swelling.

**Tunica** (L coat). Enveloping layers of membrane or tissue.

**Turbinal** (L *turbo* top). Scroll-shaped, as bone or cartilage.

**Tympanic** (L *tympanon* drum). Pertaining to the eardrum.

**Ureter** (G *oureter* ureter). The tube conducting urine from the kidney to the bladder.

**Urethra** (G *ourethra* urethra). The tube discharging the urine from the bladder externally.

**Uro-**[1] (G *ouron* urine). Combining form relating to urine (e.g., urogenital, urorectal).

**Uro-**[2] (G *oura* tail). Combining form relating to the tail (e.g., uromorphic, urochord).

**Uterus** (L womb). The muscular organ of gestation.

**Utriculus** (L small bag). The larger of the two inner-ear sacs.

**Vagus** (L wandering). Tenth cranial nerve.

**Velum** (L curtain). Veil-like structure, usually membranous.

**Ventral** (L *venter* belly). Relating to the lower or abdominal body surface; anterior (cf. Dorsal).

**Ventricle** (L *venter*). A small cavity of the heart or brain.

**Ventro-** (L *venter*). Combining form denoting ventral direction (cf. Dorso-).

**Vermis** (L worm). Median lobe of the cerebellum.

**Vesicle** (L *vesica* bladder). A bladder-like air space in tissues; small cavity containing fluid.

**Vesicular** (L *vesica*). Sac-like; composed of vesicle-like cavities.

**Vestibule** (L *vestibulum* antechamber). A cavity leading into another cavity or canal.

**Vibrissa** (L nostril hair). Long, course facial hair.

**Visceral** (L *viscus* internal organ). Relating to internal body organs.

**Vitreous** (L *vitrum* glass). Transparent.

**Vomer** (L ploughshare). Thin bony plate between nasal cavities.

**Xipho-** (G *xiphos* sword). Combining form relating to the xiphoid process of the sternum.

**Zygomatic** (G *zygon* yoke). Relating to the cheekbone.

# Selected Reading

It is hoped that this selected list of books and journal articles will help those who wish to initiate an in-depth study of chordate anatomy.

## GENERAL

Alexander, R. McN. *The Chordates*. Cambridge University Press, London, 1975.

Colbert, E. H. *Evolution of the Vertebrates*, 2nd ed. Wiley, New York, 1969.

DeBeer, G. R. *Vertebrate Zoology*, rev. ed. Macmillan, New York, 1953.

Gans, C. *Biomechanics: An Approach to Vertebrate Biology*. Lippincott, Philadelphia, 1974.

Goodrich, E. S. *Studies on the Structure and Development of Vertebrates*. Dover, New York, 1958.

Halstead, L. B. *The Pattern of Vertebrate Evolution*. W. H. Freeman, San Francisco, 1968.

Hildebrand, M. *Analysis of Vertebrate Structure*. Wiley, New York, 1974.

Jollie, M. *Chordate Morphology*. Rheinhold, New York, 1962.

Parker, T. J., W. A. Haswell, and A. J. Marshall *A Textbook of Zoology*, Vol. 2, 7th ed. Macmillan, London, 1962.

Romer, A. S. and T. S. Parsons *The Vertebrate Body*, 6th ed. Saunders, Philadelphia, 1986.

Stahl, B. J. *Vertebrate History: Problems in Evolution*. McGraw-Hill, New York, 1974.

Wake, M. H. *Hyman's Comparative Vertebrate Anatomy*, 2nd ed. University of Chicago Press, 1979.

Webster, D., and M. Webster *Comparative Vertebrate Morphology*. Academic Press, New York, 1974.

Young, J. Z. *The Life of Vertebrates*. Oxford University Press, New York, 1962.

## TAXONOMIC CLASSES

### Protochordates

Barrington, E. J. W. *The Biology of Hemichordata and Protochordata*. W. H. Freeman, San Francisco, 1965.

Barrington, E. J. W., and R. P. S. Jeffries, eds. Protochordates. In *Symp. Zool. Soc.*, No. 36, 1975.

Berril, N. J. *The Tunicata*. The Ray Society, London, 1950.

Hyman, L. H. *The Invertebrates: Smaller Coelomate Groups*. Vol. 5. McGraw-Hill, New York, 1959.

### Fishes

Alexander, R. M. *Functional Design in Fishes*, 3rd ed. Hutchinson, London, 1974.

Daniel, J. F. *The Elasmobranch Fishes*. University of California Press, Berkeley, 1934.

Gilbert, P. W., R. F. Mathewson and D. P. Rall, eds. *Sharks, Skates, and Rays*. Johns Hopkins University Press, Baltimore, 1967.

Hardesty, M. W. and I. C. Potter, eds., *The Biology of Lampreys*, 2 vols. Academic Press, New York, 1971-72.

Marshall, N. B. *The Life of Fishes*. Weidenfeld and Nicolson, London, 1965.

Northcutt, R.G., ed. Recent advances in the biology of sharks. *Am. Zool.* 27: 207–515, 1977.

### Amphibians

Francis, E. B. *The Anatomy of the Salamander*. Oxford University Press, London, 1934.

Noble, G. K. *The Biology of the Amphibia*. Dover, New York, 1954.

Schmalhausen, I. I. *The Origin of Terrestrial Vertebrates*. Academic Press, New York, 1968.

### Reptiles

Bellaus, A. *The Life of Reptiles*, 2 vols. Universe Books, New York, 1970.

Gans, C. ed. *Biology of the Reptilia*, 3 vols. Academic Press, New York, 1969.

### Birds

Farner, D. S. and J. R. King, eds. *Avian Biology*, 3 vols. Academic Press, New York, 1971–73.

Koch, T. *Anatomy of the Chicken and Domestic Birds*. Iowa State University Press, Iowa City, 1973.

Welty, J. C. *The Life of Birds*, 3rd ed. Saunders, Philadelphia, 1982.

## Mammals

Bensley, B. A. *Practical Anatomy of the Rabbit*, 7th ed. Blakeston, Philadelphia, 1946.

Crouch, J. E. *Text-Atlas of Cat Anatomy*. Lea Febiger, Philadelphia, 1969.

Ewer, R. F. *The Carnivores*. Cornell University Press, Ithaca, New York, 1973.

Field, H. E. and M. E. Taylor *An Atlas of Cat Anatomy*. University of Chicago Press, 1950.

Getty, R., ed., *Sisson and Grossman's Anatomy of the Domestic Animals*, 5th ed. Saunders, Philadelphia, 1975.

Greene, E. C. *Anatomy of the Rat*, Hafner, New York, 1959.

Griffiths, M. *The Biology of the Monotremes*. Academic Press, New York, 1978.

Hartman, C. G., and W. L. Straus, Jr., eds. *The Anatomy of the Rhesus Monkey*, Hafner, New York, 1961.

Leach, W. J. *Functional Anatomy: Mammalian and Comparative*. McGraw-Hill, New York, 1961.

Miller, M. E., G. C. Chrestensen, and H. E. Evans *Anatomy of the Dog*. Saunders, Philadelphia, 1964.

Reighard, J., and H. S. Jennings *Anatomy of the Cat*. Holt, New York, 1935.

Sisson, S., and J. D. Grossman *The Anatomy of the Domestic Animals*, 4th ed. Saunders, Philadelphia. 1953.

Slijper, E. J. *Whales*. Hutchinson, London, 1962.

Stromsten, F. A. *Davidson's Mammalian Anatomy*, rev. 7th ed. Blakiston, Philadelphia, 1947.

Taylor, W. T., and R. J. Weber *Functional Mammalian Anatomy*. Van Nostrand, New York, 1951.

Wimsatt, W. A., ed. *Biology of Bats*, Academic Press, New York, 1970.

# PHYSIOLOGIC SYSTEMS

## Integumentary System

Heath, G. W. The siphon sacs of the smooth dogfish and spiny dogfish. *Anat. Rec.*, **125**:562, 1956.

Maderson, P. F. A. When? Why? and How? Some speculations on the evolution of the vertebrate integument. *Am. Zool.*, **12**:159–171, 1972.

Nursall, J. R. Swimming and the origin of paired appendages. *Am. Zool.*, **2**:127–141, 1962.

Parkes, K. C. Speculations on the origins of feathers. *Living Bird*, **5**:77–86, 1966.

## Skeletal System

Davis, D. D. Origin of the mammalian feeding mechanism. *Am. Zool.*, **1**:229–234, 1961.

Denison, R. H. The early history of the vertebrate calcified skeleton. *Clin. Orthopaed.*, **31**:141–152, 1963.

Gregory, W. K. The pelvis from fish to man; a study in paleomorphology. *Am. Naturalist*, **69**:193–210, 1935.

Jarvik, E. On the origin of girdles and paired fins. *Israel J. Zool.*, **14**:141–172, 1965.

Pautard, F. G. E. Calcium, phosphorus, and the origin of backbones. *New Scientist*, **12**:364–366, 1961.

## Muscular System

Barclay, O. C. The mechanics of amphibian locomotion. *J. Exptl. Biol.*, **23**:177–203, 1946.

Edgeworth F. H. On the development of the coraco-branchiales and cucullaris in *Scyllium canicula*. *J. Anat.*, **60**:298–308, 1926.

Hughes, G. M., and C. M. Ballintijn The muscular basis of the respiratory pumps in the dogfish (*Scylionhinus canicula*). *J. Exp. Biol*, **43**:363–383, 1965.

Romer, A. S. Pectoral limb musculature and shoulder–girdle structure in fish and tetrapods. *Anat. Rec.*, **27**:119–143, 1924.

## Digestive System

Barrington E. J. W. The supposed pancreatic organs of *Petromyzon fluviatilis* and *Myxine glutinosa*. *Quart. J. Micro. Sc.*, **85**:391–417, 1945.

Gibbs, S. P. The anatomy and development of the buccal glands of the lake lamprey (*Petromyzon marinus L.*) and the histochemistry of their secretion. *J. Morph.*, **98**:429–470, 1956.

## Respiratory System

Atz, J. W. Narial breathing in fishes and the evolution of internal nares. *Quart. Rev. Bio.*, **27**:366–377, 1952.

Bertman, G. The vertebrate nose; remarks on its structure and functional adaptation and evolution. *Evolution*, **23**:131–152, 1969.

Cox, C. B. Cutaneous respiration and the origin of modern Amphibia. *Proc. Linn Soc. Lond.*, **178**:37–47, 1967.

## Circulatory System

Barnett, C. H., R. J. Harrison, and J. D. W. Tomlinson Variations in the venous systems of mammals. *Biol. Rev.*, **33**:442–487, 1958.

De Graaf, A. R. Investigations into the distribution of blood in the heart and aortic arches of *Xenopus laevis* (Daud). *J. Exptl. Biol.*, **34**:143–172, 1957.

Foxon, G. E. H. Problems of double circulation in vertebrates. *Biol. Rev.*, **30**:196–228, 1955.

O'Donoghue, C. H., and E. B. Abbot The blood vascular system of the spiny dogfish, *Squalus acanthias* Linne, and *Squalus sucklii* Gill. *Trans. Roy Soc. Edinburgh*, **55**:823–894, 1928.

Satchell, G. H. The reflex coordination of the heart beat with the respiration in the dogfish. *J. Exptl. Biol*, **37**:719–731, 1960.

Turner, S. C. A comparative account of the development of the heart of a newt and a frog. *Acta Zool.*, **48**:43–57, 1967.

## Urogenital System

Chase, S. W. The mesonephros and urogenital ducts of *Necturus maculosus*. Rafinesque. *J. Morph.*, **37**:457–532, 1923.

Fraser, E. A., and L. Albert The development of the vertebrate excretory system. *Biol. Rev.*, **25**:159–187, 1950.

Hisaw, F. L., and L. Albert Observations on the reproduction of the spiny dogfish, *Squalus acanthias*. *Biol. Bul.*, **92**:187–199, 1947.

## Sensory System

Allison, A. C. The morphology of the olfactory system in vertebrates. *Biol. Rev.*, **28**:195–244, 1953.

Detwiler, S. R. The eye and its structural adaptations. *Am. Scientist*, **44**:45–72, 1956.

Dijkgraaf, S. The functioning and significance of the lateral-line organs. *Biol. Rev.*, **38**:51–105, 1963.

Murry, R. W. Evidence for a mechanoreceptive function of the ampullae of Lorenzini. *Nature*, **179**:106–107, 1957.

Reed, H. D. The morphology of the sound-transmitting apparatus in caudate amphibia and its phylogenetic significance. *J. Morph.*, **33**:325–387, 1920.

Van Bergeijk, W. A. The evolution of the sense of hearing in vertebrates. *Am. Zool.*, **6**:371–377, 1966.

Weaver, E. G. The ear and hearing in the frog *Rana pipiens*. *J. Morph.*, **141**:461–477, 1973.

## Nervous System

Ebbesson, S. O. E. New insights into the organization of the shark brain. *Comp. Biochem. Physiol.*, **42A**:121–129, 1972.

Flood, P. R. A pecular mode of muscular innervation in amphioxus: Light and electron-microscopic studies of the so-called ventral roots. *J. Comp. Neurol.*, **126**:181–217, 1968.

Lindström, T. On the cranial nerves of the cyclostomes, with special reference to *N. trigeminus*. *Acta Zool.*, **30**:315–458, 1949.

Norris, H. W., and S. P. Hughes The cranial, occipital, and anterior spinal nerves of the dogfish, *Squalus acanthias*. *J. Comp. Neurol.*, **31**:293–402.

Young, J. Z. The autonomic nervous system of selachians. *Quart. J. Micro. Sc.*, **75**:571–624, 1933.

## SHEEP BRAIN

Briggs, E. A. *Anatomy of the Sheep's Brain*. Angus and Robertson, Sydney, 1933.

Igarashi, S., and Kamiya, J. *Atlas of The Vertebrate Brain. Morphologic Evolution from Cyclostomes to Mammals*. University Park Press, Baltimore, 1972.

Northcuff, R. G. *Atlas of the Sheep Brain*, 2nd ed. Stipes, Champaign, Ill., 1966.

Yoshikawa, T. *Atlas of the Brains of Domestic Animals*. Pennsylvania State University Press, University Park, Pa., 1968.

Wilkie, J. *The Dissection and Study of the Sheep's Brain*. Oxford University Press, London, 1937.

# Index

This index consists of an alphabetically arranged list of terms that appear in the headings or in bold face in the body of the text. The animal form to which the term is applied is identified in a subentry only if the term is used for more than one form. If the term is used for only one form there is no subentry, but the form can be determined from the indicated page number, since the inclusive page numbers for protochordates are 3 to 15, for the lamprey, 19 to 27, for the dogfish shark, 31 to 100, for the mud puppy, 103 to 149, and for the cat, 153 to 264.

*Continued*

*Continued*

# Supplementary Figures

This section of the manual consists of five tear-out figures that are designed primarily to enhance your understanding of the circulatory system of the cat. They have been incorporated into the manual because the circulatory system is among the more complex and therefore challenging areas for dissection in the comparative anatomy laboratory. They illustrate the following:

The supplementary figures are on one side of the page and the backs are blank. The pages are perforated near the book spine. Hence, the figures can be removed and placed alongside the pages of related text to facilitate dissection or review. Write your name on each torn-out page and store them at the site of the relevant text pages.

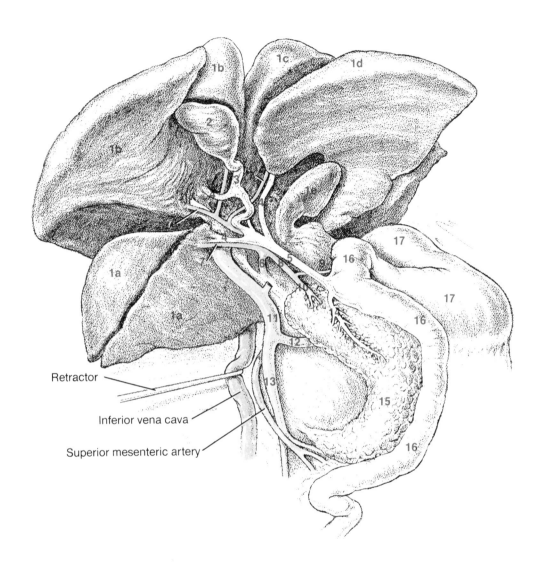

Retractor

Inferior vena cava

Superior mesenteric artery

**FIGURE S–1**

Liver, gall bladder, and associated vessels. (Lymphatic vessels and nerve fibers will be present, although not shown here.)

1 Lobes of liver
  a Right lateral
  b Right median
  c Left median
  d Left lateral
  e Caudate
2 Gall bladder
3 Cystic duct
4 Hepatic ducts
5 Common bile duct
6 Hepatic artery

7 Branches of hepatic artery
8 Gastroduodenal artery
9 Right gastroepiploic artery
10 Superior pancreaticoduodenal artery
11 Hepatic portal vein
12 Gastrosplenic vein
13 Superior mesenteric vein
14 Pancreatic duct (duct of Wirsung)
15 Head of pancreas
16 Duodenum
17 Stomach

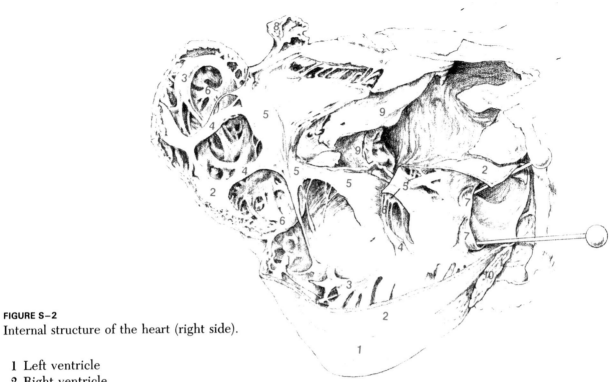

**FIGURE S–2**
Internal structure of the heart (right side).

1 Left ventricle
2 Right ventricle
3 Columna, or trabeculae carneae
4 Papillary muscle
5 Atrioventricular valve cusp
6 Chordae tendineae
7 Semilunar valve flap at base of
  pulmonary aorta (pin is in the sinus)
8 Auricula of right atrium
9 Right atrium
10 Auricula of left atrium

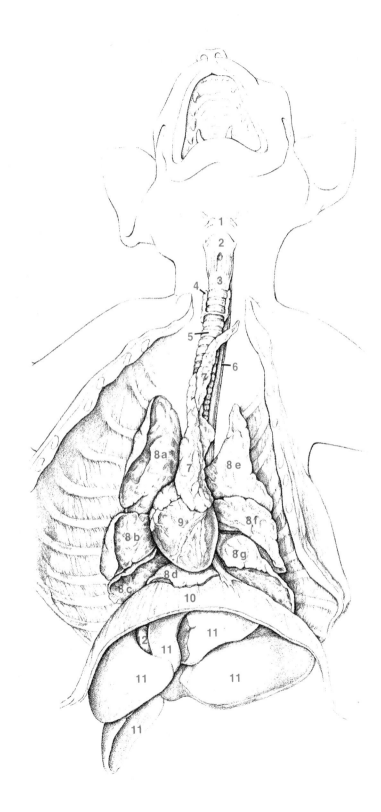

**FIGURE S–3**
Thoracic viscera.

1 Hyoid bone
2 Thyroid cartilage
3 Cricoid cartilage
4 Thyroid gland
5 Trachea
6 Esophagus
7 Thymus gland
8 Lungs
  a Right, anterior lobe
  b Right, middle lobe
  c Right, posterior lobe
  d Right, mediastinal lobe
  e Left, anterior lobe
  f Left, middle lobe
  g Left, posterior lobe
 9 Heart
10 Diaphragm
11 Liver
12 Gall bladder

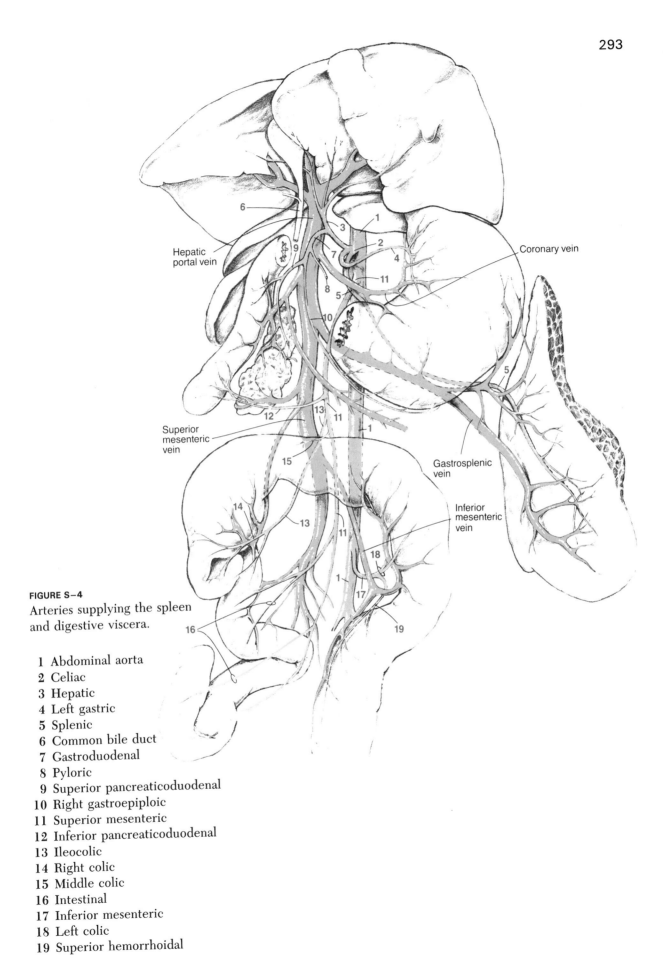

Hepatic portal vein

Coronary vein

Superior mesenteric vein

Gastrosplenic vein

Inferior mesenteric vein

**FIGURE S–4**
Arteries supplying the spleen and digestive viscera.

1 Abdominal aorta
2 Celiac
3 Hepatic
4 Left gastric
5 Splenic
6 Common bile duct
7 Gastroduodenal
8 Pyloric
9 Superior pancreaticoduodenal
10 Right gastroepiploic
11 Superior mesenteric
12 Inferior pancreaticoduodenal
13 Ileocolic
14 Right colic
15 Middle colic
16 Intestinal
17 Inferior mesenteric
18 Left colic
19 Superior hemorrhoidal

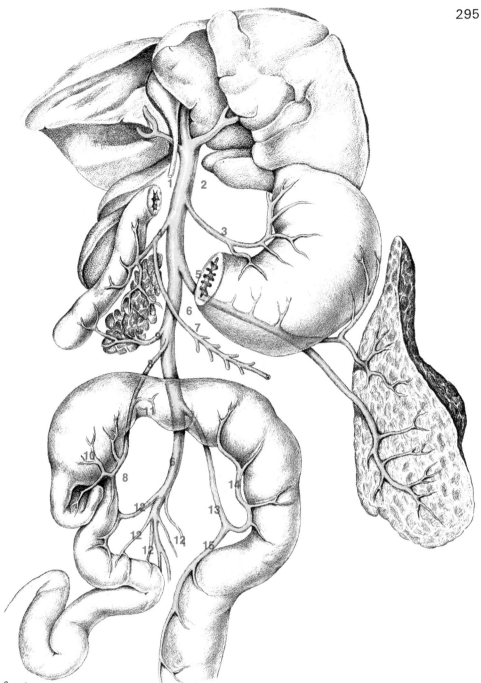

**FIGURE S–5**

Hepatic portal system of veins.

1 Common bile duct
2 Hepatic portal vein
3 Coronary
4 Superior pancreaticoduodenal
5 Gastrosplenic
6 Superior mesenteric
7 Gastroepiploic
8 Ileocolic
9 Inferior pancreaticoduodenal
10 Right colic
11 Middle colic
12 Intestinals
13 Inferior mesenteric
14 Left colic
15 Superior hemorrhoidal